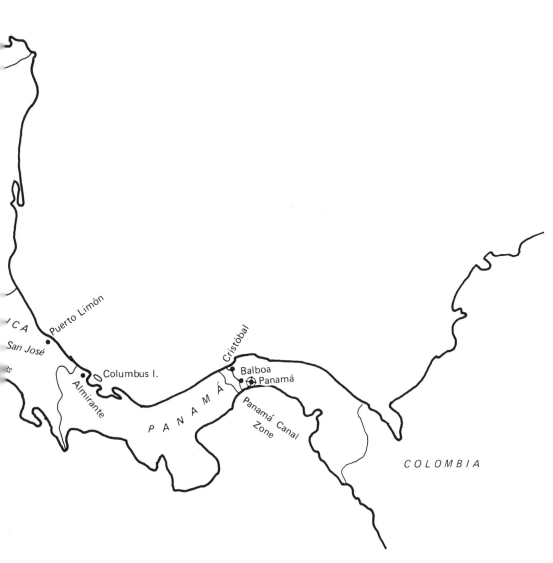

Blue Mts.

JAMAICA

Kingston

CARIBBEAN SEA

Puerto Limón

San José

ICA

Columbus I.

Almirante

Cristóbal

Balboa

Panamá

P A N A M Á

Panamá Canal
Zone

COLOMBIA

The Imperative Call

The Imperative Call

A Naturalist's Quest in Temperate and Tropical America

Alexander F. Skutch

A University of Florida Book

UNIVERSITY PRESSES OF FLORIDA
FAMU/FAU/FIU/FSU/UCF/UF/UNF/USF/UWF
Gainesville 1979

Library of Congress Cataloging in Publication Data

Skutch, Alexander F   1904–
    The imperative call.

    "A University of Florida book."
    Includes index.
    1. Skutch, Alexander F., 1904–   2. Birds—
Central America.   3. Natural history—Central
America.   4. Ornithologists—United States—
Biography.   I. Title.
QL31.S53A33   574'.092'4 [B]   79–14701
ISBN 0–8130–0579–5

University Presses of Florida is the central agency for scholarly publishing
of the State of Florida's university system. Its offices are located at 15 NW
15th Street, Gainesville, FL 32603. Works published by University
Presses of Florida are evaluated and selected for publication by a faculty
editorial committee of any one of Florida's nine public universities:
Florida A&M University (Tallahassee), Florida Atlantic University (Boca
Raton), Florida International University (Miami), Florida State University (Tallahassee), University of Central Florida (Orlando), University of
Florida (Gainesville), University of North Florida (Jacksonville), Univer-
sity of South Florida (Tampa), University of West Florida (Pensacola).

To
EUGENE EISENMANN

who generously shares his vast knowledge
of Neotropical ornithology with all
who seek his help

For the river calls and the road calls,
and oh! the call of a bird!

Gerald Gould

Contents

Foreword

Two voices summon men with a call so imperative that few who hear clearly can resist. One is the voice of religion, which bids us abandon all mundane pursuits and seek holiness, God, and life everlasting. The other is the voice of nature, which invites us to fill our spirits with its beauty and wonder and challenges us to disclose some of its closely hidden secrets. Obeying either of these voices, we may neglect nearly everything that prudent men esteem and strenuously seek: wealth, security, solid comfort, and social status. We may even abandon family, friends, and homeland to follow the call into a wilderness where perils lurk.

Although these two voices sound so different, they have much in common. Both entreat us to cultivate something vaster and more enduring than men and their creations; to care more about what we do and experience than what we possess. Perhaps, if we could hear aright, we would recognize that the two voices are in fact one, calling us in different tones to release our spirit from workaday pettiness and permit it to expand widely into realms of mystery and wonder. Following either of these voices, men have too often been guilty of cruelty and revolting excesses—for from what human endeavor can human wickedness and folly be excluded? But if we follow with a pure heart, humble mind, and compassionate spirit, they may lead us to things more precious than gold.

In this book I tell something of what happened to one person who could not resist the imperative call, especially as it issues in varied tones from the throats of birds. Of how he heard it in his formative years. Of how, not without many a nostalgic pang, he was drawn away from the homeland that he loved to pass many years studying nature in tropical America; of the travels that ensued in the United States and the Caribbean countries; of the more interesting places where he dwelt and worked and the more exciting adventures that befell him; of the procession of the seasons that he lived through in tropical lands; of some of the stately trees, lovely flowers, fascinating birds, and curious animals of other kinds that he saw.

Although this is primarily a book of natural history rather than an autobiography, I have included enough autobiographical details to give continuity to the narrative and reveal the motives of my undertakings. After all, when one is dedicated to his work, when he seeks knowledge or illumination in any field, the story of his quest constitutes the more essential part of the story of his life.

In a book like this, one has space for only some of the highlights of his most absorbing studies. Those who have read the full reports of some of these studies may wish to know the circumstances under which they were made. Others, who may desire more information on some of the subjects treated herein, must be referred to the books and articles in journals where the studies are reported in detail. To help them find these accounts, I may mention that A Naturalist in Costa Rica (University of Florida Press, 1971) contains, in an appendix, a complete bibliography of my writings published up to 1970. That companion volume of the present book continues the story of my experiences as a naturalist after I went to Costa Rica in 1935.

Acknowledgment.—A substantial part of chapter 23 first appeared, with a different title, in the September 1954 issue of Natural History Magazine, copyright by The American Museum of Natural History. The author is grateful to the publisher for permission to use this material.

1

A Child's Eden

Despite our vaunted human freedom to choose, in many of the matters that most affect our success and happiness, choice is denied us. Among these are the place and time of our birth, our parents and the genes they transmit to us, the guardians of our infancy and childhood, and our earliest teachers, who may influence our character and outlook more than the learned professors whom, at a later age, we deliberately select to instruct us. And even the career we choose, our life's work, seems to be determined for us by heredity and early influences, to a degree that we may be reluctant to admit.

Although I was not consulted about the time and place of my entry into this world, I am not sure that, with my present knowledge of history and geography, I would choose a date or a locality very different from those that were chosen for me. In the year 1904, I entered a fairly peaceful, prosperous, and hopeful world. It is true that the Russians and Japanese were then fighting stubbornly for control of northeastern Asia; but that was on the

Headpiece: Waiting for mother: a young raccoon photographed in a flashlight's beam.

opposite side of a planet which, for all practical purposes, was very much larger than it has since become. Four decades after Appomattox, the scars, economic and moral, which the disastrous War between the States had left on the defeated South were slowly disappearing. The recent Spanish-American War, which wrested from a tottering power almost the last remnants of a once-gigantic empire, was a minor conflict.

The country of which I automatically became a citizen by entering with a wail instead of a passport was victorious, strong, and confident. Modern science and technology, by many recent inventions and growing control of disease, had made human life more pleasant and secure without yet threatening its extinction, swiftly by a nuclear holocaust, or more gradually by poisoning the environment. And for the undertaking to which, after some preliminary vacillation, I decided to dedicate my life, I could hardly have chosen a more propitious date to begin my earthly career. It permitted me to start my study of tropical nature at a period when modern sanitation and transportation had made tropical America fairly healthful and accessible, but soaring populations had not yet begun to destroy its wildness and beauty at an ever-accelerated pace.

The house in which I was born, one in a long, solidly built row of two-story brick dwellings near what was then the northern edge of the growing city of Baltimore, was not such as one who loves wild nature would choose for his habitation. Years later, I was told that I entered it so reluctantly that, for a while, the physician who ushered me in despaired of saving both mother and infant, although finally he brought both through the ordeal without lasting lesions. But I have no recollection of anything that happened to me in that modest home where my parents settled after their marriage; I remember only its façade, from having often passed it while I was growing up. The great advantage of my birthplace was its proximity to some of the pleasantest country in the world, the gently rolling piedmont region of Maryland. To these green hills my family migrated every summer, to escape urban heat that now, after a long residence in the tropics, I find scarcely endurable. At first we summered in a rented cottage; but when I was about three years old my father bought "Pomona," a scant half-mile northwest of the village of Pikesville.

The white stucco house was large and ancient. The walls of the oldest part, at the front, were impressively thick; but those of the rooms that my father added at the rear were of modern thinness. The house stood on a hill, in the midst of wide lawns shaded by lindens, maples, tulip-poplars, and other trees. Beyond were the orchards, cultivated fields, pastures, and wood lots of the sixty-acre farm. From the toll road that led far up into the county, our entrance lane passed along the hillside opposite the house, swept around in a wide circle, dipped to cross a brook by a wooden bridge, and climbed obliquely up the hill to pass in front of the dwelling. Thence it continued, by a straighter and more often used "back road," out to the turnpike where the electric cars from the city passed.

In the house were many rooms, many pictures on the walls, and rows and rows of books in glass-fronted oaken cases; for my father loved fine engravings and finely printed books, and while still young had the income to indulge his expensive tastes. Nearby stood a large stone stable, where our one horse lived in winter and the carriages were kept. More distant was the huge barn, built in two levels against a steep slope, so that wagons could be hauled into the upper story to unload hay directly into the lofts, from which, through convenient trap doors, it was dropped for the animals who occupied the rows of stalls on the lower level. Behind the barn was a smaller wooden building, with space for farm machinery below a granary which, when empty, made an attractive playhouse. When we first went to live at Pomona, these farm buildings and most of the fields were rented to a farmer, who dwelt in a smaller house across the lawn from ours. My father went almost every day to the city, where he worked, and had neither time nor knowledge to manage a farm. At first, we spent only the summer months there; but before I started to school, the big townhouse that my mother had inherited from her parents was sold, and we lived the year around in the country.

Mr. Shipley, the farmer, was my first employer. While I was still quite small, he paid me five cents for picking the striped Colorado potato beetles off his potato plants. Doubtless I would have been distressed by the fate of those poor bugs after I delivered them to him in a can, as even then I had a compassionate heart for every living thing; but I do not recall that my childish thoughts reached beyond the nickel that was my wage. Mr. Shipley, or perhaps one of the hired men, was unintentionally responsible for one of the most distressing accidents that ever befell me. A large drum, filled with manure from the stable, was placed in the two-wheeled horse cart to be hauled to the garden; and of course the child who was watching wanted to go along. So I was lifted up and stood holding the rim of the drum, enjoying the ride, until the cart jolted over a rough place, the drum jumped and came down on my big toe, and strong arms carried me, bleeding profusely, back to the house.

The most exciting event of a year filled with excitements for a small boy was certainly the arrival of the thresher. The wheat had been harvested by a two-wheeled, horse-drawn machine with an attachment like a small windmill, that threw the yellow stalks against a sliding, toothed knife. It had been tied into bundles or "shocks" and transported in the hay wagon to the big barn, to await the arrival of the thresher. After a while, news would come that this outfit, making its way slowly from farm to farm, was approaching our own. At last the steam engine rolled impressively down our lane, pulling a train consisting of the big threshing machine, the smaller bailer, and a water barrel mounted on wheels, to supply the engine's boiler. Beside the barn the train stopped, and the threshing machine was attached by a broad belt to a wheel on the side of the engine. Then, all day long, men tirelessly threw bundles of wheat into the insatiable maw of the machine, which on one side delivered an endless stream of grain that was caught in burlap sacks, while on the other, from a long, wide, projecting tube, it blew out the straw and chaff into a mountainous pile. After the grain had all been threshed, the bailer was attached to the engine. Children watched in fascination while a massive L-shaped arm, swinging rhythmically up and down, packed the straw into the box, from which it emerged in heavy, wire-bound bales. At the end of such a day, one small boy was so weary and thirsty from watching in the summer heat that he was long in falling asleep.

To ride in the wagon that hauled the grain to the railroad was a great privilege. At Roslyn station on the Western Maryland Railroad, about a mile from the farm, a big, empty boxcar stood on the siding, ready for our wheat. As sack after sack of the flowing grain was emptied, the heaps at the ends of the car mounted higher. Finally, the doors were closed, and the car waited for a passing freight train to pick it up and haul it down to the grain elevators at the port of Baltimore. To watch the trains roar by, each trailing a long plume of smoke, while I waved to the engineer and passengers, who sometimes waved back, was one of the delights of my childhood. It was fascinating to see the express train, which did not even slow down at this small way station, stretch out a slender steel arm and snatch the waiting mailbag from a tower on the platform, to speed it toward the mountains of western Maryland, or down into the city. I have never outgrown my boyish interest in railroads, and to this day would rather travel by train than any other way.

In the pastures grew many chestnut trees, still unstricken by the disastrous Japanese blight that finally exterminated them over large areas of eastern North America. When the prickly pods opened with the first frosts, strewing the nuts widely over the ground, we searched for them amid fallen leaves as eagerly as though they were scattered golden coins. How delicious they were, roasted over the open fire on chill autumn evenings! Hickory trees were also abundant in the pastures, and a native walnut tree grew behind the house, but their nuts were less assiduously sought.

Along the brook that flowed in front of the house was a wet meadow where buttercups and blue-eyed grass bloomed, frogs croaked, and lurking garter snakes sent queer sensations coursing through a small boy's flesh. Beyond the brook and the meadow rose a steep, wooded hillside, where beneath the trees grew the most cherished wildflowers in the vicinity. In early spring, before the shade became dense, the Trailing Arbutus* opened its fragrant, pinkish florets among the dead leaves that littered the stony slope. Later, the

---

*When the name of an animal or plant is capitalized, the scientific equivalent will be found in the index.

Bird-foot Violet, which for reasons unknown to me we called "Johnny-jump-up," displayed its lovely blossoms, with silken petals of two colors, violet and deep lilac-purple, amid finely divided leaves. Largest and rarest of all were the Stemless Lady's-slipper Orchids, with sac-like, pink lips on leafless scapes that sprang directly from the ground. Eagerly we children gathered these treasures and carried them home in warm little hands to wilt in vases, when, no doubt, we should have left them to adorn their native woodland.

Nearly four years older than my sister, and still more widely separated in age from my two brothers, I would have been a rather solitary child if I had not had so many pets, both native and introduced. On the hill above the house stood a water tank, on a steel tower that to a small boy rose impressively high. To climb the vertical ladder to the platform surrounding the tank was a daring feat, of which I was proud. In a box that housed a measuring gauge on the tank's side, a Gray Squirrel built her nest, of dry leaves and other trash, amid which she gave birth to her young. With greater indulgence of a childish whim than foresight, my father had the baby squirrels brought to the house, where they were nourished with milk from a medicine dropper until they could take more solid food. Turned loose in the playroom, they found a congenial abode in a table drawer filled with clothes patterns made of thin paper, which they tore to shreds to make a nest. Since women's styles are so swiftly outmoded, I suppose that the loss was slight.

More companionable than these squirrels were five downy Indian-runner ducklings. Waddling behind me, single file, they would follow me for long distances, if I walked slowly. Perhaps they would have continued to be my faithful attendants until they grew up, if my cousins, having heard that ducks like water, had not put them under the heavy stream that gushed from a fire plug at a corner of the house, with fatal effects.

My father, who was more generous than prudent, often brought surprises from the city. For my birthday, he came with a trio of bantam chickens—a resplendent rooster and two demure little hens with pencilled plumage. For minutes at a time I would gaze admiringly at these living treasures that I could call my own. The death of one of these hens prompted my first poetic effort, an elegiac outpouring of heartfelt grief. If my experience is typical, as children of six or seven we feel joy and sorrow, we love and even hate, as intensely as we ever can. From this age onward, mental growth increases our ability to reason and plan but hardly our capacity to feel. Perhaps, indeed, our emotional responses are attenuated as we form the habit of reacting practically to the crises in our lives, of trying to understand their causes or foresee their consequences. Anyone who recalls how intensely he could feel while his intellect was still undeveloped will not conclude that animals feel less than we do because they are less intelligent.

A pair of Green Peafowl brought to the farm about this time turned out to be wild. The hen soon vanished; and the peacock, perhaps in search of a mate, took to wandering widely. Reports would reach us that he had been seen, or perhaps caught in the barnyard, on some distant farm; and the hired man would hitch the horse in the dogcart and go in search of him, with me as the only passenger. I loved those rides through the green countryside, even if we returned without the resplendent truant. Finally, perhaps convinced that no peahen was to be found for miles around, the peacock decided to stay with us. He elected to roost on the porch roof of the now deserted tenant's house, whence at dawn we heard his loud *kay-ho*. What finally became of him, I cannot recall. Likewise, I cannot remember what happened to my Guinea-fowl, which always remained so aloof that I received scant satisfaction from owning them. What impressed me most was the hardness of their eggshells.

As was natural, a small boy's relations with the free native birds were less intimate. I was told that I could catch them by putting salt on their tails, but they never permitted this familiarity! From my parents and nurse, who was country bred, I learned the names of the more conspicuous kinds, the Robin, the Blue Jay, the Catbird, the Baltimore Oriole, the Crow, the Whip-poor-will, the Bobwhite, and a few others; but I doubt that anyone on the farm could identify the lesser kinds, such as the wood warblers, the vireos, and the small flycatchers. I gazed in wonder at the Ruby-throated Hummingbird hovering on invisible wings before the hollyhocks in the flower garden, little dreaming that some day I would study many members of this glittering family in tropical lands where they abound. The Wood Thrush's liquid song, to which my father called my attention one evening as the setting sun burst forth brilliantly after a shower, is one of the most vivid memories of my early childhood.

Fortunately, I avoided the egg-collecting stage through which budding naturalists used to pass; but after I could wander over the farm alone, I gathered molted feathers. The blue-and-white plumes of the Blue Jay, and the flicker's flight feathers with bright yellow shafts, became cherished possessions. With proper guidance, this feather-gathering might have become the starting point of a growing knowledge of birds. The feathers might have been placed in a notebook, with the name and a picture of the bird that grew them; but no one in the household knew enough to guide me. Even today, I cannot resist picking up and saving bright molted feathers, such as those of parrots and trogons; but I am thankful that I never coveted entire birdskins, which can be obtained only by killing the birds.

In the fall, when nuts were ripening, my father hunted the Gray Squirrels that abounded in the wooded pastures, often in company with friends from the city. Proud of his marksmanship, he always used a .22 rifle; but his guests, who sometimes included clergymen, brought shotguns. One summer I was given a toy gun, and like a great hunter, sallied forth to prove my prowess. Following the long, curving entrance lane, I had almost reached the turnpike when a Turkey Vulture soared overhead and I pointed the harmless gun at it, although I had been told that these scavengers were protected by law. At that moment, one of the strangest apparitions I have ever seen ran down the highway: a bareheaded, red-faced man, clad in a long, black robe. Pausing between the stone pillars of the entrance gate, he stared at me until I turned and ran homeward in terror, never stopping until I reached the house and the comforting arms of my aunt, who happened to be the first person I encountered there. To my childish imagination, this extraordinary man, possibly an escaped lunatic, was the Law in person, come to chastise my transgression. In any case, I have never again pointed a gun at a vulture or any other feathered creature, except, on two occasions, hawks that attacked birds that I was studying in Panamá.

My father loved the country although, city bred, his knowledge of nature was not extensive. As soon as we could follow, he took his children on long walks, over the farm and along the surrounding byways, where in those days few motorcars passed. On Sunday afternoons, we were often accompanied by my cousins, and perhaps a few young friends from the city, all of whom called him "Uncle Bob." Often we trudged wearily homeward carrying wildflowers, or box tortoises to be released in the garden, or, in the fall, leafy twigs aglow with autumnal color.

Even my first years at school did not take me out of the open country. On the other side of the highway, a short distance north of our entrance, Miss Mary Livingston had started an academy for girls on a country estate; and boys were admitted to the lower grades. Here, soon after my seventh birthday, I was sent, and did well in my studies because I liked my teacher, Miss Waters. She read sad old tales so feelingly that they brought moisture into young eyes, and she decorated our copybooks with gold stars that were the mark of excellence. After my first year at this school, it was moved several miles farther from the city, to a site near Garrison Forest; and I went back and forth in the electric cars, in company with several schoolmates. Happily, I had the same teacher in the second grade as in the first. After that, I was sent to Park School, then situated beside Druid Hill Park in north Baltimore. I did not like it so well as my first school, as it was less rural and, moreover, I had to get up earlier and take a very long ride in the slow streetcar with its frequent stops.

One day, while I was in class at this second school, I was called out into the hall to meet Miss Waters. She asked me a few questions and then, as we were alone, she bent over and kissed me. She turned to walk down the steps, and I never saw her again. As I grew older and more thoughtful, to remember that kiss made me happy. It suggested that, despite occasional stubbornness, I was the kind of child that older people could love, the kind of pupil who makes teaching a pleasant rather than a tedious occupation.

One afternoon, when I was about twelve, I returned from school to a desolated farm. My father's affairs had not been going well. During the two preceding summers, we had lived in small neighboring cottages, while the big house at Pomona was rented to a family from the city. Without his children's knowledge, my father's business had failed and he had gone through bankruptcy. In the morning, while I was at school in the city, the farm and its chattels had been sold at auction; the horse and cows had already been led away. When I learned what had happened, I wept; and to console me, Father gave me a gold watch that had been his. Thus early I learned the harsh consequences of insolvency, a lesson that I have never forgotten, preferring to deny myself all but the most essential things rather than incur debts.

Soon after this, we went to live in a hotel in the city, while we awaited the completion of the house that was being built for us on its outskirts. Years passed before I again set foot on the farm about which my earliest memories cluster. Although still beautiful, it seemed to have shrunk to a fraction of its remembered size.

2

The Green Hills of Maryland

Before the winter's end, our new home was finished and we moved in. Designed in accordance with the wishes of my father, whose taste was far better than his business acumen, the house was much smaller than the one we had left, but it was attractive, comfortable, and quite large enough for a family of six. This dwelling in the New England colonial style was my home for the remainder of my boyhood and my early manhood; to it I returned from my early journeys to study nature, sometimes triumphantly, sometimes dejected by the failure of a cherished project.

The house was on Clark's Lane, beyond the northern limit of the growing city of Baltimore, which soon afterward engulfed it. The half-acre lot on which it stood was shaded by white and pin oak, ash, hickory, and sour gum—large, old trees that rose above the high-peaked roof. At the back was space for a flower garden and a small vegetable garden, where I worked and exercised my growing limbs. In front of the house, on the other side of the

Headpiece: The author's boyhood home on Clark's Lane, on the outskirts of Baltimore.

unpaved lane, stretched a wide, open field where buttercups and daisies bloomed. Only a vacant lot separated us from a broad avenue, along which passed the electric cars that ran from the heart of the city far up into the country. But in the opposite direction, only two other houses lay between us and an extensive tract of old second-growth woodland, through which flowed a brook of clear water. Here, amid the broad-leafed woods, were outcrops of gray rocks where I could sit in utter solitude while I read Homer or Keats or Wordsworth, or tried to solve some of the perplexities of a youth who instinctively adopted the Socratic principle that an unexamined life is not worth living, and who accordingly could not accept all the ways of his elders. Among other things, after reading Shelley's poetry and essays, I could not continue to eat the flesh of slaughtered animals, a refusal that brought me into conflict with my father and the family doctor, my uncle, who predicted, falsely, that my health would suffer from a deficient diet.

On summer mornings, in the interval between my graduation from high school and entry into college, I earned a little pocket money by taking to a narrow opening in this woods half a dozen small boys, to play games, hear stories, learn a little about nature, then wade in the brook, before they were returned to parents glad to have them thus amused during the long and sometimes tedious summer vacation. Beyond the woods, to the north and east, farms and more woods stretched as far as I cared to walk.

The advantage of our new location was its proximity to the open country that we loved; the disadvantage, the hours that we had to spend on the electric cars to reach the city where my father worked, my mother had many social engagements, and I attended school and college. On most weekdays for the next twelve years, I took two streetcars to reach my destination, often waiting a long while for their arrival. At school, I did well in English literature and composition, geography, history, geometry, and carpentry, but I repeatedly failed courses in German, French, and Latin—only when it became indispensable for entering college, doing advanced work in science, or living in a foreign country, could I learn a foreign language well enough to get by.

At the age of seventeen, I entered Johns Hopkins University with scarcely any grounding in science, as elementary chemistry was the only course given while I was in high school, and that was interrupted by changes of teachers. Until I read an outline of Comte's *Positive Philosophy*, I had no idea of the sequence in which the sciences should be studied; but I was so eager to learn about plants that in my freshman year I wished to take a course in botany. This was rather irregular, as I had not yet had the preliminary course in general biology and, moreover, a conflict with a subject obligatory to all freshmen prevented my attendance at some of the botanical lectures, although I was able to do all the laboratory work. But the professor of botany,

Duncan S. Johnson, was indulgent; I read to make up for the lost lectures, and did so well in the course that I was encouraged to continue. My first view of a one-celled alga through the microscope opened an exciting new world that I have never had time to explore as thoroughly as I wished. From my sophomore year onward, I crammed into my schedule all the mathematical and scientific courses that I could take, from calculus to the genetics of protozoa, and I even found time for a series of lectures on Indo-European philology by the renowned Professor Bloomfield. In my junior year, I wavered between the satisfying exactitude of physics and the exciting variety of biology; but in the end my love of living things held me to the latter, and I continued in botany until, soon after my twenty-fourth birthday, I received a doctor's degree.

As an undergraduate, I helped a graduate instructor to operate a bird-banding station. By handling the birds, often releasing again and again the same individual that repeatedly entered the trap for food, I became familiar with the appearance of certain common species, but I learned scarcely anything about the habits of these same birds. I vividly recall one snowy day late in spring, when all of us who attended the traps were kept busy during all of our time between classes, and until late in the evening, banding newcomers and releasing, often many times over, birds who returned for food that was so hard to find outside. A Cardinal on whom I was fitting a band bit my finger with his strong bill until he drew blood. Banding has supplied much valuable information on the movements, the longevity, and the social relations of birds; but in later years I shrank from subjecting them to the terror of being caught and the indignity of being handled.

Although my formal studies regularly took me cityward, in my free time I preferred to go in the opposite direction, on long bicycle rides over unfrequented country roads, or walks through woods and across fields. The Saturday excursions of the botany students, under Professor Johnson's stimulating guidance, took us as far as the Falls of the Potomac and were among the chief attractions of his courses.

Soon after my twelfth birthday, I became a Boy Scout, in a troop that met in the closely built up city, where I did not like to go. I did little in scouting until I transferred to a troop that had its own rustic, one-room meeting house in a patch of woods above the highway cut at Garrison, half an hour by the electric car from home. In this troop I continued until I went to Panamá, becoming successively patrol leader, treasurer, scribe, assistant scoutmaster and, during my years of graduate study, scoutmaster. After the Friday-night meeting, in a room heated on cold winter nights by a roaring wood fire in an old-fashioned chunk stove, we would all go outside for a vigorous game in which smaller and bigger boys could participate on fairly equal terms. It left us all warm, happy, and friendly when we parted.

On Saturday afternoons when there was no botanical field trip, I took the boys for a hike, which was for me a welcome change from the week's often exacting studies. Although many biologists seem satisfied to analyze or interpret natural phenomena, perhaps summing up their rigorous researches in a formula or a neat graph; all my life I have wished to experience nature in its concrete reality of form and color, sound and scent, no less than to disclose its hidden springs. These outings provided the opportunity to balance the work of classroom and laboratory with the craving I had for immediate contact with the natural world.

To help the boys to know and love the beautiful country in which they were growing up was, I thought, the greatest service that I could do for them. To me, this seemed more important than covering their arms with merit badges. After a walk along rural lanes and woodland paths, we would stop to pass tests for higher rank. Sometimes I arranged a treasure hunt, for which each scout, or pair of them, received a copy of a topographic map on which I had marked the exact spot where some small prize awaited the boy who, following the map from a distant starting point, reached it first. As the day ended, we kindled campfires in some secluded spot, and each of us cooked his own supper, to be eaten with an appetite that compensated for deficient seasoning. After washing our utensils in the neighboring stream, we gathered around the glowing embers to tell stories, of heroic feats of old, of adventure in far places, or of ghosts that made delightful shivers course along boyish spines. Sometimes, in favorable weather, we set up pup tents and slept out, but more often we walked home by moonlight or under sparkling winter stars.

To sit or sleep on the ground in the nocturnal woods, to walk in the dark along woodland trails, with never a thought of venomous snakes, scorpions, or huge, hairy spiders, was a high privilege that I did not adequately appreciate until I lived amid grander and richer forests where to do such things is imprudent. Now I hesitate to step outside the house in the night without a flashlight. Venomous snakes are not often seen here; but one who walks over a Bushmaster such as I killed at midnight beside our front porch does not live to tell the tale. Blessed the land that is free from the scourge of poisonous serpents, as was that in which we hiked and camped!

The early teens is the flowering of a fortunate boyhood, and many a human male is never more lovable than when at this age. Thus far the child's chief business has been to grow and perfect himself, and now the favorable results of this effort become manifest; the stream of life runs smooth and calm, as though hesitating to plunge into the often impetuous rapids of adolescence. One who guides the young should try to prolong this happy interval of harmonious growth; it will end soon enough in any case. To be an admirable boy is no less difficult than to be an admirable man; perhaps more difficult, for one has had less time to practice. Scarcely anything is more repugnant than the young male who apes the ways and vices of his elders, and ends by being neither a boy nor a man, but a sort of unclassifiable hybrid between these so different stages of human life.

For the most part, the boys in my troop avoided this aberration; until well into their teens, they were content to remain boys, and the more likeable for it. Most came from middle-class families; a few were laboring men's sons, but nearly all were gently bred. Since I had grown up in the troop, they continued to call me "Alec" even after I became their scoutmaster; but I cannot recall an instance of disrespect to their leader. In the intimacy of camp and long walks through the country, where conventional rules of etiquette are not always applicable and one inevitably reveals himself as refined or crude, I grew to love some of these boys like sons or younger brothers; all were older than the younger of my two brothers. For years I have heard nothing of any of them, and I sometimes wonder whether any were caught up in the maelstrom of the Second World War, and whether any seed that I tried to plant in their young minds took root and grew. Would I be glad to meet now, as fathers and grandfathers, boys who were so bright and promising? Have they been loyal to the promise of their boyhood, or have they betrayed it?

Sometimes I led the troop from the scout house to Gwynn's Falls, thence upward along its course to Red Run, which we followed as far as Gold Hill; and often, too, I took this walk alone. After crossing the railroad and passing over a belt of swampy ground on a boardwalk, one entered a low, riverside meadow, where Pin Oaks and other trees grew in open stand. Here, one bright, mild morning in early December, I found two young Red-headed Woodpeckers, still with the gray heads of immaturity, behaving in a manner new to me. They flew from tree to tree, pecking idly at the bark without appearing to eat. Finally, one descended to the wet ground beneath an oak tree, near a Yellow-shafted Flicker who was feeding there, and arose with something in its bill. Going to a fence post, it deposited the object in a crevice in the split chestnut wood, driving it in with a few blows of its bill. Then, espying a small insect flying in the balmy air, it flew up like a flycatcher to seize it on the wing.

After the bird's departure, I examined the post and found fourteen halves of the small acorns of the Pin Oak, split so that they would fit into the narrow fissure, and a single dried, black field cricket. Some of the acorn pieces were still covered by half of the prettily striped, glossy shell; others were naked; most were fresh but a few were decaying. Presently the woodpecker descended once more to the ground, picked up an earthworm, and wedged it in the cleft top of another fence post. Here I found half of a bruised worm wriggling helplessly in the vise where it had been deposited.

One of these woodpeckers then flew into an oak tree and, after pecking over the branches for a while, inserted its bill sideways beneath a flake of bark and withdrew a piece of acorn that it had evidently deposited there. It carried this to a different fence post and inserted it into a crack. This led me to examine all the posts in the fence line, and I found liberal stores of acorns in seven of them. The larger crevices contained whole nuts in their shells, the narrower ones split acorns or fragments. Many of the acorns had fallen so far down into deep fissures that the birds could not retrieve them later. In clefts in the posts I also discovered three living earthworms that had evidently been placed there earlier that same morning. While I examined the posts, the same woodpecker flew to one of them with a whole acorn and hammered it into a crack. On this visit, it removed some fragments from one crevice and transferred them to another in the same post, evidently for greater security. These stores were laid up by at least two young woodpeckers, one of whom could be distinguished by a few red feathers coming into its gray head. They made no protest when I examined their caches in full view of them.

In this same grove, a week later, I found one gray-headed woodpecker and another whose head was almost wholly red. Possibly the latter was the bird whose head was just beginning to redden eight days earlier, but more probably it was a different individual. For two hours, I watched this bird while it stuck acorns into cavities and decaying ends of stubs well up in trees, instead of in the fence posts. After storing three acorns that it gathered from the ground, it placed a fourth in a crevice on the upper side of a branch to hold it firm while it split the nut with blows of its bill and ate the meat. Then it broke a piece of bark from a tree and hammered it into a cavity where, I believe, it had earlier deposited acorns.

A month later, toward the end of January, two woodpeckers, one still gray on the head and one bright red, were busy with their stores in this grove. One tried to stick a splinter of bark into the end of a stub, found that it would not fit and took it to another stub and left it there. Later, this woodpecker broke a fragment of decayed wood from one tree and forced it into a crevice in the side of another tree. Unfortunately, I could not reach the places where the woodpecker left the bits of wood and bark; but others have described how, after filling a cavity with acorns, beech nuts, or insects, the Red-headed Woodpecker seals the opening with pieces of wood or bark, rammed in so hard that they are sometimes difficult to remove. As far as I saw, the woodpecker with nearly adult plumage sealed his stores but the apparently younger bird failed to do so. Perhaps the instinct to store awakens sooner than the instinct to conceal the stores from possible pilferers.

With an abundant supply of acorns and their skill in catching insects in mid-air, whenever they flew on mild days, these woodpeckers had no difficulty satisfying their hunger even in mid-winter, and they seemed to enjoy much leisure. By mid-morning, they were pecking lazily over the trees, shifting their stores from place to place in an apparently aimless fashion, or just loafing. How different from the little Downy Woodpecker, who spent the short winter days endlessly pecking, pecking, pecking, as though he could never find enough to eat! The Yellow-bellied Sapsucker had a better solution to the food problem: in a Mockernut Hickory in the grove where the Red-heads lived, a sapsucker had made a horizontal row of eight little holes in the bark, from which, even in January, sap flowed freely for them to lick up. Of the five kinds of woodpeckers that I found in this grove in mid-winter, the most handsome was the Red-bellied. When a male with a splendid red crown and boldly black-and-white barred back arrived, the more mature of the Red-headed Woodpeckers drove him unresistingly away.

The Acorn Woodpecker has a quite different and, I believe, unique method of laying up its acorns. In California, Mexico, the pinelands of Honduras, and doubtless elsewhere in its wide range that extends to Colombia, it drills in the thick bark of a tree, a dead trunk, or a service pole, a hole just big enough to hold a single acorn snugly. A large trunk may be covered with hundreds or thousands of acorns, each stored singly in its own niche. But such is not the invariable practice of this jolly, sociable woodpecker: in the mountains of Guatemala and Costa Rica, I have watched it inserting whole or fragmented acorns into clefts and crannies of trees, or beneath lichens, without carving special receptacles for them, just as the Red-headed Woodpeckers did. They also shifted their supplies from place to place in the same seemingly aimless fashion; but, as far as I saw, they did not try to seal them in. In these countries, I never saw Acorn Woodpeckers store their acorns in the manner so well known in California. Other kinds of woodpeckers, including the Red-bellied, the Hairy, and the Lewis in North America, and the Great Spotted Woodpecker in Europe, store various kinds of food at least occasionally; but in the tropics, where fresh food is more abundant throughout the year, I have known only the Acorn Woodpecker to do so.

On a morning when I watched the Red-headed Woodpeckers harvesting acorns, I saw a male White-breasted Nuthatch pick some small edible object from a cleft in the bark, carry it briefly, and, evidently not being hungry, tuck it into another crevice. Then he broke off loose fragments of bark and stuffed them into the cranny to cover his cache. And on the same morning, a Blue Jay deposited an acorn on the ground and carefully covered it with dead leaves. The woodpeckers, the jays and crows, the titmice, and the nuthatches are the four avian families that seem most frequently to practice food storage; but the New Zealand Robin is also said to do so. Even the nestlings of crows and jays try to hide in their nest unwanted bits of food that their parents have brought to them, all quite instinctively, without any adult example. Among mammals, the vegetarian rodents are the greatest harvesters and food storers.

A short distance above the grove where the woodpeckers stored acorns, Red Run flows into Gywnn's Falls. Along the valley of this small stream ran a pleasant lane that we often followed. Near the bridge that carried the road across the brook was a steep hillside where shrubby Mountain Laurel grew densely beneath small trees. In winter, when deciduous woods are bare and brown, this heath retains its bright green leaves, which reflect in a myriad gleams the mild sunshine that falls upon their glossy faces through the naked boughs above, lending a festive touch to the sere landscape. I was happy to discover how the laurel's resting leaf buds are protected from winter's harshness. They are not enclosed in overlapping scales, like the winter buds of most northern trees and shrubs, nor are they surrounded, until the old leaves fall, by the hollowed base of the petiole in whose axil they arise, as in the sumac. But the base of the Mountain Laurel's petiole bends upward against the stem, from which a cushion of tissue grows out to press against it. Snugly enclosed between stem and petiole, the naked bud is shielded from winter's blasts; as though one tried to keep his fingers warm by placing them in his armpits and pressing his arms close against his sides. Lacking this protection, the apical bud of each shoot does not even survive the summer in which it is formed; by autumn it has withered away, so that next spring the growth of the branch must be continued by a lateral bud.

Upward along the Run, we came to a patch of low, wet ground where Skunk Cabbage flourished in a copse of young trees. In the tropics, I have known many aroids, some growing profusely, high on great trees and others standing erect, with thick stems higher than my head; but none that I have seen has a more handsome spathe than this aberrant member of a heat-loving family. It may be that the inflorescences of the tropical aroids do not win such admiration because they must at all times compete for attention with an exuberant flora, whereas our pungent cabbage blooms at a season when any floral display excites delighted surprise; but I believe that even in competition with its tropical cousins, it would win a prize for elegance.

As early as January, in a mild winter, the Skunk Cabbage's hooded spathes push up among brown, dead leaves, or perhaps beside the bright emerald of a patch of delicate fern moss, or near an expanse of gray-green bog moss, close by some purling, ice-rimmed woodland brook. They display the hues of the choicest hardwoods. Here is one beautifully streaked and mottled with cherry color and green, inside and out, its polished surface showing every detail to perfection. Nearby grows another almost wholly green, with only scattered splashes of cherry. Over there is a hooded shell that seems to have been carved by a master hand from Honduran mahogany, then stained and polished with loving care. Each is a little woodland shrine, such as in foreign lands peasants erect to the Virgin or their patron saint, whose offering is the golden pollen liberally shed and contrasting beautifully with the deep tints of the hood's smooth interior—the year's first offering to the goddess of fertility.

Beside each spathe stands the foliage bud, the furled leaves already peeping forth from the pointed tips of the protecting scales, waiting for the first breath of spring to spread their broad, ribbed blades. Unlike some people, I do not find the odor of these leaves offensive. It is a wholesome, pungent, vital aroma, expressive of the robust constitution of a plant that often blooms amid winter's snow, with no resemblance to the sickly odors of corruption.

Higher up the valley, a north-facing cliff of ancient gneiss rose close beside the road, its rough, shaded face overgrown with Common Polypody Fern and some of the most charming of mosses. Most abundant was the Apple Moss, named *Bartramia* for one of North America's pioneer botanists. Its plump little globose capsules are asymmetric or lopsided; and in early May, when they were ripening, I noticed that all had grown more on the side toward the cliff, so that the mouth of each receptacle, through which the spores would soon be expelled, was directed outward, toward the free air. Only the capsules of some moss plants that grew vertically downward beneath outjutting ledges were turned inward. Here and there amid the Apple Moss grew the odd little *Webera*, with flattened capsules resembling grains of wheat, on such short stalks that they seemed to lie upon the tiny green leaves. These capsules, too, all pointed outward toward the light. It is reported that the first drops of a summer shower, striking the flat top of the still-dry capsule, send the spores through the nozzle-like opening in little puffs, to be borne away by gusts of wind. Perhaps the footfalls of small insects crawling over the moss have the same effect. On shady, moist ground at the foot of the cliff, a third moss, the Nodding Bryum, formed a verdant carpet; and its slender, long-stalked capsules, too, all bowed northward, toward the light.

At a large old farmhouse, the lane ended. Above this point, the valley, narrowing between wooded slopes, inclined northward. Pushing upward along the Run through catbriar and other tangled growth, we finally reached the serpentine barrens called "Soldiers' Delight," possibly because at some early date troops had camped in this open area. Here the low, rounded hills supported a vegetation quite different from that of all the surrounding region. The chief cover of the rough, stony ground was a Beard Grass that grew in tussocks, tinting the harsh slopes green in spring and early summer, brown in winter. The low trees that grew scattered in this inhospitable soil were chiefly Scrub Pine, stunted oaks, and junipers. Amid the tussocks of grass were modest flowering herbs peculiar to this sterile soil, or at least more abundant here than on the more fertile surrounding land. In midsummer, I found here Green Milkweed, Whorled Milkwort, and Pineweed, a wiry St. John's-wort with scale-like leaves and tiny, yellow flowers. In autumn, Rough-stemmed Goldenrod, Knotweed, and purple-flowered Spiked Blazing Star flourished on the stony slopes. But the most peculiar plant on the serpentine was *Talinum teretifolium*, a dwarf purslane that no one seems to have named in his native tongue, which grew nowhere else in the vicinity. In midsummer, it held small, rose-colored blossoms above leaves that resembled succulent pine needles. This modest herb was as resistant to desiccation as a desert plant. Placed in my study before a window with a southern exposure, it survived for months without a drop of water, even producing tiny new leaves.

An ardent naturalist might continue all his life to meet new and beautiful living forms in his home county. In my first year at college, I bought a used copy of *Gray's New Manual of Botany;* and every year, from spring to fall, I tried to learn the names of every flowering plant and fern that I could find on my usually solitary rambles through the country. In my last winter among the hills of Maryland, when I taught botany at Johns Hopkins during Professor Johnson's absence, I found unsuspected treasures of delicate loveliness on ground that I had trodden many times before. The winter woods were never dreary when benumbed fingers could pluck such daintiness from the frozen soil. This was the season to gather delicate liverworts from moist hollows in the woods and from spongy decaying logs in swamps. The smallest of the plants had to be carried back to the laboratory and viewed through a microscope to appreciate their varied shapes.

Even when the dominant vegetation was deep in winter sleep, these humbler, more primitive plants were busily preparing to propagate their kind, taking advantage of the light that reached them while trees were leafless. Some of these tiny hepatics bore, on the upturned ends of their slender stems, one- or two-celled gemmules, each capable of producing a new plant. Others were maturing spores in capsules carefully shielded from drying winds. Beside a meandering woodland rivulet, one January day, I found the hepatic *Calypogeia Sullivantii,* a delicate growth with minute, two-toothed leaves. Its young, green spore cases were protected in thick pouches that had pushed down into the ground beneath the parent stem and anchored themselves there by sending long, thread-like rhizoids into the surrounding soil. In these subterranean pockets, the spore cases stood upright, ready to stretch up into the air and shed their myriad spores at the first touch of spring's warmth. Other liverworts bore their spore capsules above ground, in little puckered pouches, tied at the mouth and almost bursting with their contents.

As I wandered along a wooded ravine, beside a clear brook, a pleasant, spicy fragrance sometimes betrayed the presence of the most robust of the local liverworts. As I followed the scent, I would find, on the rocky or earthen bank, carpets of the broad, branching, deep green ribbons of *Conocephalum conicum*. After the first inspiration, or at best the second, the evanescent aroma vanishes. I stoop and press my nose against the luxuriant growth of the liverwort, but the fragrance will not return; it is too exquisite a sensation, suggestive of mossy banks and dripping rocks overgrown with ferns, long to endure. I pluck a bit of the plant, crush it between my fingers, and perceive a strong, spicy fragrance, not unpleasant, but quite different from that which I desire to renew. Yet perhaps it is caused by the same substance in higher concentration; I recall that the faintest trace of formaldehyde is agreeably aromatic, far different from the choking pungency of a strong whiff of the corrosive gas. So subtle is the fragrance which this hepatic diffuses through the still air of a wooded dell that some of my students are unable to detect it.

All my life, my allegiance has been divided between plumage and foliage, sometimes birds, sometimes plants, claiming most of my attention. After we moved from the country to the suburbs, I built a small pigeon house and stocked it with long-legged white Maltese, rich chestnut Carneaux, and a few pigeons of the common variety. As the enclosed birds multiplied, to feed them became too costly; and, moreover, my interests were changing. Finally, I gave my pigeons to a neighbor's boy, whose pigeons lived in a spacious loft and flew free to seek food. I converted my pigeon house into a carpenter shop. As I grew older, free animals leading their natural lives, without human interference, drew me away from household pets and other dependent creatures; until today I am reluctant to make a pensioner of any animal, and have no non-human dependents except horses at large in a wide pasture and a few chickens roaming free.

Of the birds that nested around the home of my later boyhood and early manhood, a pair of Gray Catbirds received the most careful study. In mid-May, I found their nest amid the slender, thorny shoots of the barberry hedge. The outer shell of small sticks was partly lined with dead leaves and scraps of paper. Later, some long pieces of string were coiled into the interior, and a lining of fibrous rootlets completed the structure. In it three pretty greenish-blue eggs were laid on successive days, by a female who could be distinguished from her mate by a slightly deformed bill. In most pairs of Catbirds, the sexes can be distinguished only by their voices.

Sitting concealed in a blind, I watched the nest all of one morning and all of the following afternoon. During the thirteen hours, the female alone incubated the three eggs. Although she left her nest twenty-five times, she was absent for only three to twelve minutes at a stretch, and she kept her eggs covered for seventy-eight per cent of the day, a good record of constancy. While she warmed her eggs, her mate sang from an ash tree some yards away, where he could look over the nest. His song was subdued and sweet, seldom marred by the catcalls and other harsh, grating notes that Catbirds too frequently interject into their prolonged medley. Whenever the songster saw his mate leave the eggs for food, he advanced from the ash tree to a hawthorn nearer the nest, to guard it faithfully until, her hunger satisfied, she returned. Then he flew to a more distant perch and resumed his singing.

Of the many birds that I have studied, these Catbirds were boldest in the defense of their progeny. If I touched the nest while the male was guarding, he hopped close around me with spread tail and drooping, quivering wings, mewing loudly. His more valiant mate struck my offending hand with her feet, while she complained with loud, whining notes.

After thirteen days of incubation the nestlings hatched, and the male joined his mate in bringing caterpillars, spiders, miscellaneous insects, and berries to stuff into their insatiable yellow mouths; but only the female brooded them. Now the parents defended their family with more intense zeal. When I placed a hand on the nest, the female not only struck it with her feet but pecked it fiercely with her deformed bill. Once, to see how long she would continue, I left my hand over her nestlings until she alighted on the back and showered it with blows that drew a little blood. Meanwhile, her mate buffeted the back of my head. The mother bird seemed to understand how the human hand works. When I held my palm upward, she would not stand upon it, as she did upon the back, but delivered her pecks from the side, where she was in less danger of being caught by the sudden closure of my fingers.

In certain other situations, these zealous parents showed less understanding. To see what they would do, I covered the nest with a large dock leaf during their absence, then disappeared into my blind. Arriving with food for their nestlings, the Catbirds hopped all around the nest, baffled. Finally, the female tugged at the leaf with her crooked bill, shifting it about an inch; but neither she nor her mate made a sustained effort to remove it, a feat not beyond their strength. They did not even try to uncover their nestlings when a white handkerchief was laid over them; but the female managed to slip a raspberry to one of them when, rising up eagerly for food, they pushed up its edge. A pair of White-breasted Blue Mockingbirds in the Guatemalan highlands, to whom I gave the same "intelligence tests," did much better, completely removing both a leaf and a handkerchief that I had placed over their nestlings. Yet they were far more timid than the Catbirds, remaining at a respectful distance while I was near their nest. I admired the mockingbirds' heads and the Catbirds' hearts.

The mother Catbird brooded her nestlings every night until the youngest was eight days old and fairly well feathered. On their final two nights in the nest, the young slept without a coverlet. When they left at the age of eleven days, they could barely fly. In the fall, the Catbirds would travel southward, perhaps as far as Central America, whither I, too, would migrate before long. Sometimes I heard their nasal mew issuing from the impenetrable thicket that covered an abandoned banana plantation in the Caribbean lowlands of that great isthmus, or from a dense stand of giant herbs with leaves large enough to serve as an umbrella for one caught in a sudden tropical shower— riotous growths utterly different from the trim suburban gardens where many of these birds were hatched. Sometimes, by squeaking with my lips applied to the back of my hand, in the manner familiar to bird-watchers, I could draw the inquisitive Catbird to the edge of the thicket, where we could see each other. Peering out, he vibrates his wings, flirts his tail, pivots from side to side, while he regards me with bright, inquisitive eyes. His curiosity satisfied, he retires once more into the densely tangled thicket, where I cannot follow, to consort with untravelled tropical birds, who rarely respond to squeaking as he does.

It was not easy to break away from the green hills and vales where I learned to love nature, from family and friends, books and cherished associations. But far to the south grander forests, brighter and more varied birds, a thousand secrets of nature waiting to be uncovered by dedicated naturalists, were calling with voices too imperative to be resisted. Go I must, no matter how hard the parting. But I did not sever ties all at a single stroke. Each successive visit to the tropics was longer; each return to a rapidly changing homeland found it more altered and foreign; until finally I made a home in a tropical valley, and many years passed between journeys north. But before this happened, I took other journeys about which I wish to tell.

Clustered shoots of the Sea Lungwort (*Mertensia maritima*) emerging from a shingle beach in early June.

Buried, dissected stems of the Sea Lungwort (*Mertensia maritima*), a mass of interconnected strands bearing dormant buds in large knots. Right: Bases of young shoots of the present season. Left: Base of the dissected taproot.

3

The Rockbound Coast of Maine

In the summer following my first year in college, I served as a junior counsellor at a boys' camp in central Vermont. My group consisted of boys under twelve; my duties, of a little elementary teaching, helping to direct games, accompanying the boys on long walks through the surrounding hills, and guarding them in the water. On free days, I took long excursions with an older counselor, walking or "hitchhiking"—begging rides from passing motorists. Solitary drivers often carried us for long distances, seeming glad to have somebody to talk to.

On this first visit to New England, I became familiar with its rugged hills, its quiet villages with white churches of simple dignity, a few of its historic spots, including Fort Ticonderoga, and some of its industries. Among the most impressive sights was a marble quarry, or perhaps more properly mine, near Rutland. As the excavation deepened, it expanded stepwise in all directions, with the result that the bottom was much broader than the nar-

Headpiece: Seaweeds, chiefly wrack (*Fucus vesiculosus*), growing on intertidal rocks at Otter Cliffs at the end of March. Ice still covers the cliff at upper right.

row opening at the top. It seemed that one of the great pyramids of Egypt
might have fitted into the immense empty chamber with gleaming white
walls. When the camp closed, we hitchhiked to Boston and thence to New
York. My conventional mother was shocked when she learned that I had
visited Harvard University and museums in the intellectual city of Boston,
where she had relatives, in a sweater and khaki trousers instead of a proper
suit. This was the last time I hitchhiked, as I did not like asking strangers for
rides.

The following spring, Professor Johnson invited me to spend the summer
with him at the Mount Desert Island Biological Laboratory at Salisbury
Cove, on the northern shore of that picturesque island on the coast of Maine.
In late June, 1923, I went by rail to Boston, where I met him, his son
Duncan, and another student, for an excursion to Mt. Washington in his car.
The climb by woodland trails was easy enough, until we reached the elfin
forest or *Krummholz* just below timberline, a belt of dwarf trees with gnarled
trunks and twisted, interlocking branches, the tortured vanguard of the
coniferous woods on a cold, gale-swept, northern summit. As we fought our
way upward through this nearly impenetrable barrier, the stiff boughs
tugged at our clothes, threatening to tear them from our bodies. My tightly
wound spiral leggins persisted in coming loose and trailing far behind me. At
last we escaped from the uppermost stunted trees onto the rocky slopes near
the summit, where the magnificent view over the Presidential Range com-
peted for my attention with the charming alpine-arctic flora at my feet. As
soon as we had deposited our packs at the rest house, I took my *Gray's
Manual* from my knapsack and, with the professor's help, started to identify
some of the hardy little plants then just coming into bloom, while a great
mass of snow still clung to a scarped mountainside.

From the White Mountains, we proceeded in Johnson's car to the coast of
Maine, arriving at the laboratory before it was open. Until the tent in which I
was to sleep that summer became available, I spread my bed pack on a patch
of grass beside the cove and slept under the stars, found a little food at the
nearest country store, and passed the days exploring in solitude a country
quite different from that in which I grew up.

Within a week, the laboratory became active, and the professor turned his
attention to the research that I was to do under his direction. At the northern
end of the island, about the mouth of Northeast Creek, were wide salt
meadows intersected by tidal channels. Here we laid off a plot for an ecologi-
cal study of the vegetation; but before this was well started, the scene shifted
to the exposed southern coast of Mt. Desert. As we approached Otter Creek
Point through the spruce woods, I became aware of a dull roar, such as I
could only attribute to an express train; yet I knew that no railroad reached
the island. Maryland's shores are sandy beaches; I had never before been

near a rockbound coast; and I remained bewildered until the professor explained that the sound we heard was made by waves breaking against the cliffs.

Soon we emerged from the woods upon a high promontory that overlooked a vast expanse of the heaving Atlantic. Not far offshore, a bell buoy tolled as it rocked on the swell. At the foot of the cliffs lay the rocky shore, with irregular granite ledges against which the waves were breaking into floods of foaming water, a patch of great tumbled boulders, and here and there, nestled amid the rocks, pools of all sizes and shapes, isolated fragments of the vast ocean calmly reflecting the blue sky, while the tide that had filled them was low.

Through a chute-like break in the vertical cliffs, we scrambled down to the shore, and with daubs of paint on the rocks, marked off the plot that we would study intensively. With characteristic energy, the professor worked at a pace that, in my ignorance of research, I felt sure would finish in a fortnight a project that was to occupy me for three summers, while he continued to make observations for two additional summers. Our object was to study the distribution of the littoral vegetation in relation to tide levels and to discover the factors that limited the upward or downward extension of each species. We began by making a map of the hundred and fifty feet of shoreline that we had chosen for the study, from the low-water mark to the top of the cliffs that rose fifty feet or more above the ocean.

Then it was necessary to determine the elevation of all the tide pools and salient points of rock. Before this could be done, we had to establish the mean low-water mark. With considerable difficulty, we attached to wave-beaten ledges a tide stake, a long strip of wood boldly marked with feet and tenths, in black paint on a white background. Then I and a helper from the laboratory slept by turns on the springy Crowberry at the head of the cliffs, while we watched, through a lunar day of twenty-five hours, the spring tide rise and fall twice through nearly fourteen feet, and recorded, as well as we could, the time when the heaving surface of the sea reached each mark on the tall tide stake. Later, I spent a day and a night alone, making a record of the tide in the quieter water of the cove beside the laboratory. The tide stake was set well out from the shore; and to read it in the dark night I anchored my rowboat close beside it. About midnight, the incoming tide drifted a large raft of seaweed close by the little boat. How huge and menacing it looked, how like a nameless sea monster, as it floated out of the obscurity into the lantern's dim light!

Making the map and marking the elevations was a task for two, one to hold a graduated pole and the other to sight with a spirit level from the bench mark that we established on the highest point of the outer ledge. When this was done, I spent long, solitary days on this rugged coast, becoming familiar with the beautiful seaweeds that flourished there, recording their positions on the exposed ledges or in the tide pools, taking the temperature of the pools and the open sea, determining the acidity or alkalinity of the water and of the soil at the top of the cliffs, tending the water-filled bulbs of porous clay, called atmometers, that measured evaporation.

I had to be careful not to slip on rocks coated with slimy seaweed, nor to be caught on an outer ledge by the incoming tide. Once, indeed, I was nearly marooned by swiftly rising water, and with difficulty and not a little danger made my way through the surf to the inner shore. A morning of scrambling over the uneven rocks developed a sharp appetite for my lunch of sandwiches, which I ate at the top of the cliffs, reclining on a thick mat of the hardy Crowberry, amid Bunchberry and recumbent junipers, and gazing out at the empty sea. At long intervals, the little coastal steamer that plied between Rockland and Bar Harbor passed just beyond the tolling bell buoy that marked the reef; and more rarely a larger vessel, looking ghostly and unsubstantial in the distance, glided by with much of its hull below the far horizon. In the evening, I climbed up through the spruce forest to meet the car sent from the laboratory to take me back.

The mean range of the tide on this coast was a little over ten feet, much less than it became farther to the northeast, toward the Bay of Fundy, where the tide has an immense range. However, on our shore the splash of the waves and storm-driven spray made the sea's influence strongly felt at the top of the high cliffs. Here we met in miniature a problem that, in later years, I encountered in tropical lands with mountains thousands of feet high, the delimitation of vertical life zones. In countries that rise from lush tropical lowlands to peaks capped with perpetual snow, the upward or downward extension of vegetable and animal species is determined primarily by their tolerance of heat or cold, although rainfall strongly influences their altitudinal as well as their horizontal distribution.

On this rocky coast, the upward limit of certain seaweeds was set chiefly by their resistance to drying on sun-warmed rocks while tide was low. The lowest point at which certain others could thrive seemed to be determined by their need of light, of which they received less the farther below the high-tide mark they grew, because they were submerged for longer hours each day by greater depths of sea water that intercepted much of the radiant energy. And just as in mountainous regions the life zones are shifted upward or downward by the irregularities of a rugged terrain, so here they were complicated by the broken surface of the littoral rocks. Plants intolerant of drying flourished higher in shady nooks and along the sluices that drained the tide pools than on exposed bosses; while those that needed much light grew lower in exposed places than in crannies. Only on a shore, or a mountainside, that formed a perfectly even inclined plane could we expect to find life zones running perfectly parallel to the contour lines, without vertical irregularities.

On the exposed, rocky headland where we worked, the ceaseless action of the waves shifted all the vegetational belts about two feet higher than they would be on a sheltered coast. From as far downward as we could see when the waves receded at low tide up to about two feet above mean low water, the dominant plant was the kelp *Alaria esculenta*. Although intolerant of drying, this large brown alga was amazingly resistant to the pounding of the surf. Attached to solid rock by gripping holdfasts, the fronds, sometimes twelve feet long by half a foot wide, writhed and twisted as tons of surging water tugged at them, but mostly they held firm. Their smooth, thin blades were torn from margin to midrib by this perpetual lashing, as the broader blades of a banana leaf are frayed by wind. Winter ice seemed to be the great enemy of these sturdy seaweeds. On a visit to Otter Cliffs at the end of March, when sheets of snow and ice still covered much of the vertical cliff, I found only the decapitated stumps and holdfasts of the kelps that grew highest on the ledges; their blades had apparently been ground off by blocks of floating ice hurled against them by the waves.

Between the holdfasts of the kelps, the coralline alga *Melobesia lenormandi* formed a beautiful, pink, calcareous encrustation over the rocks. Intolerant of drying, it hardly grew above mean low water except in the tide pools, but it extended downward into the depths as far as we could see. Also abundant in this lowest or Sublittoral belt was the coarsely branched, filamentous red alga *Halosaccion ramentaceum*, which made miniature thickets, five or six inches high, over the rocks exposed when the tide was lowest. Among these little bushes grew several other red and brown algae and thin green sheets of sea lettuce, *Ulva lactuca*.

The next higher belt, which extended from about two to seven feet above mean low water and was designated the Lower Littoral, supported a rich variety of brown, red, and green algae, all of which were tolerant of a moderate amount of exposure, up to about seven or eight hours between high tides. Most abundant of all was the red alga *Porphyra umbilicalis*, which grew profusely not only on steeply sloping rocks but also occasionally on barnacles, and even as an epiphyte on other large algae. During low tide on bright days, the thin, overlapping fronds of this seaweed dried into tough, papery layers, much like sheets of dried gelatin, which stuck closely to the surface of the rock and to barnacles. When wetted by the first lapping waves of the rising tide, they quickly became soft and flexible and separated from each other, the better to withstand the pounding of the heavier waves that soon poured over them. Prominent among the dozen other algae that grew chiefly in this belt were the wrack *Fucus furcatus;* the spongy, felted tufts of the green alga *Spongomorpha spinescens*, which seemed never to dry completely; and *Ralfsia verrucosa*, which formed brown encrustations over intertidal rocks, as the pink coralline *Melobesia* did on lower rock surfaces that were never long exposed.

The next belt that we distinguished, the Upper Littoral, extended from seven feet above mean low water up to fourteen feet, which was about two feet above the level of the highest spring tides. Unless they grew in deeply shaded nooks or in the tide pools, the plants in this zone were, on each clear day, exposed for long hours to sunshine and wind that might dry them rather thoroughly. Most prominent in this belt were the large brown sea wracks, *Fucus vesiculosus* and *Ascophyllum nodosum*, in whose branching fronds were little air bladders that kept the anchored plants floating upright when submerged. Yet it was not these sturdy seaweeds with fairly thick tissues that grew on the most exposed rocks, but rather the little, club-shaped, unicellular green alga *Codiolum longipes*, which in summer formed continuous velvety swards on rounded, seaward-facing bosses of rock. When wet, these small, crowded algae made the smooth rocks treacherously slippery; but on a sunny day they quickly dried after the receding tide left them exposed, covering the rock face with a clinging green skin mottled with white. A much darker encrustation, blackish-green in color, was composed of crowded threads of the blue-green alga *Calothrix scopulorum*, which often covered square yards of the high, exposed intertidal rocks. On these same rocks, little patches, or sometimes more extensive sheets, of black revealed the presence of *Verrucaria striatula*, one of the few lichens that can withstand prolonged contact with seawater.

On a shore composed of rough ledges or broken strata of rocks, the receding tide does not drain off all the seawater, but much remains behind in depressions. The area that we mapped contained about two dozen of these tide pools, ranging in size from a miniature lake, twenty yards long by twelve yards wide, down to tiny pools that could be covered by a handkerchief. In some, the water stood several feet deep. Isolated by the recession of the sea, each of these pools became, temporarily, a microcosm, a little marine world with its own peculiar flora and fauna, the composition of which was determined by the size and depth of the pool, its height and the time it was separated from the ocean while the tide was low, and, to a very large extent, by the degree to which it was shaded from strong sunshine by the surrounding masses of rock. The sunny and shaded sides of the same pool contained different kinds of seaweeds. On bright days, the water in shallow, exposed pools, lying among naked expanses of heated stone, might become as much as twenty degrees Fahrenheit warmer than that of the ocean, which here was usually at about fifty degrees—much too cold for comfortable swimming—even in midsummer. As the rising tide poured into each pool, its temperature quickly fell to that of the open sea. Such great and sudden fluctuations must have prevented many kinds of delicately adjusted organisms from flourishing in the more strongly insolated pools.

For the study of the varied life of the ocean's margin, these wave-relinquished pools offer special advantages. Here, where the seaweeds were neither constantly tossed by the surf nor lying flaccid against drying rocks, their varied and often beautiful shapes, their colors, ranging from bright green through red of various shades to deep brown, were most favorably displayed. Nearly every species of alga that grew in our area was to be found in one or another of the pools; only some half-dozen kinds that appeared to be absolutely intolerant of constant submergence, such as *Codiolum* and *Calothrix*, were always absent.

In the lower pools, and even in high ones that were well shaded, grew algae that required nearly continuous submergence, so that elsewhere they were not found above the reach of moderate waves when the tide was lowest. Among these were *Alaria* and two other kinds of kelps, the encrusting pink *Melobesia*, and another calcareous alga, *Corallina officinalis*, a little pink bush with short, cylindrical branches. Other seaweeds, tolerant of continuous submergence as well as of more or less drying, grew in tide pools well above the upper limit of their occurrence on exposed rocks. Still other algae, including *Fucus, Ascophyllum,* and *Porphyra,* flourished both toward the upper edge of the intertidal zone and in tide pools at the same level. Finally, we found certain seaweeds only in the tide pools. One of the most interesting of these was the brown alga *Leathesia difformis,* which in well-lighted pools grew in the form of little, hollow, irregularly lobed spheres, attached directly to the rock or to larger algae.

Epiphytes, which perch upon other plants without drawing nourishment from their living tissues, are usually associated in our minds with the tropics; yet this habit of growth is adopted by many algae of cold waters. Some, such as the small brown alga *Elachistea fucicola,* which we found chiefly on *Fucus furcatus,* are as consistently epiphytic as many a tropical orchid or aroid, and they seem to be restricted to certain hosts, as few epiphytic seed plants are.

The sides and bottoms of favorably situated tide pools were everywhere densely covered with seaweeds. Amid this luxuriant, colorful algal vegetation lived a variety of small invertebrate animals, including spiny sea urchins, flower-like sea anemones, crabs, small molluscs, and many lesser creatures. About their margins barnacles clustered. The only one of these animals that I found time to study was the amphipod *Amphithöe rubricata,* a crustacean widespread, at higher latitudes, on both the American and European shores of the North Atlantic. Somewhat under an inch long, it is ruddy brown to olive green in color, with a white spot on the dorsal surface of each abdominal segment.

In July, the bottom of our largest tide pool was covered by a thick green turf of *Spongomorpha hystrix*. In addition to its clustered, coarse, richly branched, upright filaments, this seaweed produced more slender filaments, which traced a tortuous downward course through the principal branches, until they reached the supporting rock and attached themselves to it. Bound together by these rhizoidal threads, each *Spongomorpha* plant was a compact, spongy bush, firmly anchored to the rock, and well able to withstand the tugging of the waves that twice daily swept over the tide pool.

In these tufted green algae the amphipods built their nests. Each was a little tube, up to two inches long by about a quarter inch in diameter, open at both ends. The walls of the tube were composed of fine, colorless, silken threads, applied so thickly to the surrounding green filaments that they formed a continuous sheet. In the laboratory, I watched amphipods make these nests. If a number of the animals were placed in a dish of seawater without anything else, they all clung together in a tight knot, as though trying to cover themselves with their neighbors' bodies, although in their natural habitat they are by no means gregarious. They also attached themselves firmly to bits of seaweed or any other solid object they found in their dish. When a tuft of *Spongomorpha* was available to the amphipods in the dish, they took refuge among its branches; and each almost immediately set to work constructing a tube.

Their silk is secreted by sac-like glands, one situated at the end of each of the first two pairs of periopods (the slender legs behind the pincers) and opening by a fine pore at the very tip of the leg. The four gland-tipped legs move back and forth among the filaments, with a motion suggestive of a piano-player's hands, secreting the silk and attaching it to the alga at various points. Before long, it is evident that the branches of the *Spongomorpha* are bound together by a web of colorless fibers that encircles the spinner. After working for a while in one position, the crustacean turns a somersault and continues with its head at the opposite end of the tube. At intervals, it pivots sideward. By these movements, it applies the silk in a uniform sheet throughout the length of the tube. I watched amphipods continue spinning for an hour at a time. After twelve hours, they had recognizable nests; but several days were needed to complete a well-lined tube.

Although a single large tuft of *Spongomorpha* might contain as many as eight nests, each tube was usually inhabited by a single grown amphipod, and never by more than two. When two were present, they were nearly always a male and a female. Since I found females nearly three times as numerous as the slightly larger males, all the females could not consort with mates simultaneously. Of seventeen nests with two occupants, only one contained two females.

The amphipods' breeding season continued through July and August; even at the end of the latter month, many females still carried eggs. As in other amphipods, the eggs are laid in the female's brood-sac, which is composed of eight flat, elliptical, skin-like appendages attached to the ventral surface of her thorax between the legs, four on each side. Each flap is margined by long hairs which, becoming entangled with those of adjacent flaps, bind them all together into a pouch beneath the mother's body. In these protecting sacs, the eggs hatch and the young remain until they are colorless, transparent creatures about an eighth of an inch long, already miniatures of the adults, since they undergo no elaborate series of larval transformations. A single brood consists of from fifty to seventy-five young amphipods.

I could not learn what causes the young to leave their mother's pouch; but soon after they do so, she "shows them the door." Twice, in the laboratory, I had the good fortune to witness a female evicting her brood. Coming to the mouth of her tube, she seemed to struggle violently, kicking with her antennae and pincers. On closer examination, I noticed that young amphipods were clinging to these appendages, and she was trying to shake them off. They held on tenaciously, resisting her violent shoves. As far as I could see, the mother used her pincer-legs merely to push, never to seize or injure her offspring. Finally, chiefly with her antennae, she succeeded in shoving a bunch of the resisting youngsters over the nest's rim, leaned outward to give them an additional push with these long appendages, then withdrew into her tube. The evicted young crawled over the branches of the seaweed in which the nest was built.

On the following day, I saw this performance repeated by another amphipod. After pushing about five little ones beyond the confines of her nest, she retreated to the opposite end. A few minutes later, she carelessly stuck her head out of the door by which she had dismissed her offspring, and they attached themselves to her with alacrity. She drew them back into the tube with herself, but after a short interval reappeared at the doorway and again undertook the vigorous process of expulsion.

Young amphipods were first noticed living independently in the tide pools in mid-July. In three days or less after leaving the maternal tube, these tiny creatures start to make their own nests, which resemble those of the adults in everything but size. At first, the new tubes are only about one twenty-fifth of an inch in diameter by an eighth of an inch, or a little more, in length. Sometimes they cluster on the lip of the maternal tube, like small fumaroles about the crater of a volcano—evidence that the young do not go far before beginning to build. I tried to learn whether they occupy a succession of tubes as they mature or if, when possible, they enlarge their first tubes to keep pace with their growing bodies; I found some indication that they do the latter. Their reluctance to leave their tubes would lead them to favor this course.

In favorable conditions, these amphipods eat scarcely anything except fresh seaweeds of various species, often that in which their nest is built; but in less favorable environments, as when their tubes are surrounded by much sand or mud, they may turn carnivorous, eating sponges and small crustacea, including even young amphipods, in addition to decaying vegetation.

Since in the tide pools the amphipods were never seen outside their tubes, and indeed can hardly be forced to leave without injury, I wondered how they procured their food. Holding a shred of seaweed in a forceps at the mouth of a nest, when the tide was low, elicited no response from the inhabitant, who usually rested far down in the tube. But if I gently swayed the tuft of *Spongomorpha*, the amphipod would sometimes come to its doorway and wave its antennae in the water. If now the bit of alga were proffered, the animal would seize it in its pincers and take it into the nest. In the laboratory, a piece of seaweed presented in this fashion was sometimes held so tightly by the amphipod that with the forceps I could draw it and the whole alga in which it lived over the bottom of the dish. By quickly with-drawing the fragment of seaweed that the animal was about to grasp, I could make it lunge out of its tube, but never so far that its hindmost legs released their hold on the nest's rim; the amphipod could not be enticed to sever contact with its home.

Although I did not succeed in watching the amphipods while waves were breaking over their tide pool, from these observations I concluded that when they do so, swaying the seaweeds back and forth, the amphipod sticks its foreparts out of its doorway to seize fragments of weed that are wafted by. If it catches more of this food than it needs, it fastens the excess to the inside of its tube with silk. Later, some of this stored food may be eaten, along with the thread that covers it. In the absence of such supplies, *Amphithöe* de-vours the alga in which its home is built; indeed, this was the principal food of those that lived in the *Spongomorpha*.

Toward the end of August, the two species of *Spongomorpha* most abun-dant at Otter Cliffs began to die, both in the pools and on the exposed rocks, possibly in consequence of the unusually warm weather of this period. In the large tide pool where so many amphipods dwelt, nearly all the tufts of *Spongomorpha hystrix* became detached from the persisting holdfasts that bound them to the rocks, to float free in the water and be washed out to sea by the next high tide, perhaps along with the amphipods that inhabited them. Many of the animals, however, escaped this fate. Although the disap-pearance of the abundant *Spongomorpha* left large bare patches on the bottoms of the pools, there was no lack of other algae in the biggest tide pool; and now the surviving amphipods built their homes in these other seaweeds, where earlier I never found them. Most of the homeless amphipods chose the thin, flat sheets of the sea lettuce *Ulva* for their new abodes; and here the problem of building was quite different from that in the massed filaments of the *Spongomorpha*. However, the amphipods solved it by finding, or mak-ing, a fold in the paper-thin expanse and lining the interior with silk. Indeed,

Amphithöe is a versatile builder; when no such convenient support as the *Spongomorpha* is available, it is able to construct its tube in a great variety of situations with a great variety of materials, including loose fragments of seaweed or shells, bound together with silk. And, as I proved by experiment, amphipods taken from one type of tube were able to make a quite different type.

Among the few birds that relieved the solitude of my days on the rocky coast were a pair of Black Guillemots, who bobbed on the offshore waves, then flew up to enter a cranny in the higher cliffs to the north. Inland bred, I had never seen the nest of any seabird and was eager to examine that of the guillemots. The ledge beside which their nest was situated, near the top of the high, nearly vertical cliff, was not easy to reach; but I enlisted the support of another young student. We attached a rope to a tree at the top of the cliff, and with its aid I climbed down to the narrow shelf. As my companion followed, I warned him that he was about to set foot upon a pile of loose rubble beneath the rope; but my warning came too late. The jagged fragments tumbled down, and one struck my ankle a sharp, glancing blow, before all crashed noisily upon a lower ledge.

With a smarting ankle, I lay on the narrow shelf between sea and sky, wondering whether I would need a rescue party to haul me up the cliff. After a while, however, the pain abated, and I assured myself that no bone was broken. Arising, I tried to see the guillemots' nest, but it was inaccessibly deep in a fissure in the granite. Then, with my companion's help, I climbed to the cliff's top and hobbled through the spruce woods to our waiting car. In a few days, my ankle healed. I realize now how extremely lucky I was, not to have paid more dearly for my youthful rashness in trying to reach this nest without better preparation and some experience in cliff climbing. At least two seasoned ornithologists (Gustav Kramer and J. Stuart Rowley) have lost their lives attempting to reach nests in cliffs and caves.

The car to which I hobbled was an ancient Overland, recently bought with a hundred dollars that I had received for my twenty-first birthday, which occurred a few weeks before my graduation from college. With this somewhat dilapidated vehicle, and little experience in driving, I undertook the long journey from Baltimore to Mt. Desert Island, in company with my brother Robert. With its aid, I explored the island more widely than I had done in my two preceding summers there, and made some interesting discoveries. At the summer's end, I sold this relic for a little less than it had cost me, as repair bills had become alarming.

Here and there along the exposed southern shore of the island, especially beside the bays and inlets that deeply indented the land, were beaches of shingle—small, water-rounded stones of gray shale, mostly about the size of a walnut but sometimes much bigger, that rattled and grated together as they shifted beneath one's feet. Some of these shingle beaches fringed higher land; others were rounded bars or raised barriers between the sea and a low salt meadow, lagoon, or dammed mouth of a small stream on the landward side. On these shingle beaches I found one of the most extraordinary plants that I have ever seen, the Sea Lungwort, a relative of the Forget-me-not and the Heliotrope. Large plants of the lungwort consisted of two- or three-hundred slender stems, radiating from a central point and lying on the stony beach, which they covered with a dense mat of glaucous green leaves up to seven feet in diameter. In July and August, they were adorned with small forget-me-not-like flowers, white, rose-pink, or blue in color.

Following a habit that I had early formed, I dug into the shingle to examine the plant's subterranean parts. They resembled nothing that I had ever seen before, but reminded me of the stems of certain highly specialized tropical lianas of which I had read. The buried parts of the stems, the taproot, and the larger lateral roots were not simple structures but bundles of separate strands, united at intervals. Some of the thicker roots resembled great cables, with scores of strands twisted together. Here and there on the buried stems were lumps composed of dozens of dormant buds, awaiting the proper stimulus to grow into new shoots. How had these peculiar structures arisen? Back at the university, I searched the botanical literature, but found no mention of this peculiarity of a well-known perennial herb, widespread on northern beaches in America and Eurasia. So I resolved to discover for myself how this situation arose.

In June, 1928, instead of remaining to receive in person my doctor's degree in botany, I made my last visit to Mt. Desert to study the Sea Lungwort. On this island surrounded by the cold water of the Gulf of Maine, the tardily awakening vegetation was, in early June, in its first flush of vernal activity, with trees burgeoning and spring flowers opening. On the seashore, the young shoots of the lungwort were just sprouting from old buried plants; and seedlings were pushing up through the shingle. Those from seeds that had fallen deeply between the stones were very long, thin, and pale, like seedlings that germinate in darkness.

In a cottage by the shore, where a housewife gave me board and lodging, I set up a field laboratory, and with a razor cut thin sections of lungwort roots and shoots, stained them, and examined them through the microscope. It was soon evident that as new conducting tissues—xylem and phloem—were formed by the activity of the awakening cambium, each separate strand of them was surrounded by cork-like cells, isolating it from the older tissues of the preceding year in which it was embedded. The tissues so cut off from the plant's vital activity would die and decay, leaving the separate strands that had excited my curiosity.

The resulting cable-like structure of the lungwort fitted it admirably to withstand all the punishment, all the strains and stresses, to which it was subjected as the shingle in which it grew shifted beneath the impact of winter's storm waves, sometimes burying the plant deeply, sometimes leaving much of it exposed; just as the division of the thick woody stem of a great tropical liana into separate strands increases its resistance to the pulling and twisting it receives as the supporting trees sway in the wind or fall. In situations so utterly different as a tropical rain forest and a northern shingle beach, plants meet similar challenges by similar responses. In a few northern herbs of inland woods and fields—gentians, larkspurs, sages, corydalis, etc.—the subterranean organs split in a fashion somewhat similar to that of the Sea Lungwort, but always, as far as I could learn, on a very much smaller scale.

Among the diversions that interrupted our research on Mt. Desert were botanizing in its cold sphagnum swamps, salt meadows, coniferous woods, and numerous ponds, and climbing some of the low, granitic mountains in Lafayette National Park, which occupies the center of the delightfully diversified island. Once I helped to fight a forest fire that swept over a tract of second-growth hardwoods about a mile long by a quarter-mile wide. Twelve days after the conflagration had been halted, thanks largely to a timely shower, I returned to survey its effects. Although several other heavy rains had fallen in the interval, the fire still smoldered in the thick humus beneath the spruces at the edge of the burn. Here and there, the charred stump of an ancient tree that was probably already dead when the area was lumbered stood upon a bare granite rock, its principal roots closely following the contour of the outcrop. The overlying humus in which the tree had grown was completely consumed by the fire. The destruction of the largely organic soil had left the trunks of birches propped up on their roots six inches above the new ground level. This fire in woods composed largely of broad-leafed trees had been mostly confined to the ground and was far less spectacular than the crown fires that roar through coniferous forests, but it caused great damage, destroying humus that might take hundreds of years to replace.

Scattered rather uniformly over all the charred ground were patches of light orange, usually less than an inch in diameter, that proved to be the fungus *Pyronema omphalodes*. Twelve days after the fire, this discomycete was almost the only sign of life in the devastated area. Although *Pyronema*, as its name suggests, is a pioneer on scorched earth in many parts of the world, its soft, almost gelatinous fruiting bodies appeared strangely out of place on beds of almost pure ash, exposed to the full intensity of midsummer sunshine.

It was amazing how quickly broad-leafed trees that had not wholly succumbed to the fire asserted their vitality. The birches, as I here noticed for the first time, were particularly well fitted to regenerate promptly. The trunk of these trees is usually greatly enlarged at the very base, where it is sometimes twice as thick as it is a foot higher. The surface of this basal swelling is densely covered with crowded dormant buds, some of which spring into activity when the tree's crown is killed, and sometimes for no apparent reason. This reserve of buds is prominent on the White Birch and the Gray Birch; in the latter, the earliest of these buds appear on the collet of seedlings in their first summer, and they continue to increase in number as the tree grows. On the Sweet Birch, the basal swelling with its buds is prominent on some trees and poorly developed on others.

Twelve days after the fire, new sprouts, some already two inches high, clustered thickly around the base of Gray Birch trunks. Twenty days after the conflagration, some of these birch shoots were nine and a half inches tall. By this time most of the surviving broad-leafed trees had sprouted. I measured new shoots of the willow seven and a half inches high; White Birch, six inches; Red Maple, five and a half inches; Staghorn Sumach, four inches; Red Cherry, three inches. The tallest sprouts that I could find on the Red Oak and the American Aspen were hardly one inch tall. Probably those of the aspen were so much less advanced than those of the birches because they arose from underground roots instead of preformed buds at the base of the trunk, for both aspens and birches are fast-growing trees. In the moister soil of the lowest parts of the burnt area, deep-seated rhizomes of the Bracken had survived and promptly sent up new fronds, which three weeks after the fire were already twenty inches high. The Great Willow-herb or Fireweed, too, was sprouting through the scorched ground, from buds on its roots.

Here was a convincing demonstration of the toughness and resiliency of life. Protoplasm is a delicate substance, readily destroyed by degrees of heat or cold, and by many chemicals which hardly affect rocks and metals. Yet despite their vulnerability, living things are so tenacious that they somehow survive cataclysms that one would expect to destroy them utterly. Phoenix-like, the forest was springing up anew from its ashes. In a few centuries, if undisturbed, it would cover this devastated land with the tall conifers that once flourished there.

Returning to the burnt area in early July of the following year, I was surprised to find it overgrown with the flat, green ribbons of the liverwort *Marchantia polymorpha*. Although they grew most luxuriantly in moist depressions and beneath the shade of stumps and fallen logs, they flourished even on dry, black cinders fully exposed to the sun on the ridges, their thread-like rhizoids clinging tenaciously to charred particles and black debris, evidently drawing enough moisture even from this unpromising source. The more vigorous plants bore not only upright sexual branches but many little cupules filled with the multicellular gemmules that propagate the plant vegetatively. In places, the liverwort grew in such profusion that its crowded ribbons stood edgewise to the ground.

At the end of August, after a long spell of dry weather, the *Marchantia* in all but the moistest spots was dying and shedding its spores freely. Now the

Common Hair-cap Moss, which earlier in the summer was abundant only in moist places and depressions between outcropping rocks, had spread densely over large areas of burnt ground, replacing the liverwort. When I next visited this burn, in early June of the fourth year after the fire, two species of *Polytrichum* mosses had succeeded the liverwort almost everywhere. In the valley and on level hilltops, the Common Hair-cap flourished in close stands; while on the drier slopes, the Juniper Hair-cap grew more sparsely. Beneath the densest growths of the mosses I found remnants of dying or dead *Marchantia*; but only in certain moist and shady spots, as under the projecting edges of rocks closely crowded by *Polytrichum*, or beneath the evergreen trees at the margin of the burn, did a few of the liverworts grow freely, unencumbered by the moss. A foliose lichen, *Peltigera polydactyla*, grew amid the hair-caps.

The colonization of recently burnt land by *Marchantia polymorpha* is widespread in the North Temperate Zone, and has been reported from regions as widely separated as England and Idaho. In both of these places, the liverwort was associated with the Cord Moss rather than hair-caps, although on some burns in England the latter eventually displaced the pioneer bryophytes. But on other burns in Maine, both on Mt. Desert Island and in the interior of the state, I regularly found the hair-caps growing with *Marchantia*. In some places, the struggle between these two humble growths for possession of ground denuded by fire was vividly portrayed in all its stages.

More unexpected was the discovery of the same succession of vegetation on the summit of a tropical mountain over seven thousand feet high. Between visits to the burnt area on Mt. Desert Island, I spent a summer on the island of Jamaica and climbed its highest mountain, Blue Mountain Peak. Some years earlier, the east summit had been swept by fire. Now, in 1926, a close sward of the Juniper Hair-cap covered the top and steep northern or windward face of this summit, where for seven or eight yards downward from the top it formed a continuous carpet. Although in July the moss was brown and dry, the clouds that bathe this mountain through much of the year, along with an annual rainfall of about 168 inches, had supplied abundant moisture for its earlier development; and many of the moss plants had fruited. Among the hair-caps grew many plants of *Peltigera polydactyla*, the same lichen that I had found associated with them in Maine.

Amid the densest stands of hair-caps on the mountaintop, I detected few traces of *Marchantia*; but in the shade of fallen logs and stumps propped up on their roots, the liverwort grew luxuriantly, as it also did on the southwestern slope, beneath a scarcely penetrable stand of Bracken and lush tropical ferns higher than my head. Here on the shaded ground, the liverwort had advanced from its retreat beneath logs to grow over prostrate hair-caps, reversing the succession that had evidently occurred soon after the fire.

Although it is a well-known law of plant geography that the flora of high

altitudes in low latitudes resembles, in many ways, that of lower altitudes at
higher latitudes, the similarity of the pioneer vegetation on these burnt areas
in Jamaica and in Maine was surprising. In the woods around this scorched
summit, the trees were all quite different from those of the northern United
States. On burnt areas in humid tropical lowlands, I have seen neither
Marchantia, nor hair-cap mosses, nor the lichen *Peltigera*, although the
fungus *Pyronema* makes splashes of orange on the charred ground. Here
swiftly springing seed plants of many kinds, including herbs, shrubs, trees,
and vines, cover the ground before humbler plants can take possession.

For *Marchantia polymorpha* to spread so freely over hundreds of burnt
acres in so short a time, many thousands of spores must have fallen there
within a few months of the fire. But where did they come from? On un-
burned areas on Mt. Desert Island, this liverwort was so rare that, in three
summers, I noticed only one small, sterile patch. On Jamaica, I found it only
on the scorched summit, although other species of *Marchantia* grew lower
on the mountain. In England, it is likewise reported to be rare and sterile on
unburnt land. In greenhouses, it grows best on cinders, and it is intolerant of
shade. Evidently the profuse stands of *Marchantia* on recently burnt ground
fill the air with millions of spores, which are scattered widely by the wind
and give rise to a numerous progeny only when they fall in another burnt
area. Although, since man spread over the earth, fire has been one of the
chief destroyers of life, certain organisms have seized the opportunity for
expansion that it affords them, and their continued existence may depend
upon its recurrence.

On my last visit to the burnt areas on Mt. Desert, in early June of 1928, I
was attacked by ravenous swarms of mosquitos and little, black simulid flies,
whose combined assaults left me with a grotesquely swollen and distorted
face. For a day or so, I could not open one of my eyes. Never in the lavish
tropics have I received such severe punishment from insects as in this frugal
northern land. Although, despite such occasional discomforts, the coniferous
woods of the North are compellingly attractive to some, I never fell wholly
under their spell, doubtless because more luxuriant forests, inhabited by a
far greater variety of living things, were calling me too imperatively in the
opposite direction.

4

Mountain and Plantation in Jamaica

The tropics are the headquarters of vegetable life, which flourishes there in variety and profusion unmatched at higher latitudes. Holding that first-hand experience of the tropics is an indispensable part of a botanical education, Professor Johnson insisted that each of his graduate students spend some time there. Periodically, he led an expedition to Jamaica, then still a British colony. One of these expeditions came at the end of my first year of graduate work in his department at Johns Hopkins. As his laboratory instructor and assistant, I was chiefly responsible for packing our equipment, much of which was shipped in big cardboard boxes. With old-fashioned Yankee notions of economy, the professor insisted that each box be tied with a single long piece of string, although I would have found it easier to use shorter lengths and tie more knots.

Today, many undergraduates are sent by their colleges for a year of study in the tropics, with all expenses paid and even pocket money; but back in the

Headpiece: The author on his first voyage to the tropics. Approaching Jamaica on the S.S. *Carrillo*, June 7, 1926.

1920s, graduate students were often hard pressed to find the means to visit a tropical country. Our botanical department had a small fund for renting a house in Jamaica and certain other general expenses, but every member of the expedition had to pay his own travel and personal expenses. With my stipend of forty dollars monthly as laboratory assistant, this outlay was difficult to meet, and my fellow students were hardly more affluent. Fortunately, Dr. John R. Johnston, director of tropical research for the United Fruit Company, came to our aid with an offer of free round-trip passages on the company's steamships for three students who would undertake certain studies of the banana plant. Philip White, who later worked on tissue culture for the Rockefeller Foundation, chose the flower and its cytology; Paul Acquarone, afterward professor of botany at Akron, Ohio, selected the root for his study; while I agreed to investigate the anatomy of the banana leaf. Other members of our party were Professor John N. Couch of the University of North Carolina; Dr. Johnson's son, Duncan, then a high-school student; and Martin Curtler, a young Englishman eager to see tropical birds. We sailed from New York on June 2, 1926.

A sea voyage interposes a seemly interval between distant lands, allowing the traveller time to recover from the bustle and strain of departure before confronting a foreign country, as modern air travel does not. As the sun set amid obscuring clouds on the fourth evening after the *Carrillo* sailed from New York, we entered the Windward Passage. To the starboard, the low, scrubby hills of Cape Maisi, Cuba's eastern extremity, rose by steps or natural terraces from the shore. A lighthouse with a few cottages clustered around it and a rusting hulk, lying black and naked on the sands, all beneath a leaden sky, imparted a touch of melancholy to my first distant view of a tropical land. As we continued southward into the night, flashes of lightning played among the clouds massed in the west.

By dawn, the sky had cleared. As I went on deck soon after the sun rose out of the sea, the strength of its rays, even at this early hour, assured me that we were in the tropics. The *Carrillo* had already passed well into the lee of Jamaica, and the green island towered to our north in all its majesty. From the blue water that bathed the feet of the outermost of its hills, they rose, rank behind verdant rank, into the clouds that rolled over the highest summits. To eyes accustomed for days to the level sea, with an occasional glimpse of a low island strand, these abrupt mountains appeared tremendously lofty. The valleys visible to the voyager were mere slits carved by streams into the mountainsides, and the only level areas in sight were the narrow floodplains of rivers emergent from the hills. A fitting introduction to the tropics, at once enticing in their luxuriance and humbling in their grandeur!

Soon I noticed a dugout canoe coming from the shore through waves that seemed too high for so small a craft. In it rode a black man, with two native boys paddling. A rope ladder was dropped over the ship's side, and the man climbed up to the deck, with his blue pilot's flag tied in a cord held between his teeth. The pennant was raised to the masthead, and we proceeded under the pilot's guidance, towing the canoe at the end of a long rope, which one of the boys gave a turn around a thwart and held in his hand, while he shared a piece of sugarcane with his companion. We glided close by tiny coral islets where the sea's deep blue paled enchantingly through azure to the sparkling white of the sands. Rounding the long, protecting arm of the Palisadoes, we dropped anchor off Port Royal, for inspection by the quarantine officer. Then we continued across Kingston's broad harbor to the city, spread at the seaward edge of a broad plain that sloped gently upward to meet the mountains in the background.

As, without a tugboat, our ship maneuvered up to the dock, the enchantment of the approach was abruptly replaced by the disillusion of arrival. Before we could walk down the gangplank, our vessel was surrounded by ragged black boys in dugout canoes, singing snatches of the latest jazz songs and begging American coins from the passengers. On the wharf, everything was black with coal dust, blown off the heaped-up baskets which a heterogeneous multitude of both sexes and all ages were carrying on their heads into the hold of a British liner. Their tattered, begrimed garments hardly distinguishing man from woman, the natural darkness of their faces intensified by a thick deposit of soot, these sweating porters were a disquieting sight. Our baggage was rudely snatched away by black hands competing for small coins, although a sign on the wharf advised passengers not to tip. What a contrast between the first distant view of the green, sea-girt land and our first contact with it!

After we had settled in the high-ceilinged rooms of our hotel and sorted out our baggage, we began to explore the island. At beautiful Hope Gardens, a short ride by trolley car beyond the upper edge of the city, all of us except our leader first became familiar with tropical plants in a tropical setting, rather than in a northern greenhouse. For at least one of us, the impact of these noble growths was unforgettable. On the following day, we hired a car and drove eastward, over an unpaved but fairly smooth road that now traversed narrow stretches of coastal plain, now wound among steep foothills that jutted out to the water's edge. At intervals, we crossed brooks that meandered along channels far too wide for them, or broad watercourses that were quite dry. The sand and coarse gravel spread over the surrounding plains were evidence of the fury of the torrents that sometimes poured down from the hills, filling these broad streambeds until they overflowed, tearing away bridges, and stopping all traffic along the shore road. We were told that travellers were sometimes marooned for days between two suddenly rising streams, the mountains, and the sea.

Shielded by the central mountains from the moisture-laden trade winds, this southern coast of Jamaica was the most arid land that I had ever seen. Low, small-leafed, scrubby growth prevailed on the limestone hillsides and uncultivated areas of the plains. On more gentle slopes, and amid sand and stones on the floodplains, tall organ cacti pointed unbending arms toward the sky. These cacti were often set close together to make hedges less penetrable than barbed-wire fences. On the sides and crests of the coastal hills, the giant, leafless inflorescence stalks of agaves stood up prominently, dying as their seed pods matured. Here and there along the roadway grew a massive Silk-cotton tree, its swollen, often bottle-shaped trunk supported by plank buttresses—wing-like projections from its base, broader and higher than a man. Gumbolimbo or West Indian Birch trees, leafless at this season, rose singly or in small groups above the scrubby growth, shedding reddish-brown bark in papery sheets from smooth trunks and limbs. Among all this strange vegetation I was happy to recognize one old friend, the "life plant" *Bryophyllum calycinum,* cherished in northern greenhouses because of its curious ability to produce a new plant from each of the many notches along the edges of its fleshy, crenate leaves. Here, as in many other tropical countries, this succulent of Old World origin grew as a roadside weed.

On the more fertile plains, sugarcane thrived with irrigation. Coconut palms and bananas flourished in the moist lowlands and beside streams. About the white-walled, thatched, mud-and-wattle huts along the road stood Lignum Vitae trees, famous for the hardness and durability of their wood, and equally attractive whether, amid their glossy foliage, they bore clusters of blue flowers or of orange fruits.

Near the village of Yallahs, where the ruins of an old church bore witness to past prosperity, we visited a mangrove swamp bordering a lagoon. Although this tree was widespread on muddy tropical American shores and had been described and photographed many times, our first actual contact with so strange a growth brought our cameras into action, with no thought of saving films for future marvels. Propped above the fetid mud on arching stilt roots, the Red Mangrove simultaneously bore flowers, fruits, and viviparous embryos in all stages of development. The most advanced of these had become spindle-shaped seedlings almost a foot long, ready to slip from their collars and plummet into the mud, where they would stand upright and quickly anchor themselves by lateral roots. These aquatic banyan trees occupied a broad belt around the lagoon, to the virtual exclusion of other vascular plants. The very different habit of the Black Mangrove, which occupied a zone of slightly higher and firmer ground just landward of the Red

Mangrove, failed to arouse equal enthusiasm in young botanists. Yet with its innumerable slender pneumatophores, or breathing roots, growing upward from thicker subterranean roots and downward from the trunk and even some of the lower branches, this arboreal relative of the garden verbena was a tree quite different from any that we had ever seen.

We spent a week exploring the country around Kingston, including a visit to the splendid botanical garden at Castleton. Then our baggage was loaded on an open truck, the younger members of the party arranged themselves on top of the baggage, our leader sat beside the driver, and we set off for the Blue Mountains. After a long, steady rise over the Liguanea Plain, we began to wind up the steep, deeply eroded Port Royal Range. Some of the bends in the road were so sharp that our truck had to back once or even twice to make the turn. Passing the summit of this rather barren coastal range, we descended abruptly into the Yallahs Valley, where at a hamlet called Mavis Bank we unloaded. Advised that we needed transportation, the neighboring farmers trickled in with a dozen mules and donkeys, some with pack saddles and others outfitted for riding. After much wordy bargaining, they agreed to transport us and our impedimenta up to Abbey Green at ten shillings for each mule and six for each donkey. Suitcases and other pieces that did not fit on the pack animals were carried on human heads.

We soon learned that Jamaicans were exceedingly reluctant to carry anything in their hands. "What the hand refuseth the head taketh," is a proverb that came from the tongue of a hillsman, expressing the whole philosophy of transportation in these abrupt mountains, traversed only by trails too steep and narrow for wheeled vehicles, that we were about to explore. It was amusing to see a native girl carry a cupful of water or a bit of food wrapped in a banana leaf on the crown of her head; but we admired the splendidly upright carriage and perfect muscular coordination that this feat demanded. A common load was a five-gallon kerosene tin filled almost to the brim with water, and carried over rough paths without spilling a drop. The black man who could march all day with a heavy load balanced on his head marvelled that we could climb mountains with knapsacks on our backs. I have noticed in my travels that this method of carrying loads, which seems so easy and natural to Europeans and North Americans, is not widely favored by the inhabitants of tropical countries.

When all the loads were adjusted, we resumed our journey. Our way led across the wide, stony channel of a shrunken mountain stream, then steeply upward along narrow trails. As we rose into the zone of coffee plantations, the air became cooler, a pleasant change from Kingston's oppressive heat. My donkey was so diminutive that my feet almost touched the ground. Reluctant to impose my bulk on an animal so much shorter than myself, I chose to walk much of the way.

Toward evening we reached our destination, Abbey Green coffee planta-tion. The small planter's cottage that we had rented perched on a shelf cut into the steep southern slope of Mossman's Peak, at an altitude of about four thousand feet above sea level. Below it on the mountainside was the "fac-tory" where the coffee berries had been processed. Like the cottage, it was now unused, as this plantation had been acquired by the owner of an adjoin-ing one, who took the coffee to another factory. So in this big, barn-like building we set up our field laboratory, using our microscopes and arranging our botanical specimens on the tables where chattering women and girls had graded coffee beans, as was widely done by hand before the introduction of electronic equipment. Close by these buildings was a huge gash, deep, long, and wide, gouged in the precipitous flank of the mountain by a landslide and still nearly bare of vegetation. The torrent of muddy water from this washout had overwhelmed a coffee factory lower on the mountain.

Continuing up the slope above our cottage, the trail soon passed from the cleared land into unspoiled forest that covered the Blue Mountains above about forty-five hundred feet. Here, watered by abundant rainfall and fre-quently bathed in cloud-mist, tropical vegetation grew in lush profusion which contrasted sharply with the drought-resistant flora that we had seen on Jamaica's southern coast. Here were the epiphytic growths—the orchids and aroids and bromeliads and ferns perched upon trees, the woody lianas, the moss-draped trunks and boughs—that one expected in wet tropical woodland. Its wetness was vividly attested by bog moss growing on a steep trailside bank instead of in a swampy hollow. When even common tropical weeds were new to us, days were too short to identify, in Fawcett and Rendle's excellent *Flora of Jamaica*, all the plants that we collected.

The trees that composed this montane rain forest could not compare in size with the great oaks and other broad-leafed trees that I found at corre-sponding altitudes in the Central American forests which I later explored. Their trunks seldom exceeded a foot in diameter, and their stature was correspondingly low. Apparently, the Blue Mountain slopes were so steep, and subject to such rapid erosion, that the trees fell before they could grow very tall. Many leaned precariously downhill. Bright flowers were rare, as beneath dense tropical evergreen forest everywhere. I have no recollection of any bird in these mountain forests except the gray Rufous-throated Sol-itaire, a small thrush whose sweet, pensive whistles broke the silence of these hills. On this trip, I was more interested in plants than in birds.

In these cloud-bathed mountains, ferns grew with a profusion that I have rarely seen equalled, and never surpassed, in the heavier montane forests of continental tropical America. I promptly fell under the spell of these charm-ing plants; John Couch shared my enthusiasm; and together we scrambled over steep slopes and pushed into deep ravines, collecting as many as we could of the more than five hundred species of ferns that inhabit Jamaica. Fortunately, our party at Abbey Green was joined by Dr. William R. Maxon of the Smithsonian Institution, an authority on ferns who helped us identify our specimens.

Here the ferns had done their utmost to produce varied forms, to adapt themselves to every situation that a wet mountain forest offered, to claim a place in the sun along with the flowering trees, yet to flourish in whatever dark nook or recess was available to them in the depth of the woods. At one end of the scale were the stately tree ferns, which grew best in deep, moist ravines, where their straight, slender, branchless trunks stood twenty or thirty feet high, bearing at the summit a rosette of great compound leaves. So delicately incised and subdivided were these enormous fronds that they made a filigree silhouette against the blue tropical sky. At the opposite end of the scale were the filmy ferns, the most delicate members of a vegetable tribe renowned for its delicate grace. Except at the veins, their fronds are only a single layer of cells thick, and they flourish best in an atmosphere that is almost continuously near the saturation point.

In these wet mountain forests, we found filmy ferns in bewildering variety. Some grew in crowded colonies, draping tree trunks with a soft, thick raiment of living green; others climbed up trunks by slender rhizomes, which at intervals bore stiff, horizontal fronds of intricate design; others crept along the ground or a decaying log, holding erect their translucent leaves. Often the filmies and other diminutive ferns grew upon the trunks of the tree ferns, providing instructive contrasts. Yet other ferns, such as the prickly *Odontosoria*, scrambled vine-like into the trees. In sunny openings, the bifurcating fronds of *Dicranopteris* formed head-high, impenetrable thickets of wiry stipes.

We climbed to the summit of Mossman's Mountain, which rose behind our cottage, and twice to Blue Mountain Peak, 7,360 feet high. The baggage for our longer expedition to the Peak was loaded on mules in the corral, before anyone awoke to the fact that the animals with their bulky packs could not pass through the narrow gate. Our muleteers solved the difficulty, not by unloading the mules, but by digging up one of the gate posts, which was replaced after the exit of the baggage train! Finally, we started up the narrow but well-kept trail through the beautiful mossy forest. At the summit was a small cabin in which most of our party slept; Couch and I camped in a pup tent that we set at the edge of a precipice, whence we could overlook a welter of wild mountains to the north. We stayed on the summit for three days; and among other things I studied the succession of vegetation on the burnt east peak, of which I have already told.

Fortunately, the clouds that much of the time cluster around Blue Mountain Peak remained aloof, and we enjoyed superb views on every side.

To the south, we looked over the summits of the Port Royal Range to the blue Caribbean. Kingston, with its spacious harbor enclosed by the long, protecting arm of the Palisadoes, seemed almost to lie at our feet. To our west stretched the Blue Mountain Range, green peak beyond green peak. Walking a few hundred feet to the northern edge of the summit, we stood at the brink of an abrupt precipice and gazed over lower hills to Port Antonio's double harbor, and the Caribbean stretching northward toward Cuba. After sunset, when we donned our warmest clothing and hugged the campfire, we looked pensively down at the twinkling lights of Kingston, musing upon the contrast between the sweltering streets there and the November crispness of the air here. At night, we slept on beds of the fern *Dicranopteris*, called "hog-grass" by our porters, which made a couch softer than the hemlock boughs used by campers in the North.

In a swampy spot amid the coffee plantations below Abbey Green, we were delighted to find the Giant Horsetail. Its slender stems, four yards high, far taller than any member of this family that we knew in the North, reminded us of the time, ages ago, when its still larger relatives, the cala-mites, flourished in the Coal Measures.

After a month at Abbey Green, we packed our baggage on mules and donkeys and migrated westward to Cinchona. A friend, to whom I men-tioned this place, remarked that it sounded like a name from the phar-macopoeia, which was quite correct. This botanic garden and experimental station, on Bellevue Peak at an altitude of five thousand feet in the Blue Mountain Range, was established by the colonial government during the preceding century, when an effort was made to start plantations of the cin-chona or Peruvian bark on the island. This small tree of the madder family, the source of quinine, grows wild in the tropical Andes, chiefly from three thousand to eight thousand feet above sea level. Wasteful methods of gather-ing the bark in the rainy mountains where the tree is native were threatening to destroy what at that time was the source of the only effective remedy for malaria, and several European countries with mountainous possessions in the tropics decided to promote its culture as a plantation crop. The cinchona tree evidently grew well in the wet Blue Mountains, but the thrifty Dutch succeeded in producing quinine so much more cheaply on their Javan plan-tations that the Jamaican planters could only sell their bark at a loss. Aban-doning cinchona, the planters tried to grow tea, but the Blue Mountain climate proved to be too moist.

After these failures, Cinchona was no longer used as an experiment sta-tion, but preserved as a botanic garden. Here we lived for a week in the large bungalow that had once been the director's house, the highest dwelling in Jamaica, in the midst of trim lawns and beautiful gardens. Here oaks and Tulip-Poplars from North America and eucalyptus and grevillea trees from

Australia grew tall amid tropical palms and ferns in a magnificent mountain setting. The rows of lofty pine trees that bordered the trail leading up from the south reminded us of lands far away.

From Cinchona, our party returned to Kingston. After a few days there, the three of us who had agreed to study the banana plant for the United Fruit Company, to earn our steamship passage, were taken to Platfield Plantation in St. Mary's Parish. Acquarone and White stayed only long enough to collect material of the roots and flowers, which they preserved for future study, then hurried back to the port to join the rest of the party on the next ship sailing for New York. A banana leaf is far too large to preserve intact in a jar of alcohol or formaldehyde, and, moreover, I have always preferred to live close to whatever plant or animal I study; so I decided to remain at Platfield for the rest of the summer.

From late July until mid-September, when it was necessary to return to Baltimore for the opening of classes at Johns Hopkins, I lived alone at Platfield, in a small house that had been built for the plantation timekeeper. Here I set up a field laboratory, with monocular and binocular microscopes, slides and cover glasses, razors for cutting sections of banana leaves and stones for sharpening the razors, a small balance, and many little bottles of reagents.

The cottage was situated on a pastured hillside, not far above a Chinaman's general store, where I bought provisions. The company provided a cook to prepare my meals and clean the house, and a yard boy to cut firewood, trim the lawn, and help me in the field. Susan, my cook, did her best to keep house for the young northerner who still knew little about tropical living, and was, moreover, too engrossed in his research to pay much attention to what she did. I admired the way she got down on her hands and knees to polish the beautiful hardwood floor with a coconut husk cut in half across the fiber. But once she slipped up and gave me a dish of Akee that had been gathered at the wrong stage, or else had been improperly prepared. Although for a day I was very sick, I did not pay the supreme penalty for eating this treacherous food, as many have done. Mrs. Keith, the tall overseer's little wife, who took a motherly interest in the young scientist working on the plantation that her husband directed, gave poor Susan a terrible tongue-lashing. After my recovery, I resolved never again to eat Akee, which has higher merit as an ornamental than as a fruit tree. When the large red pod splits open to reveal the shining black seeds, half-embraced by white, dubiously edible arils, it is most attractive; but the feast is best reserved for the eye alone.

Although my assignment was only to report upon the anatomy of the leaf, I had in fact undertaken to study almost the whole banana plant. The freshly unrolled, still intact blade of a large banana leaf, four or five yards long by

over a yard in breadth, is one of the greatest continuous expanses of photo-
synthetic tissue in the whole vegetable kingdom, being exceeded in area by
the similar leaves of the Traveller's Tree, the floating leaves of the Victoria
Water Lily, and the submerged blades of certain great kelps. But the banana
leaf includes more than this broad blade that, except in sheltered spots, soon
becomes frayed in the wind.

The banana plant is often called a tree, which hardly seems an exaggera-
tion when you behold a stately plant with a trunk a foot thick at the base and
five or six yards high, with the tip of its topmost leaf waving proudly five
times a tall man's height above the ground. But this noble plant is really only
a gigantic herb, one of the most massive in the whole vegetable kingdom.
You may cut it to pieces from root to crown without finding a splinter of
wood, which is the essence of a tree. No part of it is harder than a green
apple; banana trunks, chopped up with a machete, are used as fodder for
cows and devoured in their entirety.

This smooth, massive, columnar trunk is wholly different from the trunk of
a tree; for a tree trunk is, in gross structure, simple and solid, while that of
the banana plant is composed of many separate parts. It is a false stem,
consisting of little more than the bases of the leaves, which are nearly as long
as the trunk is high, crescent-shaped when cut transversely, and full of air
spaces. These leaf bases overlap, fitting tightly together in an intricate pattern. You may strip the long leaf bases off one by one, as you would pull the
successive layers of skin from an onion; and as each is removed the trunk
becomes thinner. Finally, when you reach the center, you will find a tightly
furled young leaf, pushing its way upward in the midst of the sheathing bases
of the older leaves. Or, if the plant is mature, you will find at the heart of the
trunk the true stem, which is only a few inches thick, as white and smooth as
a rod of polished ivory, and so brittle that it will break under its own weight
without the support of the tougher and more fibrous leaf bases that surround
it. This stem produces a few foliage leaves on its enclosed length, then
pushes out from the top of the false stem in the form of a huge red bud,
bends under the bud's weight until it hangs downward, bears long, white
flowers beloved of hummingbirds and little stingless bees, and later supports the heavy bunch of fruit.

To harvest this fruit, which even for local consumption is cut while still
green, the whole noble growth is toppled over, for it is capable of bearing
only once. But surrounding it are younger shoots of various ages, that sprout
from buds in the sides of the massive, solid, subterranean rhizome of "bulb,"
until many trunks come to stand in a compact clump, called a mat. Each of
these daughter plants is capable of producing a single bunch of bananas, but
on well-attended plantations some are pruned away so that the remainder
will bear larger bunches.

From its tip to its insertion on the "bulb," a banana leaf, including its sheathing base, its stout petiole, and its expanded blade, may be twenty-five or thirty feet long. To explore every region of this great organ, I cut endless thin sections, stained them, and mounted them in water for microscopic study. One afternoon, as I sat at my microscope poring over a bit of banana leaf, a very ragged black man entered the room and greeted me with a deep bow and flourish of his battered hat, which he laid on the polished floor. Lightly touching my knee, "Sir, I beg you your trousers," he said. Although they were spotted with the brown stains that discolor the clothes of everyone who works on banana plantations, I could not then and there divest myself of this garment to oblige the genial beggar, as he seemed to expect.

The impudence of some of the young blacks in Kingston was hard for one raised in the south of the United States to accept; but I found the rural Jamaican a gentleman of engaging politeness. His punctilious good manners sometimes embarrassed the members of our party. If, neglecting the usual greeting, we hurriedly gave an order to an employee, or asked a question of a stranger, he would listen respectfully till the end, then, before replying, interpose "Good morning, boss," as though nothing had been previously said. On taking leave, he rarely neglected to wish us a safe journey or a successful undertaking. "Yessir" and "No, sir" were used almost too freely in his conversation, and one man whom I accosted had so far forgotten the meaning of "yessir" that he repeatedly answered my questions with "yessir, sir." Precise information was difficult to elicit from the country people: distances were usually "not too far," or "far enough"; quantities, "not too many," or "not too much." Some of these simple, often illiterate, people were quite sententious, repeating proverbs that expressed their philosophy of life in a nutshell. One that I still recall was "Better belly burst than good food spile." Since most were poor, they were probably rarely confronted with this harassing alternative.

Close by my cottage at Platfield stood a dense clump of tall timber bamboos that rattled and screeched whenever the wind blew. Sometimes they were visited by a flock of black Smooth-billed Anis, calling to each other in high-pitched, whining notes. At the Chinaman's store down the hill, men would sometimes argue interminably, in loud if not angry voices, about such matters as who should ride in the plantation truck. From the river that flowed at the foot of the hill arose the sound of washerwomen pounding clothes against the rocks, or beating them with a broad wooden paddle, a method of washing that achieved a measure of cleanness at the price of missing buttons and torn fabrics. But the loudest sounds that assailed my ears while I lived here came from a burning thicket of bamboos on a neighboring hillside. As the air in each compartment of the wide, hollow stems was heated by the flames, it expanded until it burst the wall with a sharp report, like that of a big Fourth-of-July firecracker.

Although to accomplish my task in the short while available, I had to apply

myself steadily to it, I found time for occasional walks through the surrounding country, to observe, collect plants, and photograph. The terrain was
broken and hilly; the modest elevation opposite my cottage was rather
grandiosely called Mt. Job. Occupied largely by small provision patches, the
country was more pleasantly diversified and picturesque than a region of
great, single-crop plantations. Coconut palms and banana plants adorned the
landscape on all sides, with here and there patches of yams, taro, and cassava. Crowning the rounded hills were groves of tall, gracefully arching
bamboos. Their plume-like shoots cast so dense a shade that scarcely any
herbage grew amid the dry, fallen leaves that carpeted the ground beneath
them. Even the most humble cabins of interwoven strips of bamboo covered
with clay, plastered white or unadorned, were attractive amid verdure and
flowers. At the nearby hamlet of Mt. Vernon, I found a thatched shed
serving as a school, while a larger, more substantial schoolhouse was being
built beside it. The schoolmaster obligingly lined up his three dozen little
black pupils for a photograph in front of their rustic classroom.

Neither on these walks nor in the Blue Mountains did I ever meet a snake,
and I spoke with natives who had never seen one. Philip Gosse, who wrote
on the natural history of Jamaica in the middle of the nineteenth century,
frequently mentions serpents, which seem formerly to have been rather
abundant on the island. But the Mongoose, introduced into Jamaica from
India in the latter half of the century, had almost exterminated them. The
advent of Kipling's Rikki-Tikki-Tavi was no unalloyed blessing to the island,
for he preyed upon native birds and domestic chickens, thereby provoking
the people to raise their hands against him.

I left Platfield in mid-September, with a sheaf of notes and many drawings
that I would elaborate into my doctoral dissertation on the anatomy of the
banana leaf. Although toward the end of my sojourn there I suffered nostalgic spells, I had proved to myself that I could work alone in far places. After a
restful voyage from Kingston to New York, I was ready for another year of
study and research at the university, where I continued to serve as Professor
Johnson's assistant.

5

Voyage to Panamá

During the two summers following my visit to Jamaica, I served as naturalist at a Boy Scout camp on the Severn River, an affluent of Chesapeake Bay south of Baltimore. Although botanical research was thereby interrupted, I believed that to help impressionable young minds to know and love nature was an undertaking at least as important as the production of another scientific paper that only a few specialists would read. Then, too, I should have an opportunity to study the natural productions of the sandy coastal plain, where the flora and fauna are somewhat different from those of the hilly piedmont region where I had grown up. My work at the camp became increasingly monotonous, especially during the second summer. Since most of the boys came for only two weeks and there was a continual influx of new campers, the same elementary instruction, the same tests of accomplishment, had to be given over and over; we could hardly advance to new subjects. As was to be expected, most of my efforts to develop a warm

Headpiece: The old church in Panamá City.

interest in nature fell upon sterile ground; but a few of the boys were enthusiastically responsive, and this was my chief reward.

After I received my doctorate in botany, my university gave me a research fellowship, which permitted me to choose my subject and the place where I would work on it. My taste of the tropics in Jamaica had made me eager for contact with the more varied life on the mainland of tropical America. I wished to continue my study of the banana plant in a spot where it grew abundantly. On my way to Mt. Desert Island at the end of May, 1928, I had visited Dr. John R. Johnston at the headquarters of the United Fruit Company in Boston. He was interested in my project and eventually arranged for me to spend half a year at a small experiment station maintained by the company near Almirante in the Caribbean lowlands of western Panamá. Finishing my paper on the Sea Lungwort delayed my departure until late November, which, as I now see it, was fortunate, for I did not reach Almirante until the long wet season of that rainy region was approaching its end. A rainy season in Caribbean Central America is sometimes difficult for a novice to endure.

On November 17, 1928, I sailed from New York on the *Toloa*, a United Fruit Company steamer bound for Cristóbal. As we voyaged away from gray November skies and green northern seas into waters that vied in intensity of blue with the sky above them, the wrench of departure was forgotten, and all misgivings about the step I was taking vanished. My daydreams were tinted with colors as bright and cheerful as those of the sky and the ocean. The flying fish, blue birds of the sea that glided away from the prow of the advancing ship, sometimes augmenting their speed by flipping the crest of a wave with their tail fins, were fit symbols of my hurrying anticipation. Life was then for me at its floodtide of promise; it seemed to be yielding everything that I might reasonably expect of it. To complete my satisfaction with the moment, in honoring my free pass the steamship office had assigned to me a stateroom with a private bath—the only time I have voyaged in such luxury! I was not then in a mood to reflect that the tides of our life are as changing as those that obey the inconstant moon.

From the books of Darwin, Wallace, Belt, Schimper, and Beebe, but chiefly from a multitude of scattered scientific articles, I had formed a vivid idea of the splendor and fascination of tropical life. During my summer on Jamaica two years earlier, I had seen some of this splendor at first hand. Now I was going to behold it on a vaster scale, amid the richer flora and fauna of the continental tropics—and to make new discoveries that would perhaps place my name not too far below those great ones. The youth in his early twenties would be craven and unworthy of encouragement if he did not cherish, in the secret places of his mind, such lofty aspirations—alas, how difficult to fulfill!

Our first port of call was Havana, where the *Toloa* spent three days unloading cargo, while passengers saw some of the sights of the city and surrounding country. Thence a three-day voyage through the Yucatán Passage and the Caribbean Sea took us to Cristóbal at the entrance to the Panamá Canal, my first landfall on the mainland of tropical America. According to my original itinerary, I was to proceed to Almirante in another ship that would call first at Puerto Colombia on the northern coast of South America; but now I learned that I could reach my destination nearly two weeks earlier by continuing on the *Toloa* to Puerto Limón, Costa Rica, to which port a launch would be sent from Almirante to bring back the wife and children of a company official stationed there. The *Toloa* would remain at Cristóbal for only one full day, and with youth's enthusiasm I set forth to see all the sights of the Isthmus of Panamá between dawn and sunset. Vain endeavor! Yet in what other part of the earth, in the days before air travel became general, could I have crowded such varied and contrasting experiences into the space of twelve hours?

The early morning journey by train across the contracted continent, through exuberant tropical vegetation, large-leafed and lush and still heavy with dew, was full of excitement for a young naturalist. For miles we rode along in sight of the Canal and the ships passing through it. Then the railroad circled the wide expanse of Gatún Lake, passing on causeways over far-reaching arms, where the skeletons of drowned trees stood gaunt and bleached in the backwaters. Many of these relics of the forest, which was flooded fourteen years earlier, bore masses of foliage that belonged to the epiphytes perched upon them, but none of their own. Beyond the lake the hills, covered with the dark verdure of tropical forest, rose crest beyond crest into the misty distance. I had never yet explored such forests.

Descending from the train at Panamá City, I took a random course, which soon brought me to a church whose squat square tower, and masonry walls adorned by plants rooted in crannies and ledges, produced an air of venerable antiquity that was not wholly dispelled by the roof of modern corrugated iron, the all-too-obvious electric wiring, and the printing shop leaning against its rear wall. Here modernity clashed so blatantly with the past that I was irresistibly drawn to examine the interior, where the uncarpeted floor bore telltale marks of the domestic fowls that wandered in through the open doors. Here I sat for nearly an hour and had my first revealing glimpse of simple people at their spontaneous devotions.

The votaries who entered in a steady trickle were mostly poor, wizened, old women of color, some carrying bundles that suggested that they had dropped in to pray on their way home from their morning shopping. Most of them brought candles, which they lighted in front of one of the many sacred images that lined the walls. A blind woman was led by a companion to a sconce where she stuck her taper. By ten o'clock in the morning, a hundred and sixteen candles burnt in the church, and, in addition, flames flickered on wicks floating on oil in a dozen drinking glasses. The image most favored by these votive lights was a badly cracked oil painting of Jesus on the cross, surrounded by a number of symbols that I could not interpret. Two little children touched themselves with holy water from a fount attached to the wall. A black man in white trousers rubbed some of this water on the brim of his straw hat. A coffee-colored woman with a neat white headdress made a cross with her finger on the floor, then reverently stooped to kiss it. And all this while, a middle-aged colored man, wearing glasses and such a conspicuous starched collar as only a simple person with scholastic yearnings could display, sat in a center pew assiduously writing with pencil in a copybook. To make himself perfectly comfortable, he had removed a shoe and rested his bare foot on a kneeling bench. Here a delightful informality pervaded an atmosphere of spontaneous, unaffected devotion.

Leaving the church, I strolled, through crooked streets full of strange sights and sounds and odors, down to the waterfront, where I first set eyes upon the ocean that Balboa discovered, here only forty miles distant from that other ocean that I had long known. Wandering still at random, I came at length to the statue of de Lesseps, the French nobleman who promoted the building of the Suez and Panamá canals. Hanging on the trees around the statue were bagworms of such stupendous size that I could not resist the impulse to carry off a few specimens, although it had not been my intention to make an entomological collection. Here science and art, history human and natural, intermixed in the most fascinating manner.

In the afternoon, I hailed a taxi, and was driven across the savanna to the ruins of Old Panamá, where I stood before the walls of the cathedral that the pirate Henry Morgan had destroyed centuries earlier. A fig, germinating in a cranny in the massive wall, had sent down roots to the ground and grown into a large tree. Begonias, orchids, aroids, and shrubby epiphytes flourished higher upon the crumbling masonry. Cattle, seeking the noontide shade in the ruined nave, had left tokens of their presence. Neither printed signs nor cicerones intruded between the pilgrim and the mysteries of time and decay. Such was the setting in which I first watched leaf-cutting ants file in long procession along well-made paths, beneath verdant burdens. As I look back across the years, I think it was here that I first succumbed to the fascination of Latin America, its romantic scenery, its tragic history, its vast variety of animal and vegetable life; I believe it was here that I began to fall under that Circe's spell which at various difficult periods in the following years I tried in vain to exorcise.

But not yet aware of the momentous thing that had befallen me, I was eager to see more. From venerable antiquity I would now turn to modern skill and efficiency, so I requested the taxi driver to take me to Miraflores Locks, where I hoped to watch a vessel pass through—the *Toloa* would not "transit." And fortune continued to favor me that day; as I stood in the rain on the wall of the lock-chamber, a British tramp was lowered slowly through the great gates to the level of the Pacific. Barely had she passed through when the chauffeur told me that if I did not go immediately I would miss the last train for Cristóbal—and likewise my ship. It was already too late to return to the station at Balboa; we speeded away in the opposite direction to catch the train at a way station. The fare that the driver asked seemed exorbitant even for the opulent days of the late 'twenties,' but there was no time to discuss his fee, for the train was approaching—and I had seen too much that day to be in a mood for haggling over money. It was my first full day in Central America!

After nightfall, we sailed for Limón, Costa Rica's Caribbean port, where we docked late on the following morning. The railroad joining the port with San José, the highland capital of the country, had suffered severely from landslides and washouts under November's torrential rains; traffic was indefinitely suspended, throwing passengers' plans into confusion. Fortunately, my destination was not the interior of Costa Rica but the port of Almirante, which we had passed, far offshore, during the night. I had scarcely finished my Thanksgiving dinner on shipboard when I was told that a launch was waiting to take three other passengers and myself down the coast, retracing in part the voyage we had just made.

The seventy-mile voyage eastward along the coast was rough, for the motor launch, jarring violently from the vibration of its engine, wallowed in the trough of the long, smooth Caribbean swell. The mother of the two little girls who had come down on the ship with me from New York was decidedly unwell; but I must confess that I selfishly exulted in that rough sea, for it seemed the fit counterpart of the rugged, seemingly uninhabited land that rose to our starboard side. A low range with jagged crests ran close to and parallel with the coast. Through rifts in the afternoon clouds, we caught glimpses of much higher mountains in the background. From the seashore to the mountaintops, all the visible terrain was darkly mantled by primeval forest. I knew that behind the coast were plantations of bananas and cacao, but neither they nor the dwellings of those who attended them were visible from the sea. As we passed along that wild coast, a few porpoises raced before the launch's bow, but our vessel was too slow to provide that keen competition in speed which they seem to enjoy; soon they tired of our leisurely progress and raced off ahead.

It was dark when we entered the Boca del Drago, which separates Columbus Island from the mainland, but the black captain of our little vessel knew the passage well. Columbus had named this strait more than four centuries earlier, when he passed through it on his last unhappy voyage. The large forested island on our port side, the bay we were now entering, and the port of Almirante, whose lights glowed faintly in the sky far to our right, were all named for the great admiral who was the first European to explore this coast. Soon we distinguished the intermittent flash of the lighthouse on Cape Toro, the northwestern point of Provision Island; then the channel lights of the port of Bocas del Toro came into view. These aids to navigation were maintained, I was informed, not by the government of Panamá, but by the underwriters who insured the vessels of the United Fruit Company, the chief users of these waters. Although the officer of the port came down to the wharf in response to the launch's whistle, he examined neither my credentials nor my baggage; I walked ashore as though I had not come from a foreign country. We were still in those carefree days before economic stress and growing distrust among nations made the passing of certain frontiers almost intolerably vexatious.

The following morning, November's last, was so bright and sparkling that the voyage across Almirante Bay remains photographed on my memory as clearly as though I had made it last month rather than five decades ago. The broad expanse of blue water, bordered to the seaward by a chain of large islands separated by narrow passages, appeared to be wholly landlocked, hemmed in by low shores completely covered by verdant forest, its continuity unbroken by any visible clearing. The forests swept up to the foothills; and these in turn were topped by higher and higher ranges, until in the background towered the commanding mountains of the continental divide, their summits wreathed by cloud wisps. Somewhere among those distant peaks stood Volcán Barú or Chiriquí, 11,265 feet high, the loftiest point in all Panamá. All these mountains were covered with verdure to their summits; neither barren rocky cliffs nor clearings made by man were visible on their flanks.

As we plowed through the tranquil water, Brown Pelicans, hulking barges of the air, glided along close to the surface or plunged below in seemingly clumsy dives. Young Frigatebirds soared easily above us, the white foreneck and breast of the immature plumage contrasting sharply with their otherwise black feathers. A remarkable fish that I have never seen elsewhere lived in this bay. It was long and slender, with a beak like a pike; when disturbed, it rose to the surface and skimmed along with great rapidity, its body standing nearly upright in the air, only its caudal fin touching the water. I saw only two, which appeared so suddenly and moved so rapidly over the surface that I could not clarify the mechanics of their aerial locomotion.

Rounding a projecting point of land, we came suddenly in view of Almirante. From this port, a single-track, narrow-gauge railroad ran inland through the banana plantations, which were served by numerous spur tracks and a network of light tramlines leading off from them. The passenger train, which came down to Almirante in the morning and returned up-country in the afternoon, had been delayed by washouts and was late in departing. When finally it left the station, it went only a short way, then backed slowly into a long, curving spur that led to the hospital maintained by the fruit company. The purpose of this deviation from the usual route was to pick up a mother and her newborn baby, and return them to their home on one of the plantations. I was delighted by the almost Arcadian informality of the proceeding, possible only in a country where Time does not rule supreme, even over railroads.

While the female passengers, chiefly English-speaking North Americans whose husbands worked for the company, admired the infant, the young mother blushed and seemed embarrassed. An elderly wag came to her relief. "What a remarkable girl!" he exclaimed, looking intently at the baby. "She has a mouth, two eyes, two ears, and even a bit of a nose. Yes, a truly remarkable girl!" But he spoiled the effect of his wisdom by repeating it over and over again, with an air of self-complacent sagacity, as each new passenger entered the first-class coach. I inferred that themes for conversation were few among the local gentry, and the passengers welcomed any talk that broke the monotony of the slow ride.

But to me that ride, through vegetation strange and lush, was neither monotonous nor too slow; I regretted that I had only eighteen miles of it. On boarding the train with my free pass, I had been formally presented by an official of the company to the conductor; and now, when we stopped at a branch line curving away through the plantations, the latter descended from his coaches to introduce me in turn to the man who awaited me. "Here he is; you can have him; I've finished with him," the conductor announced with quaint drollery, then signalled to the engineer to proceed. Doubtless my pith topee, a souvenir of my earlier visit to a British possession, betrayed to my host that I was a newcomer to Panamá. All the "old-timers" on the banana plantations wore hats of felt or straw, more suitable for the rainy climate of Caribbean Central America, where the sunshine is never of dangerous intensity.

Standing on the tracks of the branch line was a little, open, red motorcar, propelled by two cylinders attached directly to the wheels on either side. After my host and I were seated on the forward-facing bench, the man who attended the conveyance pushed it until the cylinders began to fire, then jumped on and sat with his back to us. It was his duty to run ahead and throw the switches as we approached them with reduced speed, to turn the car around by lifting one end from the rails when the direction of the journey was reversed, and to keep the machinery in good repair. Afterward, I made much use of the little red car and found travel in it a most satisfactory way to see the country. Since over this soft, deep soil railroads and tramlines were cheaper than highways to construct, this was our principal mode of conveyance, varied by journeys over the waterways, or afoot.

For a mile or so, we chugged rapidly between plantations and pastures, with the wind of our passage cooling our faces. The afternoon shadows were growing long when we dismounted from the car at the end of a concrete path, bordered by tall and slender palm trees, that led up to a bungalow set amid plants excitingly different from those around any house in which I had hitherto dwelt. This was the living quarters attached to the experiment station at which I was to pass my first half-year in Central America. I found it a delightful place, especially during the mild, bright days and balmy nights of January and February, the first two months of the dry season, when the broad-leafed Dracaena trees in the garden shed their delicious lilac scent upon the still night air. The bungalow stood a short distance from the broad, winding Changuinola Lagoon, in the midst of a carefully attended lawn adorned with palms, rosebushes, pink-flowered bougainvilleas, and a variety of other handsome shrubs, all enclosed by a hedge of hibiscus with great red blossoms. This hedge never felt a shears, for Uriah Morgan, the black West Indian "yard boy," kept it perfectly trimmed with only his machete, the long, straight bush knife that is indispensable in the tropics, and probably put to more varied uses than any other tool.

This dwelling was even better equipped than those of the overseers on the neighboring farms, all of whom were well housed. The living quarters were surrounded by a broad veranda tightly enclosed by screens entered through double doors, to keep out dangerous malarial mosquitos. Between two spacious bedrooms was the dining room, with the kitchen and another bedroom at the rear. The cook and the houseboy, who attended us in lordly style, went to their own houses after serving our supper. With three big tanks to catch the rain from the roof, we had no lack of water for our shower baths, even at the end of the dry season. An electric plant with storage batteries gave us bright illumination by night. We were connected by telephone with the towns of Almirante and Guabito and most of the plantations. Once a week, the supply train stopped in front of the house to deliver food and other necessities. When we wished, which was not often, we might ride in the evening on the little red car, through the great swamps of Silico Palms, to Guabito, the railroad headquarters beside the Sixaola River, on the Costa Rican border. Here moving pictures were shown free, once or twice a week, to company employees—which meant nearly everybody in the vicinity.

In this beautiful setting we had, in fact, most of the conveniences of a modern city with none of its clangor, bustle, and annoyances, all in the midst of wildlife that I found unusually rich and varied. Never elsewhere have I lived in greater intimacy with the great-leafed plants of the lowland clearings, with birds both aquatic and terrestrial, with quadrupeds from opossums and raccoons to alligators, which sometimes in wet weather left the lagoon to crawl through the banana groves near the house. In retrospect, it occurs to me that my later explorations might have seemed easier had I started to work in the tropics under sterner conditions; here I learned so much with so little effort, and so little deviation from the manner of living to which I had been accustomed, that undesignedly I formed standards for a favorable locality for studying tropical nature that are difficult to maintain.

In later years, while working alone in the wilder parts of Central America, glad to be covered by any roof that did not permit too much rain to leak through, content to be sheltered by four walls however rough, my thoughts sometimes reverted to the contrast between these rude surroundings and the comfort in which I dwelt during those first six months in Panamá. Sometimes, happy to be content with so little, I contemplated with philosophic satisfaction the contrast between that time and the present; but again, in the long wet seasons, when roof and walls failed miserably to keep the inside of my cabin dry, when food was hardly varied enough for health, I recalled not a little regretfully that early comfort. But the explorer, whether he seeks great things like islands, rivers, and mountains or only, as I, small things like plants and the secrets of bird life, must accustom himself to accept the vicissitudes of fortune as they are dealt out to him. To learn to wait, to suffer passively, to endure: these for an active spirit are the hardest lessons, but for all high endeavor they are basic. Until I had learned these lessons, which had not been included in my college curriculum, I was not equipped for the undertaking on which, soon after leaving Almirante, I set my heart.

A clump of the Silico Palm (*Raphia taedigera*). The tallest trunk will die after its enormous clusters of fruits have ripened.

6

Beside the Changuinola Lagoon

The primary purpose of the experiment station beside the Changuinola Lagoon was to study the banana plant, and in particular to find a way to circumvent the Panamá disease, which was attacking the plantations in Panamá and elsewhere. This disease was a great menace to the industry because, after ravaging the plants over large areas, the infection persisted for many years in the soil, making it unfit for the cultivation of the chief commercial variety, the Gros Michel. Thus it eventually caused the abandonment of whole districts, which at vast expense had been converted from primeval forest to bearing plantations, with their railroads and tramlines and telephone systems and much costly equipment, along with the port works attached to them.

The cause of the disease is a fungus, *Fusarium cubense*, whose microscopic filaments enter the roots from the ground and work upward through the trunk, turning the foliage yellow and finally killing the plant. The fruit

Headpiece: The Changuinola Lagoon in front of the experiment station, near Almirante in western Panamá. A shaded cacao plantation lines the farther shore.

61

companies had spent great sums in unsuccessful attempts to check its ravages. Because of this plague, banana growers were constantly shifting the
field of their operations, abandoning infected plantations to the swiftly
springing second growth, and sacrificing the ancient forest in fresh areas.

This experiment station beside the Changuinola Lagoon had formerly
been supervised by Mark Alfred Carleton, the unhappy wheat-dreamer of
whose triumphs and misfortunes I read, in Paul de Kruif's *Hunger Fighters*,
while dwelling in the same house that he had occupied. In an effort to find
individual banana plants immune to the Panamá disease, Carleton, I was
told, searched among the riotous growth that covered abandoned plantations
for the few that still survived and bore at times dwarfed bunches of bananas.
But these scattered survivors proved, on being tested, no more immune to
the disease than were the neighboring plants that had succumbed.

My host and companion during my six months at this station was its single
investigator, Joseph Permar, a veteran of the First World War. His was the
difficult task of breeding a new variety of banana from a parent that does not
normally produce seed. Because of its size, flavor, and especially its endur-
ance of the long journey from the tropical plantation to the consumer far
away in the North, the Gros Michel variety was, at that time, preeminently
the banana of commerce. But it is highly susceptible to the Panamá disease.
Other varieties of the banana, less suitable for marketing, are more resistant
to infection. Although when self-pollinated the commercial varieties are
always sterile, they will sometimes set a few viable seeds when cross-
pollinated by some of the less desirable varieties. Permar made hundreds of
these crosses, then carefully attended the resulting hybrid progeny through
long months to bearing age. But as he sampled, one after another, the fruits
of these hard-won and slowly maturing plants, he was forced to the disheart-
ening conclusion that none could take the place of the standard, disease-
stricken Gros Michel—indeed, many were hardly edible. His was slow,
discouraging work, demanding a degree of patience and perseverance that
deserved a greater success.

Beyond the lawn that surrounded the house grew bananas of many different kinds, gathered from the tropics of both hemispheres for study and
breeding. The northerner who knows only two or three kinds of bananas,
brought in white-hulled ships from distant ports with strange names, rarely
suspects that the varieties of bananas are as multifarious as those of apples or
roses. Some of the species and varieties at the experiment station had great,
columnar false stems and huge leaves whose tips waved ten yards above the
ground, while others were not much taller than a man. Some yielded massive clusters of fruit that hung downward from the top of the stem; others
produced bunches of much smaller fruits that stood upright in the midst of
the leaves. There were banana plants with red buds and pink buds and

crimson buds; banana plants with smooth fruits and hairy fruits; banana plants with seedless fruits, and others whose fruits were full of large, hard, black seeds. The last, which may have been the wild ancestors of the culti- vated bananas, had fruits with soft, sweet pulp no less delicious than that of the seedless kinds; but the abundance of seeds made them troublesome for people to eat. Nevertheless, they were highly attractive to birds, especially the pretty Blue Tanagers. Like so many tropical birds, the latter remained paired throughout the year. A male and female often came together to feast upon the ripe, splitting fruits of these introduced wild bananas.

When, as a boy, I read, in tales of the South Sea islands and other tropical regions, accounts of savage feasts at which plantains were eaten, I wondered vaguely how the humble weed that grew in the lawn was so altered by a tropical climate that it yielded a large and luscious fruit. Later, I learned that the plantain of the tropics is a kind of banana whose flesh is more solid at maturity, so that it is eaten cooked instead of raw. Typical plantains are much larger than the familiar kinds of bananas, and grow in looser bunches contain- ing fewer fruits. But here at the experiment station we had such a great variety of both bananas and plantains that it was hardly possible to draw the line between them. Some, like the Dominica, could be eaten raw but were better cooked. Even if we chose more technical characters to distinguish the two groups, there were borderline cases that mocked our attempts at classification. One of the interesting intermediate varieties was the apple banana, or apple plantain, so-called for its flavor. Since imported apples were expensive, our West Indian cook made excellent apple pies with this banana.

The white banana flowers, clustered beneath an uplifted red bract, yielded abundant sweet nectar eagerly sought by hummingbirds, White- tipped Brown Jays, and black, stingless meliponine bees. The tongues of these small bees were too short to reach the nectar from the mouth of the almost tubular corolla, so they bit away one of the petals to uncover the coveted liquid.

In addition to providing nourishment for a great variety of creatures from insects to man, the banana plant offers birds attractive sites for their nests, on the broad bases of the petioles, above the highest upturned hand of a bunch of fruit, in the midst of the ponderous cluster, or, for large nests such as those of the jays, in the center of the crown of foliage. In one or another of these situations, I have found nests of Ruddy Ground-Doves, Blue Tanagers, Golden-masked Tanagers, Southern House Wrens, Gray's Thrushes, Great Kiskadee Flycatchers, White-tipped Brown Jays, and other birds. The Black-cowled Oriole perforates a living banana leaf and passes fibers through it to suspend a neatly woven hammock beneath its shelter.

From January to June, I found the frail nests of the Ruddy Ground-Doves, always with two white eggs, on top of the huge, pensile bunches of bananas at the experiment station. To build a nest on such ample foundations as the banana plant affords simplifies the birds' task of construction; but despite its apparent safety, they could hardly choose a more perilous situation. The fruit is often ready to be cut before the nestlings can fly, and to harvest it the whole plant is cut down. If the bunch escapes the fruit-cutter, as it ripens it attracts raccoons, opossums, and other animals, some of which would not hesitate to add eggs or young birds to their diet of fruit.

One night in January, Permar surprised a raccoon eating half-ripe fruit in the small banana grove behind the office. At his call, I hurried out, to find that the animal had retreated to the center of the grove, where she stood facing us, her eyes glowing in the beams of our flashlights. As we stood close beside the plant where she had been feasting, she walked slowly toward us, directly into the glare of our lights. While we wondered what this unaccountable behavior could signify, she continued to approach, keeping the mat of bananas between herself and us. When I moved slightly to see her better, she came still closer and voiced a low grunt. The next thing we knew, she was fleeing at full speed toward the center of the grove, with a young raccoon, hardly half her size, following at her heels. Their bushy, ringed tails vanished together in the darkness between the banana trunks.

Now for the first time we directed the light into the plant where the raccoons had been eating. The beam disclosed a second young raccoon resting between one of the massive leafstalks and the trunk, where undoubtedly he had been the whole time, so close that I might have touched him from where I stood, but so silent that he had remained unnoticed. As he turned his black-masked, gray face into the light, he looked so innocent and appealing that I wished to lift him down, but I was restrained by the thought that his teeth and toenails were sharp and he would probably misinterpret my intentions. Wedged between the leafstalk and the stem, he appeared so uncomfortable that we prodded him gently with a stick to make him descend; but he seemed determined to wait for Mother exactly where she had left him. The mother raccoon, having courageously returned to lead one cub to safety, seemed disinclined to expose herself to apparent danger a second time and did not reappear. Before going to bed, I went to the banana grove again and found the young raccoon in the same spot where we had left him; but by morning he had vanished. Probably his mother had returned for him during the night.

Raccoons were numerous in the vicinity. One night, as I sat writing until a late hour, I heard a continuous low snarling, which sounded much like the growling of dogs engaged in a mock fight. Taking a flashlight, I went out to investigate, and found five raccoons together, milling around, snarling and growling, on a sheet of corrugated iron roofing that lay on the ground behind the tool shed. What was the occasion of their dispute I could only surmise; but they were so absorbed in it that I stood above them, watching in the flashlight beam, for a minute or more before they finally sensed my presence and walked off quietly, two by two, not precipitately as though alarmed, but calmly and deliberately. The sole remaining raccoon approached me until only a few feet away, where it paused, stood on its hind legs and sniffed as though not recognizing by sight what I was, then in a moment followed the others into the darkness.

The little, black meliponine bees, which swarmed in such great numbers to the banana flowers to gather nectar and to palm flowers to collect the yellow pollen on their hindlegs, lived in a tremendous black termites' nest in a cacao tree beside the house. They had provided it with an entrance on the side in the form of a widely flaring funnel, through which they poured in and out in an endless stream all day long. Whether they had found the termitarium already deserted by its little, soft-bodied, white builders, or had invaded it while still occupied and driven them out, I could not learn; but to judge by the good state of preservation of the structure and the experience of other naturalists, I believe that it was a case of forcible eviction. Bees are not the only animals that appropriate termitaries, for several kinds of birds, including parrots, trogons, kingfishers, jacamars, and puffbirds, carve nest chambers in them. Sometimes, too, they are invaded by black wood ants. The householder who has had the beams and boards of his dwelling consumed by termites until they are empty shells may find some slight consolation in knowing that sometimes the tables are turned, and the invaders become the invaded.

The little, black, stingless bees, belonging to the large tropical American genus *Trigona*, are not themselves the most desirable of neighbors. In their indefatigable search for waxy materials to make their combs, they chew the edges from the leaves of lemon and orange trees to obtain the suberous epidermis. Sometimes they diminish the banana-grower's profits by biting the waxy covering of the fruit, thereby detracting from its appearance and market value. Occasionally they attack the hives of domestic bees introduced from the Old World, overcome their much larger, stinging adversaries by biting off their wings, then carry off their stored food.

The most abundant waterfowl on the lagoon in front of our bungalow was the Middle American Jacana, a handsome, bantam-sized bird with a deep chestnut body, black head and neck, and a prominent yellow shield on its forehead. With picturesque metaphor, the West Indian plantation laborers called the jacanas "Christ walkers," because their long, spreading toes support them on floating aquatic vegetation and they appear to walk over the water, like Jesus on Galilee. In wet weather, these birds foraged in the pasture outside our garden and were almost as tame as chickens. When I came too near, they flew a short distance, and on alighting held their yellowish-green wings upright above their backs for a few seconds, in a statuesque attitude that displayed the peculiar yellow spur at the bend of each wing.

Scarcely less numerous, in the earlier months of the year, were the Little Blue Herons, which fed along the lagoon, and in wet weather also foraged in the low, marshy pasture behind the house. Pure white young birds flocked with slate-blue adults, which they outnumbered by about two to one, making a pleasing color contrast as they flew and wheeled together. While molting in May, the young herons became strikingly and irregularly patterned with blue and white; but by this date the number of Little Blue Herons was dwindling, as they left for distant breeding grounds. From time to time, we beheld a snow-white Common Egret standing tall and lean in the shallows, a vision to make any day memorable.

Purple Gallinules and Common Gallinules abounded on the lagoon and among the marshy bordering vegetation. The former often ate the berries of the wild plantain, *Heliconia elongata*, which flourished along the banks. The green flowers of this plant, a relative of the banana, are borne within fleshy, cup-like bracts, which stand in two overlapping rows along the upright stalk. Each cup is almost always full of rainwater, beneath which the flowers develop, and above which, as in certain aquatic plants, they push the tips of their corollas when ready for pollination. Hummingbirds, especially the long-tailed, brownish hermits, are doubtless the chief pollinators; but the flowers are self-fertilized when not visited by bird or insect. The berries develop submerged in these little aerial pools; but as they ripen, the rapid elongation of their fleshy, white stalks lifts them above the water, where their deep cobalt color catches the eyes of passing birds, who swallow them and so scatter the seeds enclosed in hard, indigestible coats.

Instead of waiting until the berries were made accessible to them by the lengthening of their stalks, the Purple Gallinules often tore apart the thick bracts to reach the nearly ripe fruits still hidden within them. These bracts were beautifully tinted with red, yellow, and green, but their beauty was surpassed by that of the lovely violet bird, who grasped the stem with long, yellow toes and tore the bracts with its strong, bright red bill. Together, bird and plant formed a picture that fulfilled my dreams of the beauty of tropical nature.

While rowing on the lagoon one day in late March, I met a Common Gallinule swimming along with six newly hatched chicks. By pulling hard on the oars, I intercepted two of the little ones from their parent and the shore. As the skiff bore down on them, they dived; but one came up again beside the boat and was readily caught. It was covered with soft, black down that shed water so well that it revealed no effect of recent submergence. The little bill already showed the yellowish tip and red base of the adults. The down on the chick's forehead was sparse, foretelling baldness. When replaced on the water, the infant gallinule turned toward the shore, and, paddling like a duckling with its webless feet, it soon reached the marginal stand of Water-Hyacinth, into which it vanished. Its parent, who had retreated up the shore, returned as soon as I withdrew and called until she had gathered her little brood about her again. She swam down the lagoon in their midst, clucking continuously to keep them together.

The lagoon, which twenty years earlier had been the main channel of the Río Changuinola, meandered for about twelve miles between plantations of bananas and cacao and much marshy, uncultivated land at its lower end. When bananas were first extensively planted in this region, the fruit was shipped in barges to the river's mouth, thence by a canal, that ran parallel to the coast, to Almirante Bay and the port of Bocas del Toro. With the development of the port of Almirante and the construction of the railroad, all this was changed; the fruit was sent directly to the ship's side by rail.

On Sundays, Permar and I often made long voyages along this lagoon, in a cayuco hollowed from a solid log or else in a rowboat, usually with Uriah Morgan as boatman. Low shores bore extensive stands of Wild Cane, sometimes forty feet tall. Steeper banks were overgrown with wild plantains (species of *Heliconia*) and shell-flowers, with giant leaves and ponderous inflorescences of many forms and colors. In still reaches floated innumerable Water-Hyacinths, lovely plants with showy clusters of large blue, yellow-eyed flowers. In places this water weed grew so profusely that it clogged the channel and impeded the passage of our vessel, making us forget its beauty while we attacked it vigorously with our machetes.

One of the shorter oxbows, leading off from the main lagoon, was completely covered with two kinds of aquatic plants. The taller, dark green Water-Hyacinth had been drifted by the wind into long, gracefully curving rows; and all the space between was occupied by the lighter green of the floating Water-Lettuce, whose hairy leaves lay close to the surface. It was like a quaint, old-fashioned formal garden, in which the Water-Hyacinth formed the flowering hedges and the Water-Lettuce the velvety lawns. The wind was the playful landscape artist who planned and executed the design. I have never elsewhere seen a water garden so attractive.

As we paddled down the quiet water of the lagoon, great tarpons broke the surface, their silvery scales glistening in the sunshine; flocks of jacanas fled cackling before us while Yellow-crowned Night-Herons perched motionless on logs a few yards away and indifferently watched us pass. Blackish Anhingas with long, snaky necks stood on branches above the shore, holding their wings spread to the sunlight. From the cane-brakes, white-faced Capuchin Monkeys peered out and barked at us, often following along through the canes to keep us in view. On the lower part of the waterway, we saw bands of black Howling Monkeys browsing upon the foliage in the tops of the guarumo or cecropia trees. The male Howlers have a swollen, bony larynx and roar with incredible volume, at dawn, when molested by other animals or by trespassing troupes of their own kind, or when anything displeases them. Especially loud and vehement are their complaints when they feel the first cold drops of a shower. Since our rain generally came from the coast to the north, these bands of Howlers between us and the sea were our weather forecasters. Their far-carrying roars of protest brought us notice of an approaching shower, if we happened to be working among the nearer banana groves, in ample time to hurry back to the house and avoid a drenching.

Although we did not often see sloths, one day in late April we noticed five Three-toed Sloths eating in trees along the bank of the lagoon, all on the same side and within a distance of about one mile. One was a female carrying a baby on her breast while she climbed down the trunk of a cecropia tree that had been defoliated to the last leaf, evidently by the animal who was abandoning it. We watched her until she vanished into the low, dense growth above which the tree stood. Nearby was another cecropia tree that had been stripped of foliage at an earlier date and was now putting forth new leaves. Only one of the five sloths was in some other kind of tree, although there was no lack of other species along the lagoon. The big, palmate leaves of the cecropia are certainly the favorite food of these animals, as they are of the Howling Monkey. Clinging by its long, hooked toenails beneath a limb, the sloth draws in a leaf with a forepaw and nibbles at its edge. It eats so slowly that a single leaf may engage it for half an hour or more. One might suppose that the tree could replace its foliage as fast as a sloth devours it.

Its hunger satisfied, the sloth often moves in to the base of the branch on which it has been eating and starts to scratch itself, everywhere from head to tail. Sometimes it spends five minutes or more scratching a single spot, with a sustained, mechanical motion of a forelimb. At times, hooking its hindfeet over a bough, the animal hangs head downward, while it scratches one arm with the claws of the other. Spider monkeys scratch as long and assiduously as sloths, but of course with more rapid movements; occasionally they hang upright by one arm and scratch it with the other. Aside from eating and scratching, the sloth's chief occupation is sleeping, which it does sitting upright at the base of a stout, nearly horizontal branch, its back resting against the trunk, its head bowed forward until its face is hidden between its arms—a featureless gray mass. The onset of a shower often causes the sloth to rest in this manner.

One of the five sloths that we saw on the same day was eating on a low branch of a cecropia tree on the bank of the lagoon. Without prompting, Uriah announced that he would capture it for me. We pushed the skiff's bow into the marshy bank; he and his helper jumped out and waded with bare feet through the grassy margin into the thicket of tall Wild Canes that covered the bank higher up. Directly beneath the sloth, they shook canes that almost touched the animal, but its only reaction was to move its head from side to side with ludicrous slowness. It remained unperturbed while the two men, with noisy blows of their machetes, cleared a space in which they could operate. Choosing a long, straight cane, Uriah tied a short stick to its end with a length of vine, thereby forming a crude hook. With the other man's aid, he raised the heavy cane and caught the end of the bough to which the sloth clung. Still the animal remained calm, as though no danger threatened. After more maneuvering, Uriah gave the sharp tug on which he depended to snap the cecropia branch. The hook slipped off without breaking it or dislodging the animal, who was, however, shaken out of its lethargy. With what, for a sloth, might be considered precipitate haste, it climbed inward to the trunk, then outward along another low limb until it reached the end and could go no farther.

To be foiled by a creature so dull-witted as a sloth would have been intolerable to Uriah, who successfully hunted such "big game" as the *tigre* or Jaguar and the *león* or Puma. Piling canes against the leaning cecropia trunk, he climbed up on them and barely managed to hook the end of the branch to which the animal now clung. A sudden jerk broke the brittle, hollow bough, which came hurtling over my head and plumped into the shallow, weed-covered water beside the skiff, the sloth still clinging to it.

"The experiment worked, sir," was Uriah's laconic comment on his success.

The sloth lay partly submerged, head buried in the grass, still clinging to the branch and making no effort to escape. When Uriah tried to lift it by a hindleg, it grasped rooted aquatic vegetation so tightly that a few skillful touches of his machete were necessary to cut it loose. Placing the sloth in the skiff's bow, Uriah tied a length of pliable vine around a leg and gave me the end to hold. This was a needless precaution because, although nowise injured by its fall, the animal lay prone on the bottom of the boat, hardly moving, much less trying to escape, during the two-hour homeward voyage.

Sitting before him on the bow thwart, I had ample leisure to contemplate the strange creature that had so unexpectedly come into my possession. He was covered everywhere, except on the face and a patch on the back, with long, coarse, flattened gray hairs, the color of "Spanish moss," and of much the same "feel." The resemblance was doubtless not wholly fortuitous, as both sloths and this bromeliad hang from the boughs of tropical trees. On the back, the long hairs were directed toward the rear and the middle, so that they pointed downward, the better to shed rain, when the sloth hung below a branch in his customary inverted position. Beneath the coarse pelage, the body was covered with a soft, close, brown underfur. On an area between the shoulders, about as large as the palm of my hand, the long, coarse hair was absent, and the underfur was particularly short, fine, and dense. This area was traversed by a broad, black, longitudinal stripe, bordered on either side by orange-buff, which merged in turn into gray, black, and brown at the edges. From a distance, this highly colored area often looks like a pit in the sloth's back. Confined to males, it appears to be a sexual recognition sign, and its presence suggests that sloths possess visual powers that their usual behavior belies.

The sloth's tail was short and stubby. The three strong nails on each foot were about two inches long. The short, close-set facial hair was light gray, contrasting with the black muzzle and the dark brown frontal ruff over his forehead, which resembled a pompadour rather than bangs, as in other species of sloths. In full daylight, the pupils of his muddy brown eyes narrowed to mere pinholes, but at night they dilated widely. Whenever the animal closed his thick eyelids, as he frequently did, the eyeballs receded into their sockets, much as in a frog. His minute earlobes were hidden amid the long hairs on the back of his head; evidently hearing played a very minor role in his life—he had remained unperturbed by all the noise we made while trying to capture him. The sloth's whole expression was one of utter stupidity and helplessness; but as I came to know him better, I found something very appealing in the quiet resignation and absolute lack of aggressiveness that this countenance suggested and his behavior confirmed. A baby Two-toed Sloth, who years later was brought to me in the Costa Rican mountains, was a much more spirited animal.

As the sloth lay on the bottom of the boat, many small azteca ants, which had swarmed out of the hollow cecropia branch, struggled among his coarse pelage, powerless to bite through the thick fur. I placed one on the back of my hand, where even the sparse, short hairs impeded its effort to bite; it floundered around for several minutes before it found a suitable spot. The ant's tiny mandibles produced a slight, stinging sensation, not nearly as painful as the caudal sting of a fire ant. On my palm, I could feel the ant's mandibles grating over the ridges of skin without penetrating it. Nevertheless, as I discovered later, when a number of these ants bite the thin skin on one's face and neck, the effect is not pleasant. I doubted that the azteca ants were responsible for the sloth's long-continued scratching.

Some naturalists have supposed that these ants serve as a garrison to defend the cecropia tree against the attacks of animals of various sorts, thereby paying for their lodging in the hollow trunk and branches and the food that the tree provides for them in the form of tiny, white protein corpuscles on the furry bases of its long petioles. But the aztecas certainly fail to protect the tree from the defoliating visits of sloths, Howling Monkeys, and a variety of insects; while Lineated Woodpeckers damage it severely by pecking into trunk and branches in order to reach and devour the ants and their pupae. Far from repelling guests with the aid of its pampered ants, the cecropia is the most hospitable tree of tropical America.

The long, grooved hairs on the sloth's neck, back, and arms were green with a minute incrusting alga. A number of small, gray moths ran over the animal's back or took short flights over it, to hide again in its abundant pelage. These insects, *Cryptoses choloepi*, were evidently permanent residents, for some stayed with the sloth for the several days that I kept him. The larvae are said to eat the algae that encrust the hairs—an interesting example of multiple symbiosis—but I could find none on this particular animal, nor any parasites. His pelage was not in the least moth-eaten.

As I carried my captive from the boat landing up to the house, he stretched long, thin arms toward the trees we passed—his first sign of volition since we captured him. Placed in a wire cage for observation, he clung all day in a corner, his head bowed forward between his arms, a picture of indifference and dejection. But at night he became more active and climbed slowly around his cage. Although he was given leaves of the cecropia and other trees, he refused to eat. We did not see him drink or excrete during the four days that I kept him.

I have never known a more spiritless, defenseless, unemotional animal. One could pick him up and handle him as one pleased; he never put up the least resistance. He submitted passively when I opened his mouth to examine the fairly complete set of teeth, necessary for grinding foliage, of this member of the Edentata. I know no other undomesticated animal, even including those much smaller, that I could have handled so freely, with such impunity. Aside from a sort of sighing, caused by deep respiration, that he made when pulled from the support to which he clung, I heard no sound from this silent creature; like other sloths that I have tested, he was quite unresponsive to noises. To remove him from his cage, however, was difficult. Once, when I tried to loosen his hold, I found his grip too strong for me. With a finger clamped between his long nails and the wire, I remained captive until Uriah came with a screwdriver to pry up his claws and release me.

When I placed the sloth on a lawn, he turned his head slowly from side to side until he sighted a tree. Then he started to walk toward it, hindlegs doubled up and forelegs stretched out, advancing one limb at a time, with a slow, unsteady motion that reminded me of a palsied old man. His joints seemed rusty. When I permitted him to climb an isolated banana plant, he ascended as far as the plant would support him, then looked around for something taller to which he could pass. Finding nothing within reach, he turned away and drowsily closed his eyes, as though overwhelmed by the problem of existence—his usual reaction to a difficult situation. But he was not incapable of learning. For photography, we set a small guava tree in the lawn. When first released here, the sloth climbed to the top, then stretched one long arm as far as it would reach toward the banana plants beyond. After he discovered that this tree was not a road to freedom, he persistently climbed down rather than up when placed in it.

My first intention was to keep the sloth a few days for observation and photography, then release him in a cecropia tree near the house. But when I searched vainly in the few available books for mention of a Three-toed Sloth with a black-and-buff patch on the back, I wondered whether he represented an undescribed species. Accordingly, I decided to retain him as a zoological specimen, for identification at the National Museum in Washington.

Since no chloroform was available, I put the sloth to sleep with ether. When I first applied the saturated cotton to his nose, he made a few feeble passes toward his head with his forelegs, which I easily warded off. Otherwise, he submitted passively to his fate. Gradually his breathing became slower and his hindfeet relaxed their grip on the cage. Although he made so little effort to preserve his life, it flowed strongly in him; three hours passed before his breathing ceased. He died as he had lived, without ever a vigorous protest, much as some venerable Buddhist mystic, certain of attaining Nirvana, might have passed away. When I saw how easily the sloth surrendered his life, I, who have a strong aversion to taking the life of any creature, felt less compunction about demanding his for science. Later, I regretted having done so. Who can tell what satisfactions a sloth finds in its apparently somnolent, almost vegetative existence? For all its seeming indifference, life may be as precious to it as to the more sprightly monkeys and birds—as to ourselves. It was sacrificed in vain, for it was identified as a male Three-toed Sloth, a species that had long been known.

When curled up asleep in a crotch, its head buried in its forearms, a sloth is not easy to detect. Its hair matches the gray bark of many a tree; it is tinged green with cryptogamic growths, as tree trunks often are; it harbors gray moths, like bark. If not confused with the tree itself, the sleeping sloth might be mistaken for a large, gray wasps' nest, such as are often found on tropical trees. Although well camouflaged when resting so, when actively eating in the open crown of a cecropia tree, a sloth is conspicuous enough, whether moving or at rest; so it is not evident what advantage it derives from slow movement. The monkeys, whose diet of arboreal foliage is quite similar to that of the sloths, are far more active and agile. Both sloths and monkeys are prey of great raptors, such as the Harpy Eagle.

Where the Lagoon joined the broad estuary of the Changuinola River, "Old Joes," as the Negroes picturesquely called the Brown Pelicans, broke the water with heavy dives for fish. From near the mouth of the river, the

Alligator Lagoon led away on a course nearly parallel to the seacoast. This was in many ways the most picturesque channel of all, for its dark water wound between stands of Silico Palms whose tremendous, plumy fronds arched gracefully over the shore on both sides. Fringing the palms was a narrow marginal thicket of the peculiar aroid *Montrichardia arborescens*, with straight, branchless stems that grew to be eighteen feet tall and four inches thick near the base. These soft, succulent stems arose from subterranean rhizomes and were covered with short spines. Standing erect like stakes, they bore arrowhead leaves that were often a foot long. These leaves were held vertically, with their faces toward the open lagoon, and with few exceptions had twisted on their stalks until they pointed to the zenith. The aroids were in turn bordered on the side of the channel by the much lower Water-Hyacinths.

The Silico Palms along the Alligator Lagoon were the outliers of extensive stands that stretched, with occasional breaks, for miles inland, and were intersected by the railroad that ran from Almirante to Guabito. They belonged to a species, *Raphia taedigera*, that grows on the Atlantic coast of tropical America from Nicaragua to Brazil, and has a close relative on the opposite shore of Africa. Some of the groves in the swamplands farther from the coast were much taller and more impressive than those that we saw along the lagoon. The columnar trunks, as much as sixteen inches thick and often fifty feet high, are ragged with the persisting bases of the older dead leaves. The outside of the trunk is so hard that it resists the machete, but the center, as in many palms, is softer and decays more quickly, leaving a hollow tube. Each living trunk is surmounted by a crown of huge pinnate fronds, as long as the trunk is high. Despite their gigantic size, these leaves, composed of hundreds of narrow, ribbon-like divisions nearly two yards long, crowded along the length of the massive stalk and standing in various planes, are as feathery and graceful as ostrich plumes, with no suggestion of the stiffness of the Cohune Palm's much smaller fronds.

Each palm trunk grows for many years without flowering. Then, having attained maturity, it makes a titanic reproductive effort and forms at its crown a huge inflorescence consisting of four or five branches, from nine to twelve feet long, which hang down between the bases of the leaves on all sides of the trunk. By attaching a sharp machete to the end of a long sapling and standing on a ladder that we made by tying poles together, with great exertion we managed to cut down part of an inflorescence. Each branch is crowded with myriad small, woody, yellowish flowers of both sexes, the staminate predominating. The latter are so closely packed on the pectinate ultimate branches that they cannot properly shed their pollen, much of which decays in and around the persisting male flowers.

The far less numerous female flowers develop into fruits of unique appearance. Quite variable in size, the largest are three or four inches long by about two inches thick. Each is wholly covered by broad, triangular scales, which point backward and are of a russet color with a high gloss. Within each fruit is a single large seed, from the side rather than the end of which the embryo emerges. When I brought some of these palm fruits back to the house, the cook asked for one to use as a darning egg. An entire ponderous cluster of them weighs far more than a man can lift.

Exhausted by the vast expenditure of material needed to produce so many flowers and fruits, the palm tree dies after its single crop of seeds has ripened. But it is surrounded by numerous other stems that arise as suckers from its base, somewhat as in the banana plant. The oldest of these daughter palms are by this time nearly full grown and have already given rise to sprouts of the third generation, which in turn bear offshoots whose stems have not yet risen above the ground. All continue to grow until each in turn is ready to fruit. Thus the palm forest consists of a number of separate clumps of trees, each from a single seed and consisting of trunks of various ages.

The wet soil beneath the Silico Palms was almost free of other vegetation. In the more watery parts of the grove, we sank into the muck nearly up to our knees, but our further submergence was halted by rope-like roots that ran horizontally in great numbers at this depth. From these main roots, innumerable, thin, breathing roots grew upward through the saturated soil and projected a few inches into the air, where they branched profusely. Their vast numbers covered the surface like a sward of stiff, brown grass. Where more solid mud was exposed above the water, especially at the edge of the swamp, it was deeply indented by the cloven hoofs of the peccaries who had come to eat the thin, tough, orange-colored inner layer of the palm fruits, which was evidently palatable to them, despite its bitterness to the human tongue.

These gigantic Silico Palms form one of the most unique and impressive plant societies that I have ever seen. With mixed emotions, I wandered through dim passageways between the crowded clumps. The trunks, ragged with the decaying remnants of earlier fronds, were forbidding in spite of the graceful ferns that found there a congenial soil; and the muddy ground held one aloof. But these slight feelings of repulsion were dissipated by the majesty of the huge, plume-like leaves that interarched so high above me, lifting the spirit with a sense of awe and reverence, as in the nave of some grand cathedral. Amid such a titanic display of the forces of tropical vegetation, I felt hardly more effectual, and no more important, than the pretty little frogs, jet black marked with gold above and light blue below, that chirped all day between the bases of the palm trunks.

One of the first strong effects of my contact with tropical nature in this region of amazing biological wealth was a deep and abiding feeling of humility. At the completion of my studies at the university, where I had moved in an environment more or less circumscribed, where subjects were as a rule presented in an orderly sequence, I had begun to feel myself master of the science in which I specialized. Now, less than a year later, plunged into the midst of a vast diversity of living things, nearly all strange to me, that grew and faded in conditions which I had never before contemplated closely, with few books that helped to learn even the names of all these unfamiliar organisms, my feeling of the immensity of my ignorance, the inadequacy of my knowledge, attained almost oppressive intensity. I now knew of a certainty that, in my efforts to cultivate the tree of knowledge, I had barely scratched the surface of the ground.

Pensile nest of the Black-fronted Tody-Flycatcher (*Todirostrum cinereum*) in a *Codiaeum* shrub near Almirante, Panamá.

Swan Key, where boobies and tropicbirds nested.

7

How I Became a Bird-Watcher

As told in chapter 1, from early childhood I have been strongly attracted by feathers and the creatures that grow them, but I never happened to come into close contact with an accomplished ornithologist until I met Frank M. Chapman on Barro Colorado Island late in 1930, when I was already deeply immersed in the study of tropical birds. Neither my reading, diffuse as it was, nor direct experience, had disclosed to me the rewards of the intensive observation of individual birds, nor the stratagems by which their shyness at their nests could be circumvented. Until I went to Almirante, plants claimed far more of my attention than birds, largely in consequence of my education.

In my student days, American universities, led by Cornell and Harvard, were just beginning to give courses in ornithology, and Johns Hopkins was not among these pioneers. There, the department of zoology, directed by the distinguished professor, Herbert S. Jennings, leaned heavily toward his specialty, the Protozoa. Courses dealing with larger animals consisted

Headpiece: A Rufous-tailed Hummingbird (*Amazilia tzacatl*) incubating in her lichen-encrusted nest in a bougainvillea bush.

mainly of anatomy, which was of importance to the many students preparing for the medical school; little attention was given to behavior, the aspect of animal life that most interested me. I read what I could find about living animals, but I took no advanced courses in zoology other than those on the genetics and the physiology of the protozoa. Although deeply interested in the whole of nature, I chose to specialize in botany rather than zoology for reasons that were largely aesthetic: trees, flowers, and vegetable productions of all kinds, with their beautifully symmetric, often fragrant tissues, attracted me as strongly as the procedures, sights, and odors of the dissecting laboratory repelled. Accordingly, until I went to Panamá, birds were only a subordinate interest; but beside the Changuinola Lagoon they so forced themselves upon my attention that they soon became a major interest. Above all, one particular hummingbird showed me what could be seen and learned, what deeper understanding could be won, from the prolonged observation of a single nest.

The principal and avowed purpose of my visit to Panamá was to pursue certain technical investigations of the anatomy and physiology of that noble and paradoxical giant herb, the banana plant, which I had begun to study two years earlier in Jamaica. But in addition to the project of these formal researches, absorbing as I was sure to find them, I came to the richer continental tropics with glowing visions, no less intense because they were imaginative and ill-defined, of the color and beauty and majesty of tropical nature, with its infinite variety of living things. I came also with vague anticipations of intimate contacts with its animate life. It was "Amazilia," a sprightly, glittering green Rufous-tailed Hummingbird, who most helped me to fulfill my yearning to penetrate deeply into the lives of free animals. To her I owe an immeasurable debt of gratitude.

Indeed, Amazilia seemed to go out of her way to open my eyes to the possibilities of bird watching. While I bent over my microscope in the small wooden building beside the banana grove that served as the office and laboratory of the experiment station, examining thin slices of banana tissue, she came to build her nest in a Ramie plant just outside. On the afternoon of December 19, I happened to raise my head from my work and noticed her, resting on a leafstalk barely three yards from where I sat, separated from me only by the window screen. Her odd way of perching attracted my attention; looking more intently, I detected something light-colored almost hidden beneath her. After she flew away, I went out to examine her perch and found there a little tuft of plant down, fastened in the angle between the stem and the hairy petiole with cobweb, a length of white thread, and several hairs from the cattle that grazed outside the small enclosure.

During the following days, as I sat at my work table poring over thin bits of the banana plant, Amazilia worked steadily at her growing nest. The rite of adding a new tuft of down followed an invariable routine. Returning with the material in her long, slender bill, she would alight softly on the unfinished nest, push in the stuff where it was needed, then proceed to shape the structure. Bending down her head and pivoting around and around, she used her bill to mold the nest to the contour of her body. As she pressed the yielding down more closely to her breast, she erected the bronzy-green feathers of her crown and vibrated her folded wings, as though she thrilled in anticipation of the completed nest and the nestlings it would cradle. Then she sat facing in one direction, and from the way her body bounced up and down, I inferred that she was kneading the material with her toes, which were invisible beneath her. Sometimes she darted away and then, as though the shaping and kneading had not been done to her satisfaction, promptly returned with empty bill to resume these operations.

While building up the body of her nest, Amazilia also brought bits of green moss and fragments of gray lichens, which she attached to the outside with liberal applications of cobweb, apparently for ornamentation. It was as though the painters and decorators started work on a house before the masons finished putting up the walls. But always more down was added to her walls and pressed closely to her breast. So, as the structure grew, it became just large enough to fit snugly around the central part of her body, leaving neck and head, rump and tail, protruding beyond its rim.

As is almost universal in the hummingbird family, the female nested without help. The sexes of the Rufous-tailed Hummingbird are hardly distinguishable, but I did not once see two birds show interest in Amazilia's nest outside my window—unless a stranger tried to pilfer down from it, a frequent occurrence here where hummingbirds were numerous and downy material apparently scarce. In the first three days, Amazilia made an excellent start in her building; but during the next four days, rain fell in torrents, soaking the half-finished structure, to which the builder came seldom, accomplishing little. When drier days returned, she resumed her task with renewed ardor and continued until, twelve days after its foundations were laid, the nest had been completed and the first egg was deposited in it. The beautiful little cup now had thick walls composed of vegetable down and fine fibrous materials. The exterior was tastefully decorated with gray lichens and pensile tufts of green moss. Amazilia incubated the single egg sporadically during the following day, but she did not lay the second and last until two days after the first.

The solitary hummingbird faithfully incubated her two tiny, elongate, white eggs, leaving them only briefly to seek food and still more down and cobweb to reinforce the nest. Sixteen days after the second egg was laid, the shells opened and liberated two, minute, black creatures into the nest. A downy day-old chick, that peeps softly and looks around with bright eyes, promptly captures one's affection; a newly hatched songbird, although certainly not a thing of beauty, at least gives promise of becoming attractive; but a hummingbird that has just escaped the shell is a pathetically undeveloped grub, in spite of the fact that, counting from the start of incubation, it is three or four days older than a newborn thrush or tanager. I almost pitied Amazilia as she viewed the outcome of long days of patient sitting: two grotesque, blind, squirming mites, with a mere bump of a bill, hardly longer than that of a newly hatched Whip-poor-will, and dark skin that was naked except for two lines of short, tawny down along the middle of the back. At intervals, these almost embryonic creatures reared up spasmodically, widely opening yellow-lined mouths in voiceless calls for food. Exhausted by this effort, each sank, head drooping, into the bottom of the nest, and sometimes rolled over onto its back, revealing much of its internal equipment through the bare, transparent skin of its abdomen. The transformation of such a creature into a feathered gem with vibrating wings is hardly less wonderful, if less abrupt, than the metamorphosis of a creeping caterpillar into a butterfly with painted wings.

During the first week after her eggs hatched, Amazilia divided her time between brooding, seeking food, and feeding her offspring, which she did in the manner customary in her family, by regurgitation. Perching upright on her nest's rim, she thrust her rapier-like bill into the nearest gaping mouth, pushing it down until it threatened to pierce the nestling's entrails. Then, with convulsive movements of her body, she pumped tiny insects and spiders mixed with nectar into her infant's throat. As a rule, she fed both nestlings each time she came, and sometimes she gave food to each of them twice, alternately. After they grew older, each might be fed four times on a single maternal visit. After delivering the meal, Amazilia usually settled down to brood, repeatedly thrusting out her long, white tongue as she sat in the nest warming her babies.

With these constant ministrations, the nestlings grew apace. At the age of six days, when their beady black eyes began to peep out of their still naked heads and their bills had lengthened considerably, they quite filled the bottom of their downy cradle. By the following day, their eyes were fully open, and the tawny tips of feathers were escaping from their horny sheaths. Before retiring on that seventh night of their lives, I visited the young hummingbirds and found that their mother was not covering them, as she had always done on past nights. When I came nearer, my flashlight's beam revealed a tragic scene. The nestlings were dead, with small, brown Fire-Ants swarming over their bodies. The ants, whose sting is painful to men, filed up and down the Ramie stem, flowed over the nest, and devoured their victims' flesh. I was convinced that they had attacked the young hummingbirds while still alive, for I had seen Amazilia feed them in the afternoon of the same day, when they were too full of life and vigor to have died so soon without an external agent. On another occasion, I saw still smaller ants attack living nestlings of the Scarlet-rumped Black Tanager.

All through the night and the following morning, the ants continued to march up and down the Ramie stem and to swarm over the nestlings' bodies, devouring them. In mid-afternoon, nemesis visited the assassins in the form of a small, yellow and black Banded Wasp, which hovered over the nest, darting down swiftly to attack an ant. In rapid succession, it mangled one after another with its mandibles, leaving each victim doubled up and helpless, vainly waving its legs but unable to perform coordinated movements, until the nest's rim was covered with struggling, disabled ants. After each vicious forward dart, in which an ant was nipped and left writhing, the wasp recoiled so swiftly that the other ants could not grapple with it, only to renew its fatal attacks in an unexpected quarter. After a few minutes of this carnage, the wasp would circle around at a distance, as though to rest before renewing the unequal conflict. But the ants continued to file up and down the stem and flow over the corpses; their heedless multitudes were too much for the prowess of the lone wasp, which, like many wasps, was doubtless also a flesh-eater, eager to feast on the carrion or to carry it to its clay-walled nest. After a half hour of this strife, it flew away. On the following morning, many mangled ants still floundered helplessly on Amazilia's nest.

In this interval, the hummingbird sometimes visited her despoiled nest, but she prudently desisted from alighting on it. The ants continued to swarm over her nestlings until their flesh was wholly consumed and only a heap of tiny bones remained. Then they withdrew.

Some weeks later, another Rufous-tailed Hummingbird found the deserted nest and pulled it apart, to carry the down to the nest she was building on the other side of the laboratory. When she had finished the work of demolition, all that remained of the nest in the Ramie was its base, a shallow cup composed of fibers and fine bits of grass, with scarcely any lining, all darkened and discolored by the weather. Then, six weeks after the massacre of the innocents, I found a single hummingbird's egg resting on the hard, compacted base. Fresh pieces of grass had been added to the walls around it, and new lichens decorated the exterior. Even after the second egg was laid, the hummingbird continued to build up and line the old structure, until it appeared as solid and comfortable as when new. I suspect, from hummingbirds' known attachment to their nest sites, that Amazilia had returned to try again to raise a family in front of my window.

For the first few days after this new set of eggs hatched, everything went well; but even before the nestlings opened their eyes, the first of a long series of tribulations visited them. The leaf on which the nest rested had at last died and become detached, and the nest, firmly bound to the leaning stem with fibers and cobweb, pivoted around to the lower side, where it hung at an angle that threatened to spill out its tiny occupants. I straightened the nest and fastened it in place with pins.

Soon leaves above the nest fell, exposing the naked nestlings to the full glare of the afternoon sun. They sat with necks stretched upward, mouths widely gaping, and glassy, staring eyes. I saw their mother perch on the nest's rim with wings outspread, trying to shield them from the sun's rays; but her position was wrong; her shadow fell to one side of the nestlings and they continued to gasp. Later, she covered them on the nest; but one nestling, pushing up its head between her wing and body, continued to pant. Fearing that they might succumb to the strong insolation, I tried to arrange a sunshade. I had not yet learned how tough and resistant young hummingbirds are.

When the elder nestling was a week old, its body began to bristle with pinfeathers and its eyes opened. By the following day, the tips of the true feathers began to protrude from their sheaths. That night, for the first time, Amazilia failed to sleep over her offspring, which surprised me, although I later learned that such early cessation of nocturnal brooding is usual in hummingbirds of warm lowlands.

The growing nestlings were already rather crowded in their narrow cradle, and next day I noticed that the walls were splitting from the rim downward. A few hours later, I looked up from the microscope just in time to witness a lively scene. A whole side of the old nest had broken away, dropping one of the nestlings to the ground. The other, slipping down, struggled desperately to crawl up on what was left of its ruined home. Clinging by its feet, it tried to support and raise itself by hooking its short bill over the remnant of the structure. Its excited mother hovered above it, holding a tuft of down in her bill, as though she attempted to salvage at least a trifle from the general debacle. I watched breathlessly until the squirming nestling hung by a single foot, then hurried out to rescue the dangling birdling, and lift the other, uninjured, from the ground. Meanwhile, Amazilia retired to the electric wire overhead and continued her distressed twittering.

From a circular piece of stiff paper, I fashioned a shallow cone, which I tied to the Ramie stem as near the site of the fallen nest as I could, then lined it with soft cotton. Next, I placed the nestlings in their substitute nest, which, after a little hesitation, was accepted by their mother, who soon returned to feed them.

Now I received a striking demonstration of the accuracy and persistence of a bird's sense of location. Arriving with food, Amazilia, from force of habit, poised first exactly where the fallen nest had been, then, after a few seconds, dropped to the paper nest a few inches lower. Even after she had visited her nestlings several times in their new abode, she continued to hover first before their former location, neglecting for the moment their present nest in plain view.

I marvelled at the vitality of the young hummingbirds. Exposure to tropical sunshine, a four-foot fall, repeated necessary handling by fingers much larger than themselves, had not injured them. Now a more severe ordeal awaited them: twenty-four hours of rain with hardly an intermission, including some beating tropical downpours. Amazilia had definitely ceased to cover her offspring, and the sparse remaining foliage of the Ramie afforded slight protection. After a night of this severe punishment, I watched the nestlings through much of the dreary day. When the heavy downpours came, and the big drops pelted ceaselessly upon them, the two-week-old hummingbirds sat in the improvised nest with eyes closed and bills pointing straight upward, shaking their heads from side to side when struck by a particularly large raindrop. Their budding plumage gave little comfort, and the cool rain soaked them to the skin.

Amazilia faithfully attended the nestlings throughout the rainy day. In the intervals between the hardest showers she came, her black bill dusted with white pollen from the banana flowers that she had been probing, perched on the rim of the paper cup, and fed her drenched offspring. Often one or the other, or sometimes both, of the nestlings refused to accept nourishment, and then she gently touched its bill once or twice with hers, as though coaxing it to take food. Often, however, it was too cold and miserable to respond. After each visit, she ran over the plumage of the youngsters, or part of the nest, with her tongue, an activity that I never witnessed in dry weather. From the movements of her throat, I concluded that she was sucking up some of the excess water, as some kinds of wasps do when their brood cells become wet.

With such unfailing care, the unfledged birdlings pulled through the ordeal. Then I began to understand the reason for the great abundance and wide distribution of their species, which from Mexico to Ecuador, and from sea level up to about five thousand feet, is one of the most abundant hummingbirds in gardens, plantations, and other deforested country. The nest of these hummingbirds appears to sacrifice utility to beauty; in a region where a substantial proportion of the birds build some sort of covered structure to protect the occupants from burning sun and beating rain, theirs is open to the sky. Moreover, it is too small comfortably to accommodate the two nestlings until they can fly, with the result that their growing bodies sometimes burst it asunder and they fall. The Rufous-tailed Hummingbirds' success as a species springs from the amazing toughness of the tiny nestlings, coupled with the indefatigable attentions of their devoted mothers.

Before they left the paper nest, the fledglings acquired plumage much like that of the adults, although their colors were not so bright, and brown tufts of down still adhered to the tips of the green feathers, making them appear rather rough. Two days before their departure, they would ruffle up their feathers when I touched them, and attack my finger with bills still considerably shorter than that of the adult. While I examined the nest, the first young hummingbird flew from it, aged twenty-one days. Its wings spread and began to whirr; it rose into the air and, uttering a low twitter, darted off until it was lost to view amid the banana plants. Its very first flight revealed power and control. The second fledgling abandoned the nest two days later, aged twenty-two days. Amazilia continued to feed them by regurgitation for a number of days after their departure, but I did not learn exactly how long. In later years, I found that hummingbirds of this and other species feed their young for as much as five or six weeks after they quit the nest.

Although Rufous-tailed Hummingbirds and Ruddy Ground-Doves were already nesting in December and January, not until March did I begin to find many nests of other kinds of birds on the grounds of the experiment station. On April 1, I wrote in my notes: "I am now convinced that the birds here, at nine degrees north latitude, have a definite nesting season, and it coincides with the return of spring." This surprised me. At that time, little was known about the breeding seasons of tropical birds; I had supposed that in a region with a year-long growing season, where low temperatures and severe drought are unknown, birds might raise their broods equally well in any month. In later years, I and others have gathered abundant evidence that throughout Central America the great majority of the birds have a limited breeding season between March and July or August, an interval that includes the end of the dry season (where this occurs) and the early months of the rainy season. Where the dry season is short or practically absent, nesting begins earlier than where it is prolonged. But some birds can be found nesting in almost every month; and certain groups, such as hummingbirds and hawks, have peculiar breeding seasons that do not coincide with that of the great majority.

I had no time to search in distant forests for the nests of rare birds, but some of the common, widespread species that lived in the garden claimed most of the hours that I could spare from my botanical work. Among them were the Black-fronted Tody-Flycatchers, charming feathered mites, no larger than a hummingbird or a kinglet, grayish above and bright yellow below, with a relatively long, flat, black bill, and a black, white-tipped tail that they wagged amusingly from side to side, as they flitted through the foliage catching tiny insects. Their pale yellow eyes were curiously suffused with deep red above the pupil. They uttered low, ticking notes, and the male and female, who remained together at all seasons, answered each other with clear, resonant, little trills.

Like many of the flycatchers of tropical America, tody-flycatchers build hanging nests with a side entrance, rather than a conventional open cup.

Theirs is an untidy structure, composed of vegetable fibers, fine grasses, scraps of papery bark, fragments of weeds, small withered flowers, seed down, or whatever the locality affords, all bound together with much cobweb. Within this ragged, pensile mass is a small, rounded chamber, lined with fine fibers, horsehairs, feathers, or sometimes many withered flowers. In this cozy nursery the eggs, the nestlings, and the attendant female are sheltered from tropical sunshine and showers. Since the nest hangs from a slender twig of tree or shrub or vine, one would suppose that its occupants would also be fairly safe. Unhappily, this is not true; a considerable proportion of the nests are pillaged by tree snakes that remove the occupants without damaging the structure and mammals or predatory birds that tear it open.

I watched a pair of tody-flycatchers build a nest amid the red, yellow, and green foliage of a *Codiaeum* bush on the lawn in front of the bungalow. Both sexes worked, taking rather equal shares; but they proceeded in such a leisurely fashion that, although their structure was already well-started when found on February 25, they were still building a month later. The first egg was not deposited in the feather-lined chamber until March 28. A neighboring pair, who began to build about March 7, took even longer, not laying their first egg until April 17. Subsequent stages of the nesting proceeded at the same slow pace. The two or three tiny, white eggs, incubated by the female alone, hatched seventeen or eighteen days after the last was laid— which is about fifty per cent longer than the larger eggs of a sparrow or a thrush. Because of repeated misfortunes, I have never succeeded in determining the nestling period of the Black-fronted Tody-Flycatcher, but it is probably from nineteen to twenty-one days, as in the related Slate-headed Tody-Flycatcher—about twice as long as that of many northern passerine birds of much greater size.

These tody-flycatchers took about ten weeks to build their nest and raise their two or three young to the age of flight. Many a bigger bird of northern lands raises a larger brood in less than half this time. When I studied this flycatcher, scarcely any information on how long it took tropical birds to rear their families was available, and the extreme slowness of the process, in certain species, surprised me greatly. Other flycatchers with hanging nests take even longer. The Royal Flycatcher, for example, requires twenty-two or twenty-three days for incubation, and the young remain in the long, hanging nest for about the same interval. Few birds of high latitudes, exposed to the hazards of migration or else to the ordeal of a severe winter, could balance their annual mortality if they took so long to raise such a small family.

One of the most abundant birds about the experiment station was the Scarlet-rumped Black Tanager, sometimes called the Velvet Tanager. Slightly larger than a House or English Sparrow, the male is velvety black with a brilliant scarlet rump; the female is clad in nondescript shades of grayish-brown and yellowish-olive. These fruit-eaters travelled through bushy growth in straggling flocks, in which males in fully adult plumage were much less numerous than individuals in female plumage. Early in March, a female built a nest in a small lemon tree between the house and the laboratory. As is customary in this tanager, she worked alone, but her mate often accompanied her as she flew back and forth bringing material. When finished, her nest was an ample open cup, composed of coarse grass, weeds, and dry strips of banana leaves. The outside was adorned with a fillet of a small polypody fern that creeps over trees, and the inside was lined with slender, wiry, dry stalks of an unidentified flower.

After completing her nest, the female tanager laid two eggs, which were pale blue, irregularly blotched with brown on the large end. She alone incubated them, and after twelve days they hatched. Then her mate helped her to bring insects and fruits and place them in the nestlings' uplifted, gaping, red mouths. Both parents were very shy and refused to go to their nest when they saw me watching. Everything went well with this family until the evening of the nestlings' third day, when the parents' great excitement drew my attention to it. Looking into the cup, I found it swarming with small, blackish ants, which filed up and down the supporting branch. One young tanager lay on the grass near the base of the lemon tree, but the other had quite vanished.

I soaked the nest with alcohol to remove the ants, then encircled the supporting limb with adhesive tape, sticky side out, to prevent more of the insects from climbing up. After the nest dried, I replaced the surviving nestling; but since its mother did not return to brood it through the night and it was becoming chilled, I took it into the house and covered it with cotton. Replaced in its nest at daybreak, it was fed by the parents. But the ants returned, and at midday I found the nestling hanging, dead, on a lower branch. A black and yellow Banded Wasp—the same kind that attacked the ants on Amazilia's nest—was eating its flesh. These social wasps dwelt in a clay urn in the lemon tree where the tangers nested.

About the time the tanagers' eggs hatched, a Vermilion-crowned Flycatcher started to build in the same lemon tree. After the ants killed the nestlings, the tanagers' nest remained deserted. Now the yellow-breasted flycatchers claimed this empty bowl for themselves. Transferring the material of her incipient structure to the tanagers' nest, and bringing many straws and weed stems from a distance, the female flycatcher converted the original cup into a covered structure with a wide, round doorway in the side, such as Vermilion-crowned Flycatchers always build. By this procedure, she saved herself some work and revealed a measure of mental flexibility. It seemed an intelligent action. I have not again known a Vermilion-crowned Flycatcher to effect such a conversion, but I have watched the related Gray-capped Flycatchers transform a variety of open nests into the oven-shaped structures that they habitually build.

After completing her nest in early April, the Vermilion-crowned Flycatcher laid, at two-day intervals, three pearly white eggs speckled and blotched, especially on the thicker end, with shades of brown and tan. Only the female incubated, sitting in her domed chamber facing outward, with her tail raised against the wall at the back. After fifteen days two of the eggs hatched, and then the male helped his mate to fill the nestlings' yellow mouths with small insects. When two weeks old, the young flycatchers were well feathered, with olive upper plumage, yellow underparts, and dark heads marked with white superciliary stripes, just like their parents. Whenever anyone approached their nest, the adults complained with high-pitched, plaintive notes that sounded like *cheeep* and *cheee*, the vowels long drawn out. Hearing this alarm, the nestlings, if facing outward, crouched until their heads touched the bottom of the nest, thereby making their white headbands less visible. If the young faced the other way, they crowded against the wall at the rear.

As I walked from the house to the laboratory on a rainy morning in mid-May, my attention was arrested by a great commotion among the birds, chiefly the Vermilion-crowned Flycatchers, Gray-capped Flycatchers, and Gray's Thrushes. A large black and white hawk flew from the Vermilion-crowns' nest to another small tree a few paces away. As the big raptor rested there, the air around him was alive with Vermilion-crowned Flycatchers calling querulously *cheee cheee cheeep*, Gray-caps scolding with a staccato *whip, whip, whip*, and thrushes complaining with loud, melancholy cries, so different from their lilting song. In its yellow talons the hawk clutched the two young Vermilion-crowned Flycatchers, now fully feathered, along with the straws that had covered their nest, all crushed together in one cruel embrace. The nest was a roofless ruin.

When I approached, the raptor flew to a low branch of a fig tree that grew on the bank near the laboratory. The flycatchers and thrushes pursued it, darting so close to its head that it sometimes ducked. Peering through my field glasses in the rain, I distinguished the white-striped head of one fledgling flycatcher, the yellow breast of the other, showing through the straw from the nest in which they were shrouded. The hawk perched quite still, occasionally glancing disdainfully at the small birds who threatened it, sometimes erecting its white, black-tipped crown feathers, but making no move to escape or to devour its prey. After a while, all the small birds withdrew except the thrushes, three or four of whom kept up a spirited offensive, darting by within an inch of the enemy's head, if not actually grazing it, and ever and again repeating their loud complaints.

The sight of the hawk perching there, clutching so complacently the fledglings from the nest that I had watched with such interest for so long, the fruit of so much devoted parental care, distressed me and angered me. Among my baggage was a revolver which my father had insisted that I take for personal protection in what he believed to be a wild and dangerous region. Now, for the first time, I loaded the weapon, and from a distance of about ten yards fired five shots at the raptor. The surprising result was, not that I missed—I had never practiced with a revolver—but that the big bird sat calmly watching me while I emptied the magazine at him, and the jacanas in the neighboring marsh set up a frightened cackling. An hour passed, and still the hawk perched in the same spot, holding its untasted victims, about which flesh-eating insects were already buzzing.

The marauding hawk was still resting motionless on the same branch when Uriah, the great huntsman, arrived with his shotgun. He fired once, and the bird fell like a dead weight into the tangle of bushes and vines below the tree. Taking his machete, Uriah cut his way down the bank, and presently returned holding the bird by its wings, one of which was broken, although I detected no trace of blood on its black and white plumage. Dropped on the lawn, the raptor lay motionless, gazing up at us unflinchingly with hard yellow eyes that revealed neither pain nor fear nor anger. At my suggestion, Uriah delivered the coup de grâce with his machete.

"There now," he said, feeling the strong talons to make sure that the life had quite ebbed out of them, "you won't kill no more chickens, no more little birds in their nests."

From a photograph that I took of the corpse, the culprit was later identified as an immature Black Hawk-Eagle, a raptor that early in the year often soars in the blue high above Central American forests, repeating a far-carrying, melodious cry. Possibly the bird's immaturity was responsible for its strange behavior, its failure to devour its prey or to flee from danger.

Doubtless conservationists, who often appear to have more fellow-feeling for the fierce predator than for the milder creatures on which it preys, will condemn me for trying to shoot the hawk and permitting Uriah to kill it. Others will say that I was unfaithful to the ancient Indian ethic of *ahimsa* or universal harmlessness, of which I approve and have long tried earnestly, if falteringly, to follow. But on this planet where the superabundance of life inevitably results in multilateral conflict and the widespread destruction of living things, it is often difficult for even the most compassionate of men to decide what course to pursue. The detached monk or wandering ascetic, living on alms without economic problems and centering his thoughts on supermundane things, may pass amidst the strife of nature without becoming involved in it. The householder or agriculturist who has property or crops to protect may be ruined if he fails to oppose the creatures that threaten to destroy them. The naturalist or nature-lover who becomes deeply interested in free animals sometimes faces perplexing dilemmas. One who encourages birds or other animals to frequent his garden for food or shelter seems morally obligated to protect them, even at the price of taking severe measures against their enemies. One who dwells in the midst of a fairly harmonious association of birds, as I did at the experiment station, will resist the disruption of this concord by an invader. With ants, snakes, and other undetected nest pillagers, the birds around the house were having a sufficiently hard time raising their families; most of their attempts to do so failed. As an alien who arrived to disrupt a peaceful society, the hawk was removed by the only means at our disposal. Although at that time I had not thought as long and deeply about such problems as I have since done, I believe that in similar circumstances I would still take the same course.

The Gray's Thrushes who so vigorously attacked the hawk-eagle had several nests in the vicinity. One was built on the broad leaf of a Panama-hat Plant growing at a corner of the front porch. The builder had gathered much mud from the lawn, along with pieces of the small plants that grew there, including grass, the chickweed *Drymaria cordata*, the spiderwort *Tradescantia*, and the fern-ally *Selaginella*. In the wet weather then prevailing, these herbs rooted in the hardened mud of the nest's middle layer, sending up green, leafy sprouts all around it. In the midst of this attractive aerial garden, three blue eggs, heavily splashed all over with rusty brown, lay on the fibrous lining. These plain brown thrushes, about the size of the American Robin, were sometimes called "tropical robins" by North Americans resident in Panamá. Although similar, their long-continued, melodious song is sweeter and more varied than that of their northern cousin.

At intervals, I accompanied Permar to Columbus Island, where he went to inspect a property attached to the experiment station. Standing on the high northern extremity of the island, we looked across the Caribbean Sea to Swan Key, a dark mass rising above the blue water a mile or so away. In March, a resident of the island told us that boobies were beginning to nest on this key; but so many things kept me busy at the experiment station that June arrived before I could realize my long-standing desire to visit an islet where seabirds nest.

At seven o'clock on the morning of June 1, our launch left the port of Bocas del Toro and turned westward along the outer side of Columbus Island. As we rounded Norte Point, the isolated limestone mass of Swan Key, rising one hundred and eighty feet above the sea, loomed ahead of us like a grim, medieval castle. Soon we detected a smaller islet close beside it, and as our vessel drew nearer, a third diminutive islet rose above the water between the first two. Finally, as we chugged still nearer, it became evident that these three unequal islands were all parts of the same ragged mass of rock, joined by low isthmuses that had been hidden below the horizon.

We dropped anchor in the lee of the key and went ashore in the skiff, which we pulled up on one of the few stretches of sandy beach, between rocky walls that rose from the water's edge. As we stepped ashore, the air above us was full of Brown Boobies in mature and juvenal plumage, Magnificent Frigatebirds, Red-billed Tropicbirds, and Gray-breasted Martins, all circling about in wonderment over this invasion of their island home.

I began my exploration of the islet by climbing laboriously to the top of the main pinnacle for a general survey. Among the lush vegetation clinging to the steep slopes was a big, shrubby nettle that blocked my way, and stung my hands as I brushed past it. Reaching the fairly flat, table-like summit, I found it covered with a scrap of lofty forest much like that on the neighboring mainland. Here pigeons nested in low shrubs, and leaf-cutting ants travelled their well-worn trails, waving irregular pieces of green leaf like parasols above their backs. Through gaps in the high canopy, I caught glimpses of frigatebirds, boobies, and tropicbirds circling above the treetops. A few young boobies rested on nests at the top of the steep seaward face of this elevation, but most were on the two lower, treeless rock masses over which I looked. So I scrambled down again to visit the heart of the colony.

The coral limestone that composed these humps had weathered most irregularly, leaving sharp, angular projections with roughly pitted surfaces. Deep fissures in the rock were so completely screened by a profusion of fleshy *Cissus* vines, which everywhere covered the surface, that I fell into several before I was aware of their presence. Rock crabs, of colors too varied to describe, in patterns of bewildering intricacy, scrambled over the cliffs among the nesting birds, and hermit crabs scuttled along in their borrowed shells. Crested Gray Basilisk Lizards, whose long toes on the hind feet enable them to run swiftly over the water's surface in bipedal fashion, were abundant on the more open parts of the islet. Coconut palms grew in the soil that had collected at the base of the principal rock mass, some of them rooted twenty feet above the sea. These engaged the boat crew, while I gave attention to the birds.

The Brown Boobies nested in shallow, soil-filled depressions in the rock, or on a level part of its surface. Most of the "nests" were no more than bare patches where the birds sat amid surrounding greenery; some were marked by a few sticks placed around the spot where the eggs were laid. A single booby, who held a stick in its bill as I approached, incubated a solitary egg that rested on the bare ground. Another booby stood above two newly hatched young, gazing at me so intently that it seemed unaware that it was standing on its nestlings with broad, webbed feet, apparently without injuring them. Although Brown Boobies commonly lay two eggs, they rarely raise two young. These were the only recently hatched nestlings and likewise the only twins that I saw; every other nest contained a single older nestling. Prolonged exposure to sunshine might have been fatal to these tender twins, so their parent bravely remained shading them, while all the other adults circled overhead or plunged arrow-like into the sea for fish, leaving their older offspring unguarded. Since male and female boobies look alike, I could not learn the sex of this devoted parent.

Half-grown boobies were covered with soft, thick, pure white down. The bare skin around their eyes and at the base of the bill was lead-colored; their feet were dirty yellow. Nestlings somewhat older had sprouted wing and tail feathers, which were dark gray and contrasted strongly with the white down that continued to clothe their bodies. Still older nestlings, who had acquired their juvenal plumage, were gray almost everywhere, but the gray was of a lighter shade on the lower breast and abdomen. White and dark gray nestlings from adjacent nests sparred at each other when driven too close together by my approach. Many of the dark gray young were already on the wing and wheeled above the island in company with the adults, from whom they differed most conspicuously by having gray rather than white abdomens.

As I approached a nestling booby, it uttered loud, hoarse cries, while it backed with spread wings to the farther side of the circle of bare ground that represented its nest. As I drew nearer, its cry grew louder, until it emitted a continuous *ah-h-h* with its bill wide open, revealing the rudimentary pouch between the branches of its lower mandible. It did not rely for protection wholly on vain vocal protests, but flapping widely spread wings, defended itself with vigorous blows of its bill. It bit my hand, my shoe, or anything else I placed within reach. When one of the nestlings, retreating backward from me, came too near another, it was greeted with pecks and bites, which it repaid with the same coin. This made me feel less like an intruder; I saw that I was treated just like another booby.

The Red-billed Tropicbirds nested on ledges and in sheltered niches in the pitted cliffs. Many of their young were nearly fledged. Taken in hand, the young birds uttered piercing screams, at the same time trying to inflict bites that it was wise to avoid, and struggling violently to escape. "Feathered bulldogs," the British vice-consul at Bocas del Toro called them. Before I was aware of what was happening, the boatmen had removed a number of young tropicbirds from their nests and deposited them by the boat landing, where they set up a deafening uproar. I commanded the men to return these nestlings to their homes. They were accordingly dispersed, but I had no way to ascertain whether each was taken to the niche whence it came, and regretted that I had become so deeply interested in the birds that I failed to keep a closer watch over the activities of my boat crew. It is lamentable that, with the most innocent intentions, we so often disrupt the lives of weaker creatures.

I found one full-grown tropicbird sitting in a miniature cavern in a rocky outcrop. It hissed and retreated deeper into the cavity as I approached. It did not strike when I felt beneath it, but continued its hissing expirations. When I tried to lift it gently, it advanced with a low cry toward the mouth of the crevice. Then it disgorged three fish of moderate size and flew off, leaving an empty nest. Was it about to lay an egg, so late in the season?

I had wished to pass a night on the key, but was told that I would find no level spot for sleeping. Actually, I could have camped very well in the woods on the summit, if proper arrangements had been made. Reluctantly, I tore myself away from the boobies, the tropicbirds, and all the fascinating life of a coral islet. To one who delights in birds, a visit to a colony of sea fowl is one of the most satisfying of adventures.

In this chapter, I have recorded only some of the more outstanding of my experiences with birds during my six months in western Panamá— experiences that yielded growing insight into their characters and patterns of life. Fortunately, while I was at Almirante, I learned of the recent publication of Mrs. Sturgis's *Field Guide to Birds of the Panama Canal Zone*, which helped me to learn the names of the local birds, although I searched in vain for descriptions of some of the common Central American species that reach the Almirante Bay region but do not range eastward to the Canal Zone.

Meanwhile, with the mammals, in which I was equally interested, my contacts were more fleeting and less satisfying. Our nocturnal adventure with the mother raccoon, who advanced in the face of apparent danger to rescue her cubs, could not fail to leave a warm glow in my breast; she was no less devoted to her offspring than Amazilia or any of the other birds. But I never, to my knowledge, saw her again; she had no discoverable nest to bind her to a single spot where she could be visited day after day, as did the birds; my encounter with her was an isolated episode, not a continued and growing association. And so it was with most of the other furred creatures that I chanced to meet; my contacts with them were all too transitory. They were shy; or they were active only at night, when it was difficult to see what they did; or they raised their families in dark burrows where I could not watch them. Only the clans of black Howling Monkeys, numerous along the lower stretches of the Changuinola Lagoon, seemed to be sufficiently stationary, open, and diurnal to permit sustained watching. But the nearest were a long way from my abode, and I could not visit them often. These monkeys have since been studied more thoroughly than other Central American mammals.

So it happened that, when I left Almirante to go northward in June, I was convinced that by studying birds I could best satisfy my growing curiosity about the life, especially the psychic life, of warm-blooded animals other than man; so it was that I became a student of birds.

8

Lancetilla

On my way northward from Almirante in June, I visited Guatemala City and Antigua, then spent ten days at the Lancetilla Experiment Station near Tela, on the northern coast of Honduras. During the months following my return home in July, I was busy preparing my Panamanian studies for publication and reading as widely as I could about birds and other things in which I had become interested on my recent trip. A renewal of my research fellowship permitted me to continue my studies, and on Professor Johnson's prompting I visited England and Germany. But in Europe I did not find any research that absorbed me, and without intense interest in my subject I could not do good work. Accordingly, to the distress of my sponsors at the university, I returned to the United States, and from there I arranged to go back to Central America to continue my study of the banana plant in a region where it thrives, and to learn more about tropical birds. The experiment station beside the Changuinola Lagoon was about to be abandoned, but I was welcomed at the Lancetilla Experiment Station of the same company.

Headpiece: Lancetilla Experiment Station, amid the steep, forested foothills of Caribbean Honduras, in 1930.

Although Lancetilla is five hundred miles north of Almirante, it is, climatically and biologically, in the same zone. From Panamá to British Honduras, the same lofty, dense, ever-verdant forests covered the Caribbean lowlands of Central America, nourished by the same heavy rainfall and high humidity. In Honduras and Guatemala the dry season begins a little later, may last a little longer, and be somewhat more severe than in Costa Rica and Panamá, but except in certain narrow valleys in the rain-shadow of the mountains, it is never so prolonged that a considerable proportion of the trees become simultaneously leafless, as on the Pacific side of the great Central American isthmus.

Along all this humid Caribbean coast, the population was, in the early part of this century, sparse and scattered; the few small towns existed, in the main, because they were ports essential to the commerce of the interior, or because they were centers of the banana industry. This fruit was the principal product of the region; when in any district its cultivation waned as a result of the ravages of the Panamá disease, most of the plantation lands were recaptured by the wilderness, while the human population dwindled. Everywhere in these hot, humid lowlands, which until recently were scourged by malaria, the African thrives better than the European or the American Indian; accordingly, the population becomes increasingly black. Many species of lowland plants range from end to end of Caribbean Central America, and the bird life, especially from Almirante Bay northward, is rather uniform. The lowlands of southern Central America are richer in species of birds, possibly also in individuals, than those of Honduras and Guatemala, but many interesting forms are restricted to these northern republics.

Late in April, 1930, I landed at the palm-shaded Caribbean port of Tela, where a single, long, wooden pier jutted out from a wide crescent of sandy beach. This banana-exporting port was situated on the open coast; when in the winter months fierce northers stirred up an angry sea, ships were unable to dock. The town was divided by the narrow Tela River into two contrasting sections: "Old Tela," the rather squalid native quarter on the right; and, on the opposite side, "New Tela," where the fruit company had built modern bungalows for its foreign employees, a hospital, a clubhouse, and other structures. More picturesque than either of these were the Carib villages of San Juan and Triunfo de la Cruz, situated on the coast at nearly equal distances on opposite sides of the port. Each was about an hour's easy walk from Tela along the broad, sandy beach.

These Caribs traced their origin to that fierce anthropophagous tribe which Colombus found in the Lesser Antilles, from whose name, by an etymological quirk, the word "cannibal" is derived. Only the male members of the conquered tribes were thrown into the cooking pots; the female war captives became concubines and household slaves of the sea-robbers. These unfortunate women preserved their mother tongue and transmitted it to their daughters, with the result that in Carib households two languages were spoken, one, it is said, understood by the females alone—a great convenience at times, I fancy. Until late in the eighteenth century, remnants of this bold race continued to be so troublesome on some of the Lesser Antilles that the British government finally dealt with them in the same drastic fashion that it had earlier used with the French colonists in Acadia: it bundled them on shipboard and deported them to the Bay Islands in the Gulf of Honduras. Thence they gradually migrated along the mainland coast to Nicaragua and British Honduras, now Belize.

These Caribs had so largely intermarried with Negroes that in the features and complexions of the inhabitants of San Juan and Triunfo I could detect little trace of Indian ancestry. Although Negroes in outward aspect, they clung proudly to the once-feared name. Despite the unpleasant etymological connotations of their name, I found them agreeable people, whose smiles revealed splendid sets of glistening teeth. Their villages, built upon low, sandy ground close to the shore, were scrupulously clean, in pleasing contrast to such "native" settlements as Old Tela. The houses were made almost wholly of the noble Cohune or Manaca Palm, abundant along this coast. All were thatched with its huge, pinnate fronds. In some, these same fronds, neatly woven into mats, formed the walls. But the more substantial dwellings were walled with slabs split from the hard outer layer of the palm trunk, placed with the smooth side outward. Houses built of such slabs neatly fitted together, with the corners, windows, and doorways framed with sawed boards, were tight and most picturesque. Most of the buildings had no floor except the hard-packed, sandy soil.

These villages differed strikingly from purely Negro settlements in the absence from the dooryards of the flowers and ornamental shrubbery in which the West Indian Negro delights. Possibly the sandy soil was unfavorable for flower gardens. Nevertheless, with their neat palm cottages and prevailing cleanliness, these villages were most attractive. Could I have changed the color of my skin, I might have found it in my heart to tarry in Triunfo, the prettier of the two settlements, where I saw evidence of neither wealth nor poverty, but all seemed to enjoy enough of the good things that nature so freely provided on this fertile coast. The only product of the machine age that had conspicuously invaded this Arcadian setting was the sewing machine—one of the least upsetting of modern inventions.

Despite the predominant admixture of Negro blood, these Caribs preserved their ancestral attachment to the sea. I found men engaged in carving cayucas from solid tree trunks; in these narrow vessels, they would venture far out on the open ocean to fish or catch turtles. At a period when boys in the North were making model airplanes, the Carib lads built model sailships, some large and excellently finished, with intricate rigging. It took all my tact to persuade a boy named Claudio to pose for a photograph with his model schooner *Mariposa*, but after I had made an ally of him, I had no difficulty obtaining additional subjects for my camera. A promise of a copy of the photograph was a greater incentive to pose than a cash reward. Claudio wrote his name and address in my notebook in so clear and firm a hand that I had no doubt that penmanship was taught with more care in the little thatched schoolhouse of Triunfo than in the large, elaborately equipped schools in my own country.

From the port of Tela, a light tramline wound up the narrow valley of the Tela River, between the precipitous mountain slopes characteristic of the Honduran littoral. It passed through nearly impenetrable second-growth thickets, choked with vines and the huge leaves of wild plantains and shell-flowers, among which stood giant cannas that in April raised long spikes of scarlet blossoms higher than I could reach. In the moister parts of the valley floor grew willow trees with shapely, rounded crowns and dainty foliage that seemed out of place amid this riot of tremendous leaves. Here and there, the bush was interrupted by weedy pastures in which little White-collared Seed-eaters swarmed. Graceful Turquoise-browed Motmots perching on the telephone wire dived into thickets in front of the advancing motor tramcar, and jet black tanagers surprised one by turning a brilliant scarlet rump as they fled. Four miles inland, the tramline ended at the extensive orchards and plantations of the Lancetilla Experiment Station, where Dr. Wilson Popenoe had gathered a large collection of tropical fruit trees and shrubs for propagation and distribution among the neighboring republics.

I was given a room in the airy staff house, which stood upon a hill, overlooking the plantations, the office in their midst, and beyond that a grove of tall and forbiddingly armed bamboos. Through the midst of this grove, where Spotted-breasted Wrens sang sweetly throughout the year, flowed a rivulet prettily overarched by the enormous yet gracefully plumed sprays of the bamboos. On the other side of the office, beside the meteorological station, was a cluster of eucalyptus trees that had thrived wonderfully in the moist lowland heat and shot up sixty feet in four years. Beside these grew Ilang-ilang trees, whose wide-spreading, green flowers poured an unrivalled fragrance upon the evening air. This grove of introduced trees was a favorite resort of the Gray's Thrushes, which abounded in the plantation, and in the spring months greeted every dawn with a volume of sweet melody that saturated the air. They roosted by night in another avenue of lofty bamboos that screened the snake pit, where the poisonous reptiles of the region were kept for the extraction of their venom, to be sent north for use in the preparation of anti-venin serum.

The staff house, which was my comfortable home for more than half a year, was fronted by a long, screened veranda that at one end projected outward to form an L and provided a place where we could rest exposed to the breeze. This arrangement for our comfort was disastrous to hummingbirds. Birds flying toward the projection could see trees through the enclosing screens, and since in all their experience they could fly directly to what they saw, they often dashed at full speed into the fabric of fine wires. The momentum of their flight drove their slender bills into the meshes with such force that they were difficult to extract. Sometimes, after recovering from the shock of sudden impact, the hummingbird managed to liberate itself by its own struggles. More often, it hung there until somebody in the house noticed it and gently pushed its bill outward. But at times nobody arrived soon enough to save the poor captive from a slow and horrible death; for hanging so in the sunshine, it succumbed in a few hours and was found, too late, a lifeless corpse.

Among the several kinds of hummingbirds that got caught in the screen were some that do not live in the open and were only crossing the clearing between the woods on either side. One lovely hummingbird, clad in ultramarine, green, and deepest purple, whose lifeless body I found hanging in the screen, was of a kind so rare that I never saw another.

Birds do not dash blindly against the side of a house. Of all the many yards of screen across the front of this building, they habitually struck only those parts through which they could see trees or other vegetation beyond. I have lived as long in other houses with an even greater area of screen, about which hummingbirds were at least as numerous as here, without seeing a single bird trapped, because there were no screened projections, or, if present, they were in dark, shady places. But hummingbirds are not the only feathered creatures that try to fly through wire netting. I saw a photograph of a toucan who had thrust his great beak through a house screen and hung there until captured.

Around the head of the Lancetilla Valley, mountains rose steeply to heights of fifteen hundred or two thousand feet. At the time of my visit, these hills were nearly everywhere covered by magnificent primeval forest. Looking over the roof of this forest from the cleared valley, we beheld, amid the vast expanse of green of varied hues, isolated patches of red or pink or yellow, each the crown of a single tree in full bloom. But as we wandered beneath the forest canopy, we rarely encountered any flower, except the small, inconspicuous ones of palms and aroids, and those bright fallen blossoms that often littered the ground beneath a single tree. The source of these was usually so far above us that, looking upward, it was impossible to tell whether they came from the tree itself or from some great woody vine that spread over its uppermost boughs.

The most abundant large palm in this forest was the Cohune or Manaca, of which the houses in the Carib villages on the coasts, and many others, were largely constructed. Its massive, columnar trunk grew to be sixty or seventy feet high and eighteen inches thick. Each bore at its summit a rosette of huge, stiff, flat, pinnate fronds. In these mountain forests we also found small trees of *Dracaena americana*, one of the few monocotyledons whose trunk and branches are thickened by a cambium, as regularly occurs in woody dicotyledons.

Contrary to a prevailing impression, it was not difficult to walk through this lofty rain forest. The arboreal canopy was so high and dense that little light filtered through to the ground, which accordingly supported a rather sparse undergrowth, through which we could move with comparative ease. It was otherwise in the dense thickets that sprang up on the cleared floor of the valley, where every forward step was won by hard labor with the machete. The undergrowth of the high forest consisted largely of low palms, principally the Lancetilla Palm, cruelly armed with long, flattened spines which it was prudent to avoid, for their sharp points broke off in one's flesh. From these palms the valley received its name, "little lance." Beneath the palms grew very little herbaceous vegetation. The ground was covered with brown fallen leaves, which in the dry season crackled loudly as we walked over them, bringing memories of tramps through autumnal woods in the North. But here were no autumnal colors, and the temperature suggested August rather than October.

Conspicuous among the dark green of the palm leaves and the sombre brown of the ground litter were three brightly colored objects, and scarcely more than three. The first was the spike of large, scarlet flowers of *Aphelandra aurantiaca*, a low herb belonging to the acanthus family, which bloomed throughout the year. The second was the Morpho butterfly, whose wide wings flashed intense blue as it pursued its erratic course between the great columnar trunks. But the butterfly had a rival in a humble lizard with three golden stripes running the length of its glossy black back. As it scuttled off among the dead leaves at our approach, this little reptile dragged a bright cobalt tail as intensely colored as the Morpho's wing.

The birds in the lower levels of the forest were mostly clad in sombre hues like those amid which they dwelt; the brightly colored kinds remained high up in the canopy, where it required much tiresome neck-stretching to glimpse them. One of the most abundant birds of the forest undergrowth was the Lowland Wood-Wren, a diminutive, brown, white-breasted bird with a voice of wondrous power and sweetness. I often found the frail, globular nests that it builds in vines or saplings a yard or two above the ground and uses only as a dormitory, one wren sleeping in each little shelter. But the more substantial nests that it makes for its eggs and young are so well hidden near the ground that I rarely saw them.

A favorite walk was along the trail that led to a summit on the southeastern flank of the valley. One afternoon in May, two Guatemalan students of horticulture and I went to pass the night in a palm-thatched shelter that had been constructed on this trail, at about twelve hundred feet above sea level. As we toiled up the steep path, we were soon so bathed in perspiration that we could hardly raise our eyes to admire the splendid trees beneath which we passed, without being blinded by the acrid drops running down from our foreheads. The labor of climbing absorbed all our energy and attention. When we offered water to the two Punjabi plantation laborers who helped to carry our hammocks and food, they squatted on their heels and held a cupped hand to their mouths to catch the liquid that we poured from our canteens, just as one sees them drinking from a water-skin in pictures of their native India. On reaching the shelter, they again refreshed themselves at a water hole in a granite pocket at the bottom of a dry ravine, then hurried downward to reach the valley before dark, while we busied ourselves preparing camp.

At sunset, a delightfully refreshing breeze sprang up, blowing briskly and swaying the high treetops. It drove away the fatigue of the warm day and the steaming upward pull and was itself sufficient reward for the toilsome climb. Unfortunately, it was of short duration and died with the day. Sunset was only a rosy glow in the western sky, glimpsed through gaps in the foliage; there was no outlook from our forest camp. At dusk, a band of brownish, white-breasted Spider Monkeys, indistinctly seen in the dim treetops, approached us, doubtless impelled by curiosity about the campfire and the terrestrial bipeds who moved around it. After barking at us and chattering noisily together for a few minutes, they swung off down the mountain and left us alone.

Now most brilliant sparks, rivalling Venus in their refulgence, blew up from the depths of the forest and came straight to our fire. Many ended their luminous courses in the very flames, which to my distress quenched their brilliance in a twinkling. These fireflies of the genus *Pyrophorus* bore three distinct lights, two that were greenish, like eyes luminous in darkness, on top of the thorax, and a third, at the base of the abdomen, which was yellow and even more brilliant but was visible only when the insect flew toward us. Those that fell on their backs righted themselves with a click and a bounce into the air, like other elaterid beetles, but some more deftly turned over without employing their clicking mechanism. They were, as usual, crepuscular rather than nocturnal; after it was quite dark, we saw fewer of them.

I have seldom passed a quieter night than that in the midst of the rain forest. The only sounds were the chirping of tree crickets and other insects, the peeping of tree frogs—such sounds as one may hear any summer night in woods almost anywhere. In the eastern United States at this season, these weaker voices would have been drowned by the Whip-poor-will's interminable repetition of his name; here we had not even his local representative, the Pauraque, a bird of the cleared lands who was now singing far below us in the valley. We heard no cries of the larger mammals, nothing louder than the low, mysterious hooting of a solitary owl. Only the impressive height of the trees that towered into the darkness above us, the thickness of the woody vines that hung from them, were proof that we were camping in tropical rain forest.

I woke with the first light but was disappointed with the morning bird chorus. There were a score of voices in the choir which at this season I heard every morning in my bedroom in the cleared valley for every one that greeted me here. One voice in that meager chorus atoned for its lack of volume and variety. It was a voice that we had noticed just at sunset on the preceding evening, the clear, exquisite notes of a White-throated Thrush, resembling those of the northern Wood Thrush, which arose from the palm-covered mountainside below us. It led me down a precipitous slope into a ravine through which the tiny precursor of a river trickled over a rocky bed, interrupted here and there by little sandy pools margined by brown dead leaves. As I stood there among low palms, I heard the clear notes float down through the foliage again and again, but I caught only fleeting glimpses of the wary songster.

As the three of us descended the trail later in the morning, the noise of our progress over rustling dead leaves frightened away most of the quadrupeds and birds; we saw few except the Yellow-thighed Manakins, who were not shy. With a scarlet head contrasting sharply with his black body, this brisk, minuscule creature is one of the most common and conspicuous birds of humid forests through the length of the Caribbean lowlands of Central America. Throughout a long breeding season, each male passes most of his time on one particular, slender, horizontal branch, which, according to the region, is either high or low. Here he advertises his presence by long, thin whistles and a variety of other notes, by loud snapping noises made by beating his short wings, by jumping about, displaying his bright yellow pantaloons, and indulging in various other queer antics. He has a whole bagful of tricks. As a rule, three or four male manakins establish their display perches close together, and while awaiting the arrival of a female, they pay friendly visits to each other on intermediate branches, where they practice their stunts for their mutual diversion. But each scrupulously respects the display stations of all his rivals, even during the exciting moments when a female arrives. The obscure, olive-green female builds her frail, shallow nest, hatches her two mottled eggs, and raises her young without the least help from a mate.

On a later occasion, we descended from the palm shelter by way of the stream where I had listened to the thrush. It was a memorable experience to follow the isolated pools in the rock basins until they were threaded together by a thin trickle of running water, then to accompany this trickle downward until it grew into a prattling stream, large enough to lisp the language of running water, and at length, augmented by various tributaries, swelled to a rushing mountain torrent that surged and foamed over its rocky, precipitous bed with a brawling that drowned our voices.

The descent was precarious; leaping from boulder to boulder, we crossed and recrossed the channel to find the most practicable way, often letting ourselves down five feet at a jump. Looking downward, we beheld a succession of isolated limpid pools, often broad and deep, hemmed in by rocky walls, through a gap in which the pent-up waters found an outlet, to rush headlong between the boulders until the next basin provided another temporary resting place. When we turned to look backward and upward, the scene was surprisingly different. Now our eyes were greeted by more rock than water rising above us in the long, narrow avenue between the columnar trunks of the forest giants, all festooned and draped with heavy vines and epiphytes. Only here and there, often at long intervals, a white cascade of foaming water poured visibly between the huge boulders piled in vast disorder, to lose itself again in the stony bed, through which it rushed unseen but not unheard.

At a narrow and difficult pass in the stream, I stretched out an arm to support myself by a boulder at my side, but noticing a small snake coiled up beneath my hand, I withdrew it with alacrity. The snake was a young *Barba amarilla* or Fer-de-lance, one of the most dangerous serpents of the lowlands, slightly over a foot long, but already possessor of enough venom to endanger a man's life. My companion, Ray Stadelman, had passed that way just ahead of me, and, miraculously enough, had escaped either noticing the snake or receiving its strike. Returning at my call, he captured the reptile to take back to the snake pit, of which he was in charge.

Not far below this point, we found a nearly full-grown snake of the same kind, four feet long, resting on a boulder beside a large rock pool. As the snake-hunter advanced toward the serpent, it slipped into the pool, but with the hooked stick that he carried for this purpose, he pulled it out, and after a lively tussle succeeded in seizing it by the neck. Now he wished to place his latest captive in the little flour sack that he carried, although it was only a third the length of the reptile that it was intended to contain and, moreover, already held a dangerous occupant. As the unfortunate person elected to hold the bag, I insisted that the big *Barba amarilla* be introduced headfirst; but the snake-tamer pointed out that it would then be able to double back and might bite either of us, while—so I gathered from his explanation—if he slipped it in tail first, I alone would be exposed to its fangs after he had removed his hand from its neck. I bowed before his superior experience in these matters, but the snake jumped out on the first attempt to bag it, and only after another scuffle was it recaptured and secured in the sack.

I mention these snakes in such detail because they were, up to that time, the only two really dangerous ones that I had met in much tramping through the tropical lowlands, and the first of the kind that Stadelman had seen in several months of snake-hunting in Honduras. Although Humboldt commented upon the abundance of venomous serpents along the banks of the Orinoco River and in the humid mountains of Choco, the consensus of explorers and naturalists seems to be that they are rather seldom encountered in the humid forests of tropical America. Poisonous snakes appear to be more common, or at least more frequently seen, in certain extra-tropical regions, such as the southern United States. Nevertheless, while wandering through tropical woodland, the prudent person is never wholly forgetful of this lurking menace of sudden and horrible death. Stout shoes and thick puttees give a feeling of greater security, for most victims of snake bites are struck below the knee.

The real terrors of the American tropics, however, are not snakes, scorpions, tarantulas, and jaguars, which are by no means so abundant in nature as in a certain class of sensational literature, but such small and unimpressive but at times very abundant creatures as mosquitos, ticks, redbugs, and ants, any one of which causes in the aggregate more pain and discomfort, if not actually death, than all the famed monsters of the tropical lowlands taken together. In fairness, though, I must add that I have never suffered so severely from insects in the tropics as from the combined attacks of mosquitos and little black flies in northern woods.

Climbing the trail to the thatched shelter, one Sunday morning in October, I had stopped to listen to the frog-like song of some Rainbow-billed Toucans when I was accosted by a man coming down the path with a shotgun on his shoulder. Beneath an old blue cap set backwards on the back of his head, his brown face was covered by a bristly growth. While he was still several paces distant, I caught a whiff of acrid wood smoke that emanated from his rough clothes. "Surely," I said to myself as we exchanged *buenos dias*, "here is a squatter in the hills who knows all their secrets." He had not yet shot anything that morning and asked permission to accompany me back up the trail, which I granted somewhat reluctantly, since as a rule one man alone in the woods sees more of its animate life than two together. He lost a little glamor in my eyes when he told me that he was a carpenter who worked for "The Company" and spent only his Sundays in the bush.

Whenever I stopped to watch birds, my self-elected companion squatted on his heels, waiting in patient silence until I was ready to continue onward. He followed me along the top of a precipitous slope, where the falling away of the level of the treetops admitted more sunlight beneath the crowns of the trees that grew on the summit itself. Such a sharp ridge is an excellent place for finding birds; here one usually sees more of them than in the sombre depths of the neighboring forest. This is not only because one commands a view of the upper levels of the trees on the slopes below; birds appear to be more abundant among the lower branches and undergrowth on the summit itself. I believe it is the sunlight in these localities that attracts them; they seek relief from the forest's gloom in its lighter glades.

We followed this ridge until it dropped off abruptly on three sides, then retraced our steps and continued upward along the trail. We had not gone far when I paused to watch an unfamiliar woodpecker working its way up a tree trunk. My companion, who had been leading, stopped a few paces in advance of me and presently pointed into the branches of a low tree. I looked in the direction that he indicated but at first saw nothing out of the ordinary; he raised his gun and motioned to me to sight along the barrel, which pointed to a kind of toucan—*pico navajo*, he called it—new to me, perching so quietly that I had failed to notice it. Its plumage was rich and beautiful without being in the least conspicuous. The back and wings were olive-green, the head black except for a conspicuous yellow band that arched over the eye and continued down the side of the neck. Below, the bird was black, with a bright red patch under the tail and chestnut thighs. The long, heavy, knife-like bill was greenish-yellow above and olive below, but it exposed a red interior when the bird opened it to bite off a hard fruit growing near its perch.

As I stood delighting in this unfamiliar bird, the man beside me asked softly, "Do you want it?" and I heard the click of a shell pressed into the chamber of his shotgun. The name *Selenidera* flashed into my mind, but I did not know this bird. Perhaps the splendid toucan before me was unknown to science and I might add a wonderful new bird to the known avifauna of Central America, possibly even have it named for me—and it was a sure shot. For a moment, the collector's impulses contended in my mind with those of one who delights in free creatures for their own sake. While I hesitated, the hunter fingered his trigger and the bird's life hung in the balance. But the more generous feeling won; just in time to save the beautiful creature, I replied, "Do not shoot it," and the man lowered his weapon.

Whether new or old to science, to me, at least, this toucan was a new species, and I did not wish the pleasant memory of my first meeting with it to travel down the years disfigured by an ugly blood stain. Moreover, I could see no great advantage in knowing the name of a bird about which I knew nothing more essential. Later, should I have the good fortune to learn about its mode of life, and I could not identify it from published descriptions or the examination of museum specimens, it would be time enough to collect a specimen for identification. Finding the fruit in its bill stubborn and inedible, my first Yellow-eared Toucanet dropped it and flew off through the forest. Nearly four decades were to pass before I again met one of these small toucans, this time in the Caribbean lowlands of Costa Rica, where I watched a male, then his mate, bathe in a pool of rainwater that had collected in a hollow high up in a great tree trunk at the forest's edge.

Despite his rude exterior, my new companion had on this occasion displayed courtesy and fine feeling that won my respect. We continued upward into a region where the greater profusion of mosses on trunks and branches indicated more prevalent moisture. Ferns were also more abundant; small tree ferns now appeared, and massive horizontal limbs of the great trees were heavily draped on the sides with delicate filmy ferns, while a thick tapestry of hart's-tongue ferns hung below, and erect kinds grew luxuriantly on the upper side. Far below, through a gap in the foliage, we glimpsed the blue Caribbean, with a large island in the distance.

As we approached the summit of the range, we heard a band of Spider Monkeys eating in a tall tree. Few native woodsmen can refrain from shooting a monkey when ammunition is plentiful, and my companion of the day was not one of them. With a wild look in his eyes, he raised his gun and fired into the treetop before I could urge him to desist. Nothing fell, and with much chattering the whole troupe swung through the canopy of the forest down the farther slope. Their attention drawn by the report of the shot, a party of White-faced Monkeys, not far along the ridge to our right, peered through the leaves at us, barking, shaking the boughs, and jumping up and down like excited children, whether from anger or curiosity, I could not decide. But having fired his one shot, my hunter was satisfied; we turned back down the mountainside with no further attempt at carnage.

On the last day of October began a rainstorm that in the following forty-eight hours deluged the valley with twenty-two and a half inches of water. Half the annual rainfall of New York or Washington in two days! As the third day dawned, a slight drizzle fell from a lowering sky. After so long indoors, listening to the rain patter on the iron roof, I was eager for an outing. Ray Stadelman readily agreed to accompany me on my last expedition into the surrounding mountains. We chose a trail, new to us, that led by a roundabout route to a summit east of the valley. Streams were swollen; to cross rivulets that at other times we might have passed dry-shod on stepping-stones, we waded through water well above our knees, stemming a current that threatened to push us over. The saturated atmosphere was cool and invigorating, even a trifle chilly when we stood still; with a light, effortless pace we climbed steep slopes that earlier in the year we would have toiled up painfully, measuring every step, sweating profusely, our interest centered soddenly on the exhausting business of climbing. In the thickets, Spotted-breasted Wrens sang cheerily, despite the gloomy weather, but most other birds were in silent seclusion. Only the roar arising from the dashing rivulets below us, and the *plunk* of large drops falling from high foliage upon the large leaf of aroid or palm, broke the silence of the forest. Scarcely any plant bloomed at this wet season.

To eat our lunch of sandwiches and oranges, each of us stood beneath an enormous leaf of a terrestrial aroid (a species of *Xanthosoma*, I believe), which amply sheltered us from the renewed rainfall. The water that collected at the base of each horizontal blade provided our drink. While we ate, a pair of little Golden-masked Tanagers, lovely in blue, yellow, black, and white, alighted on a dead branch, then flew off through the shower.

The rain abating, we continued upward after our repast, soon reaching the altitude where the clouds drifted through the trees above us. Along the gradually rising ridge that we followed, Manaca palms were abundant. Their great fallen fronds blocked the trail and had to be severed with our machetes so that we could pass. The large, hard fruits littered the ground everywhere.

To our surprise, on reaching the summit we found a milpa about two acres in extent, completely surrounded by the primeval forest through which we had climbed. The maize had been ripe for some time, and the dead stalks were already leaning and falling, many still bearing fat ears of white corn. The charred remains of trunks and branches, amid which the maize had grown in this new clearing, made it as difficult to walk across the field as to pass through the litter of tops and branches that lumbermen leave in a clear cutting in northern coniferous forest. In the midst of the clearing stood a small shed, roofed and walled with Manaca fronds, where the unhusked maize was being stored on a raised platform, made of slabs from the trunk of the same palm. The owner of this milpa evidently lived far down in the valley and was not in sight; so, without asking permission, we took shelter from the recurrent rain in his shed, while we ate our remaining oranges and wished for more.

A sharp wind blew in from the mist-hidden sea a thousand feet below us, penetrating our thin garments and chilling us thoroughly. A flock of wood warblers, recently arrived from the North, flitted through the clearing. I recognized a Black-throated Green Warbler, but before I could identify the others in their dull winter attire, the driving rain and mist coated the lenses of my binoculars, obscuring vision. What a bleak scene surrounded us: dead, sere cornstalks, leaning this way and that; tangled, fallen logs; and beyond, the dark foliage of the trees on the crest, tossed by the chill sea breeze and all enveloped in gray mist. It reminded me of a stormy day in late October in my native land, but appeared even more desolate. Here was no bright-hued autumnal foliage to relieve the grayness of the prospect, and the cornstalks, instead of being stacked in neat towers amid even stubble, were falling helter-skelter amid a waste of prostrate trees.

A Broad-billed Motmot (*Electron platyrhynchum*) photographed at sunset on Barro Colorado Island. The longer central tail feather has already become racquet-shaped; the shorter one is still entire.

"Chancha," a young Collared Peccary, with her master.

9

Captives and Pets

One of the most beautiful birds in the Lancetilla Valley, indeed, one of the loveliest of all birds, is the Turquoise-browed Motmot. It is a paradox that this exquisite creature is hatched and raised in a foul hole in a bank, and emerges at last with its wonderful plumage all undefiled. It is stranger still that it finds its colors in the earth, for they are not hard and glittering, like gems and other earth-products, but as soft and delicately blended as the rainbow or the sky at sunset. Central America has many birds more brightly colored; numerous orioles, tanagers, trogons, jacamars, and macaws are far more brilliant; but the subdued beauty of the motmot is of a different, and perhaps higher, order. Although these others might be painted with enamels, only pastels could do justice to the motmot's colors: soft shades of green, chestnut, and blue, with black on the face, throat, and wings, and above each eye a broad band of turquoise, the bird's brightest color.

But perhaps the most arresting feature of the Turquoise-browed Motmot

Headpiece: A Turquoise-browed Motmot (*Eumomota superciliosa*) about 26 days old.

is its tail. The two central feathers extend far beyond the others; their shafts, instead of being bordered by vanes throughout their length, have near their ends a naked portion, which is terminated by an isolated disk of vane. These racquet-like tail feathers are characteristic of the motmots, although a few of the eight or nine species, such as the Blue-throated Green Motmot of the Guatemalan highlands, lack them. When the feathers of a motmot's tail first grow out, the shafts are bordered everywhere by the barbs that compose the vanes; but along the part that will later become naked, the barbs are loosely attached and soon break away, perhaps while the bird preens. In the Turquoise-browed Motmot, the slender, vaneless stalks are longer than in other motmots, holding the terminal disks well away from the body like two little blue and black flags, and imparting to this beautiful bird an airy grace all its own.

After I had watched Turquoise-browed Motmots dig their long, narrow tunnels in the steep, sandy river bank, and hatch their four white eggs, and attend their nestlings until they outgrew their ugly natal nakedness to become smaller replicas of their parents' loveliness, I wished to follow the process of trimming the central tail feathers into the peculiar racquet-form of the adults. Since, before this was accomplished, the young motmots would leave their burrow and fly into the thickets where I could not watch them closely, I resolved to transfer two of the brood of four to a cage, a few days before they were ready to fly.

But when I beheld through the meshes of a cage a bird I had hitherto known only wild and free, I was troubled, as though I had done an unworthy deed. If I raised these birds as dull and spiritless captives, lacking the brightness and alertness that is the heritage of the free creature alone, motmots could not continue to be to me as they had been, birds lovely, alert, and elusive to glimpse which enhanced my day. They would inevitably become stupid and commonplace; for we are unable to view the captive with the same high regard in which we hold the free creature, even when we are ourselves responsible for that captivity. To cage these fledglings on the threshold of life was contrary to the spirit in which I studied birds. Moreover, I loved freedom too dearly to dare to deny it to any other being, least of all to a bird whose element is the boundless air.

Accordingly, I decided to release my beautiful captives, if it were not already too late to return them to their parents' care, and as a field naturalist to try to see motmots trim their tails in their proper environment. Fearful lest the adults had led their other fledglings too far off to be found again, I carried the two young motmots back to the vicinity of their burrow. Here, to my great satisfaction, they heard the loud, wooden *cawak cawak* of their parents, which they answered in voices weaker and higher in pitch. From my open hands, they flew into the thicket, whither their two nest mates had preceded them, and the adults coaxed the united family deeper into the impenetrable tangle of bushes and vines. What I had possibly lost in knowledge I gained in happiness and the approval of conscience—which I value more highly—for the motmots' nesting had, after all, terminated as I wish every nesting that I watch to end, with the parents leading their fledglings away.

In later years, I never saw a Turquoise-browed Motmot with a full-grown central tail feather that did not already have the shaft denuded; the barbs become detached from a definite portion of the shaft as the feather elongates. Apparently, this early denudation is associated with the fact that the Turquoise-browed Motmot has a much greater length of naked shaft than other members of the family.

In other motmots, such as the Broad-billed, the tail feathers remain entire until they attain nearly or quite their full length; but the vanes are from the first contracted above the terminal disk, where later they will become detached. On Barro Colorado Island in January, 1931, one of these motmots appeared almost every evening, around sunset, at the edge of the clearing in which the buildings stood, announcing its arrival with a full, deep-toned *cwaa cwaa*. It had two or three favorite perches, to which it returned at each visit, and by focusing our heavy, ground-glass-plate cameras on these, Frank M. Chapman and I took a series of photographs that showed the gradual removal of the vanes from the contracted part of one of the central tail feathers. Unfortunately, this motmot never preened its feathers while in the clearing in the twilight; we could not observe the actual process of removing the barbs; nor have I witnessed it in any other motmot. Nevertheless, with perseverance and good fortune, it was possible to learn something of the manner in which motmots acquire their peculiar racquet-shaped tail feathers, without depriving them of freedom.

I have never again placed a bird in a cage. Indeed, when I behold feathered creatures in captivity, they seem to be something less than birds, to have lost some of that peculiar quality that makes birds different from all other creatures. Once, years ago in Panamá, I was taken to see an aviary stocked with a great variety of the local birds. I had eagerly anticipated the visit as an unusual opportunity to increase my familiarity with shy kinds, to behold at close range birds that I had glimpsed only far above me in the treetops. But the aviary affected me otherwise than I had expected. These poor captives were not the same birds that I had met in woodland and thicket; they seemed somehow to have been altered almost beyond recognition; they had undergone some subtle yet profound change that set them apart from those of their kind that I knew in freedom. I did not remain to see all of the aviary. I preferred a fleeting glimpse of a bird in its native habitat to an hour's close scrutiny of one behind a wire netting. As happens so often, the brief experience in proper circumstances can be more significant and instructive than prolonged study in an unnatural setting.

Nor have I tried to make pets of free birds, to teach them to eat from my hand or to permit me to touch them on their nests. Rarely a bird has suffered me to lay a finger softly on her plumage while she warmed her eggs, among them a robin in Maryland, a manakin in Panamá, a hummingbird in the Ecuadorian Andes. Others, such as a tinamou and a wood-quail in Costa Rica, have allowed me to touch them gently with the end of a short stick, but would not stay until I could make contact with a hand. Green Kingfishers and Blue-throated Green Motmots, whose parental devotion was at highest pitch while they brooded newly hatched nestlings, have permitted them-selves to be lifted from their dark subterranean burrows, rather than aban-don their progeny. In Panamá, a male Slaty Antshrike, a bird no larger than a sparrow, nipped the finger that I placed upon the frail cup that cradled his two nestlings. In Honduras, a male Groove-billed Ani buffeted the back of my head whenever I looked into his bulky nest of sticks. On the outskirts of a great city in the North, a Catbird would peck my hand when I touched her treasures in the hedge, while her mate, not to be outdone in parental zeal, attacked my head from the rear. I have welcomed these close physical con-tacts with free birds, but I have not sought them. I could not feel flattered when incubating birds passively allowed me to touch them; they were doubt-less merely relying on their dull coloration and immobility to escape detec-tion.

I have watched small birds such as gnatcatchers, honeycreepers, and castlebuilders work at their nests while I stood unconcealed hardly more than arm's-length away; and scores of other birds have carried on all their domestic occupations while I stood in plain view at no great distance. These displays of confidence have been gratifying to me, but I have not considered them essential to my studies. When the birds have refused to approach their nests in sight of the dreaded human form, I have hidden this objectionable presence in a little wigwam of brown cloth, which is much simpler to set up, and more convenient for observation, than any screen composed of foliage or other woodland material. Except in a few rare and inexplicable instances, this unobtrusive hide—as the British call such blinds—has been speedily accepted by the birds as part of the harmless inanimate surroundings of their nests; comfortably seated in it, I have watched for long hours, seeing yet unseen.

Only in the cases of woodpeckers and a few other birds have I made an effort to overcome their natural shyness at the nest by gradually accustoming them to my presence. Since I wished to make repeated observations of their nests, which they also use as dormitories, over long periods, this course seemed easier than to set up a blind before each visit. As is well known to bird watchers, one can, by the exercise of vast patience and restraint, accustom many shy birds to the close proximity of people. But when I find a new kind of nest, I am so eager to learn the habits of its attendants that long periods of probationary watching would be irksome to me—to say nothing of the danger of causing the birds to desert while trying to accustom them to my presence. By concealing my body, I can the sooner cultivate spiritual intimacy with the birds, and this is what matters.

In Costa Rica, a pair of Southern House Wrens were my nearest and most intimately watched neighbors for a year and a half. But even at the end of this long interval, they would not enter the nest in the gourd that I had given them if I stood near it. This persistent shyness of man, for which they doubtless had good reasons, did not trouble me, although at times it started a train of melancholy reflections on the destructive habits of my kind. As long as I watched them through the widely open window of my cabin, the wrens seemed oblivious of my presence, and I learned more about them than if I had devoted my time to training them to take food from my lips, which would have been an achievement to make my neighbors gape, but of no great significance to the serious student of avian behavior. Moreover, it would be unfair deliberately to divest birds of their fear of me. They might not distinguish other featherless bipeds from myself. It would be most imprudent for them to place unlimited trust in all who bear the human form, and I could not undertake to guard them against every man.

At Lancetilla, we had some interesting mammalian acquaintances, both temporary and permanent. When the squatters in the valley learned that the men at the research station would pay good prices for animals, they brought various kinds for sale: anteaters, coatis, young peccaries, agoutis, and especially snakes, for which there was endless demand to supply the snake pit. One evening, a native arrived with two half-grown Nine-banded Armadillos tied up in a sugar sack. He offered them at a dollar each. I wished to keep one for a few days, to learn something of its habits. When I had taken the first armadillo at his price, the man was so eager to dispose of the second that he persuaded me to buy it for twenty-five cents. Both were males and evidently brothers; the four young of a normal armadillo litter are always of the same sex, as they are derived from a single fertilized egg.

When I lifted the pair from the sack, they attempted neither to bite nor to scratch; they only tried to push away my hands with their hind feet. Since their claws were sharp, I had to be careful of them. Placed on the floor, the young animals moved along at a leisurely gait, investigating with pink, pointed noses whatever they encountered, until they came to a corner of the room, into which they pressed headfirst as hard as they could and remained there quietly. When I laid my hand on an armadillo's rump, he drew up his hind legs and kicked backward with both together, like a little mule.

We placed these armadillos in a cage, where each pressed his head into a separate corner and so passed the night. During their second night with us, they surprised me by climbing the wire sides of the cage to the top, whence, finding escape blocked in this direction, they jumped heavily to the floor, a drop of one yard. From their clumsy appearance, I had never expected such agility. For the rest, they slept day and night, either with their heads doubled beneath their breasts, or stretched at full length with muzzles resting on extended forefeet.

The armadillos disdained bread and raw potatoes but were very eager for earthworms, of which it was impossible to dig enough to satisfy them. They located their food wholly by the sense of smell and utterly ignored the worms until they caught their scent. I placed a worm on the floor, where it wriggled around and left some telltale traces of itself, then held it a few inches above the spot. When I brought up the armadillo, he sniffed at the exact point where the worm had lain, eagerly pressing the floor with his long, mobile snout, but he ignored the actual worm dangling a few inches in front of his beady black eyes, until his nose happened to brush past it. Whether I tried this experiment by day or by night, the result was the same. To reverse the situation, I held a worm concealed in my loosely closed hand before an armadillo's nose; he stuck in his slender snout and tried to scratch it free. The armadillos drank by rapidly protruding the long, pointed tongue into the water and then withdrawing it into the mouth. Now I understood why, when I stood quietly in front of their free kindred as they foraged in the banana plantation at night, they sometimes nuzzled along the ground until they collided with my feet, then sniffed at my shoes before turning stolidly away.

After keeping them for two days, I released the young armadillos at night in the path in front of the house. They went wandering in single file down the ditch beside the walk, like two little beagles following a trail. When they came to a soft place in the bank, they rooted with their noses, and sometimes dug a little with their forefeet. Finally, they left the ditch and entered the stubble of the newly cut grass beyond. Through this they moved slowly, almost hidden from view, nuzzling at the roots and under piles of dead grass, often sniffing audibly, hunting their evening meal. Their experience of captivity had not taught them wariness, for one wandered out into the path again, right into the glare of my flashlight, almost bumping into my feet without showing the least concern. Full-grown armadillos are a trifle more alert than these youngsters, but even they are often decidedly obtuse.

Nine-banded Armadillos are largely nocturnal, but rarely one sees them abroad in full daylight. Recently, as I passed through second-growth woods in the middle of a bright morning, I noticed a young armadillo, less than half grown, in the path ahead of me. It was rooting in the wet ground, sometimes pushing its whole head beneath fallen leaves and making its long snout very muddy. It continued its eager search for food while I approached and stood only a few yards away. Finally, noticing me, it ambled, still rooting at intervals, down the slope below the path and entered a newly dug burrow, with a pile of freshly stirred earth in front. When I reached the burrow, the mouth was already blocked with a large mass of dead leaves, so that I could not look in. Returning after an interval, I found the little animal standing on its hindfeet in the burrow's mouth, holding its forefeet up while it looked around with its little beady eyes and sniffed the air. When I made a noise, it slipped inside its burrow. Evidently it engaged in this unusual diurnal activity because it had devoted a good part of the preceding night to digging its new burrow, and daybreak found it still hungry.

A more responsive and engaging animal was Chancha, a young Collared Peccary who had been taken from the mountains while still a suckling. Separated from her mother, all her filial affection was transferred to her purchaser, the snake-keeper, whom, when permitted, she followed everywhere, trotting along at his heels like a well-trained little dog. When he ran, she galloped along behind, usually managing to keep up. If she became separated from him by some obstruction that she did not know how to circumvent, or if he entered a building and closed the door in her face, she squealed loudly and imperiously, with more than a trace of indignation in her voice, until united with him once more. She loved to be lifted to her master's lap and petted like a favorite cat. While he could do almost anything to her with impunity, any teasing by another person was promptly resented with squeals and angry snaps. Since her teeth had begun to grow, these efforts to preserve her dignity commanded respect. She was nourished with milk and endless bananas, and thrived wonderfully.

Her attachment to her master was thicker than water. When permitted, she trotted behind him in the evenings to the swimming pool in the river. Already on her first visit to the water, she did not hesitate to swim after him the moment he entered the pool, and she followed wherever he went. Her squeals on being separated from him in the water were usually transformed into an incoherent *glub glub*, for she swallowed much of the river while attempting to utter them. If her master threw her from himself into the water, as soon as she came to the surface she turned her course toward him and swam with only her head emerged, gurgling ludicrously. Upon reaching her goal, if he happened to be standing in deep water, she tried to climb upon his shoulder. Once, when she lost her way and became entangled in some half-floating brush near the bank, her cries were piteous to hear. If permitted to stay in the water too long, she became obviously distended with the quantity she swallowed, then crawled out on the shore and curled up to sleep on her master's clothes until he arrived to claim them.

White-lipped Peccaries are the most dreaded animals of the Central American forests. They travel in large bands, sometimes, it is said, containing "hundreds" of individuals, and it is further averred that they will attack and destroy the man they happen to meet, unless he can escape by climbing a tree. Even the Jaguar, hunters affirm, will not attack a White-lipped Peccary that remains in the herd, although it sometimes seizes an isolated individual by surprise. I have never chanced to encounter these peccaries in the forest, for they are active chiefly by night, while my days have always been so full that I have done little wandering through the nocturnal forest. Frequently I have met small parties, containing not more than eight or ten individuals, of the milder-tempered, largely diurnal Collared Peccaries, who usually have fled with excited grunts as I approached, leaving their strong musty odor on the air. Captured young, they make affectionate and engaging pets, somewhat porcine in their habits, and are sometimes seen in the dooryards of isolated forest dwellings.

Some years after I left Lancetilla, while residing with a family of refugees from Nazi intolerance, among the rain-drenched forest that clothed the welter of low hills at the eastern base of the equatorial Andes, I witnessed in a touching manner the unhappy fate that awaits most creatures snatched from their natural haunts to serve as man's playthings. A missionary, returning from among the savage Indians along the Río Napo, brought a half-grown Coatimundi, an animal related to the raccoon, from which it is distinguished by its greater size, longer and more slender muzzle, longer and usually less prominently ringed tail, and diurnal rather than nocturnal activity. When he resumed his journey to Quito, the missionary left this denizen of the oriental forests, who had found great favor in the eyes of Frau Jansen.

The coati was called Guachi, the Indian name for the species. She had a gray face; the foreparts of her body as far as the middle were brown; her hindquarters and her feet were black; and her long tail was ringed with black and light gray in alternating bands. A most engaging little quadruped, she immediately became everyone's favorite, to be petted more than is good for any creature. It was amusing to watch her turn her long, slender, flexible snout to one side or the other, or wrinkle it upward, to sniff a hand or anything else that approached her. To eat or drink from a dish, she would give a sharp upward turn to her little proboscis, which projected well beyond the end of her narrow lower jaw and would have been very much in the way at mealtime if it had not been so flexible.

An expert climber, Guachi easily scrambled up the supporting posts of the rustic house to the rafters, where she roved tirelessly about, searching for the cockroaches, spiders, and other small creatures that lurked in crevices and in the thatch of palm fronds. She assiduously inserted the black tip of her pointed snout into all the crannies and wormholes in the boards where such small animals might be hiding, and when her nose gave notice of the presence of any creature, she dug it out with the long claws of her forefeet and promptly devoured it. Whether the coati discovered her victims by the sense of smell or that of touch I could not decide. One would suppose that this method of hunting insects would expose the sensitive nasal extremity to occasional bites and stings, perhaps even to the venomous injection of a scorpion; but our animal seemed never to suffer any inconvenience of this sort while foraging among the rafters.

When Guachi tired of hunting in the upper regions of the house, she would descend, headfirst, by way of a post or upright projecting batten, into one of the rooms (the house lacked ceilings to complicate her downward journey) and seek a soft spot for her nap. She liked best of all to snuggle down on some sitting person's lap, or beside one in bed, or, if such warmth-giving contact with mankind were not at the moment available, she would lie upon a blanket or cushion. Here she would turn her long head under her breast, cover her face with her forelegs, and soon fall asleep. Her notes, uttered chiefly when someone played with her and when she was not permitted to have her way, were small, rapid, shrill, and bird-like.

But alas, the ways of coatis and other forest creatures are not those of man. Since all their ancestors have for countless generations dwelt among the trees, how could they know how to behave in a house? Guachi soon began to commit serious misdemeanors that in the end caused her banishment. In her tireless search for more and more insects, she tore pieces of palm leaf from the thatch, making holes which allowed the roof to leak; she dug into a roll of paper intended for botanical specimens. She found and devoured forbidden dainties; and since her bodily functions were governed by no inhibitions, while hunting in the upper levels she sometimes released a stream of water that fell upon a bed, upon papers, or upon other spots where such impure rain was decidedly unwelcome.

When these various nuisances became unbearable to the owners of the house, the poor creature was harnessed with a cord and tied in a corner of the kitchen. Here I did not visit her, for it distresses me to see a wild creature in captivity. A day or two later, she was given away to some visitor who, I surmised, would laugh at her pretty ways and indulge her pranks until they ceased to amuse, then shut her up in a cage or pass her unwanted little body on to someone else. Guachi had begun her life among humans in an Indian hut on the Río Napo, the abode of many children. Here, doubtless, she had been happiest; for these Indians, having few dainty or costly articles in their forest dwellings, could best entertain her as an equal.

Not long after her banishment, I learned that Guachi had died. When, I asked myself, will well-meaning people take to heart that wise maxim of one of the most perceptive of nature writers, W. H. Hudson: "Neither persecute nor pet"? After all, our petting is, in many instances, only another mode of persecution, more agreeable to ourselves than to the animals who are forced to receive our caresses. We cannot teach a coati or a monkey or a parrot to lead the life of a man, but if we continue our cosseting long enough, we can make it impossible for the poor creature ever afterward to lead the normal existence of its kind. It should not surprise us that animals who have long been confined frequently refuse to leave their cages even when the door is open; deprived of the opportunity to learn from their elders while young, habituated to an enervating dependence, they must feel their inadequacy to fend for themselves in the manner of their free relations.

Sometimes I have bought a captive bird or other animal from a local inhabitant, kept it a few hours or at most a day or two so that it might earn the price of its freedom by showing me some of its habits, then returned it to its native environment. But I have not made a practice of ransoming these poor prisoners, for to do so would create a market, to supply which other animals would be captured. Usually, when I happen to see these unhappy captives, I have considered it the wisest, if not the easiest, course to look the other way and pass sadly onward.

One of the most disagreeable circumstances of my travels in the tropics has been the sight of so many native birds in captivity, frequently in cages so small that they could hardly turn around. Sometimes the prisoner is kept night after night, uncovered, beside a continuously burning light—electric current, unfortunately, often being sold by the number of bulbs in the house rather than by meter. Latin-American countries lag behind certain northern nations in the protection of wildlife. Some of these republics have enacted good laws on the subject, but they tend to be enforced laxly, if at all. Fortunately, relatively few people in these countries can afford highly efficient modern firearms; for tropical birds, as a rule, reproduce more slowly than those at higher latitudes where winter takes a heavy toll of life, and they could not long survive the annual carnage to which so-called game birds are subjected in lands industrially more advanced. It is especially distressing to see in a cage a migratory bird such as the Rose-breasted Grosbeak or the Indigo Bunting, whose capture is prohibited by law in the countries where they are hatched. North Americans should insist that their fellow countrymen be treated with respect on their travels.

Despite my great interest in the habits of all living creatures, I have not for many years kept any kind of pet. A monkey moving with ease and grace among the treetops is a pleasure to watch; a monkey in the house is to me an unnatural abomination; in a cage, it is among the most pitiable objects that I know. Knowing parrots of many kinds in their native habitats, where they are among the most alert and farthest-ranging of birds, I find it painful to look at poor, dull, phlegmatic Polly on her perch in the house. The harsh voices of parrots, which have a certain wild, exultant harmony as they float afar over tropical mountain and valley, make only an intolerable pandemonium in the confined space of an aviary. Always I have endeavored, sometimes at considerable discomfort and expense, to know the various creatures in their natural environments; for habitat and inhabitant are complementary and one cannot be understood apart from the other.

Were I to be confined to some great city, I do not doubt that I should find more pleasure and instruction in watching squirrels, starlings, and pigeons leading more or less normal lives in the parks and about the towers, than in staring at all the exotic animals gathered from the ends of the earth and held captive in the zoo. In favor of the zoological garden, it may be said that it arouses an interest in animals and their protection by urban dwellers who otherwise might hardly be aware of their existence; but only well-supported zoos, which keep their exhibits in surroundings as natural as art can devise, can justify themselves in the eyes of friends of animals. But even better than the best of zoological gardens for bringing awareness and knowledge of animals to city dwellers are, in my opinion, excellent moving pictures of wildlife. Such cinemas can be made without detriment to their free actors; they can be made available to more people at less expense than zoos, which only wealthy cities can afford; they can be preserved for future generations with no outlay for feeding and maintenance.

I have not even kept a dog. As a boy, I was master of a constant succession of canine friends, loved in life and lamented in death. But in the course of my travels, I have seen too much of the seamy side of canine nature; too often I have been obliged to share narrow quarters with cringing, bone-bare, ill-tempered curs, mangy, flea-bitten, and offensive to the nose. Then, too, as my interest in wild nature deepened, I gradually learned that a dog's habits diverge too widely from my own. The dog, despite some amiable traits, possesses one defect that makes him intolerable to me as a companion in field and woodland: he will never willingly leave any other creature in peace, but must constantly be barking at unoffending people, frightening the horses, worrying the cows, chasing the squirrels up trees, making the lizards scuttle into hiding, putting all the ground birds to flight. Until advancing age shortens his wind and stiffens his limbs, he cannot easily be taught gentlemanly behavior out of doors, however well bred he may be in the house. I know no other animal that is half so great a ruffian in the open. The dog is the friend of man, but like many of the companions of our childhood, he is a friend that we may outgrow.

I have derived the most lasting satisfaction from associating with untamed animals in their native habitats, even when, unable to differentiate my intentions from those of other men, they have feared and tried to elude me. These only, free and self-sufficient and submissive to no master, are my equals—or, possibly, my superiors. And I hold that a man does best to associate with his equals.

The Tela River, winding through the Caribbean foothills of Honduras, at low water. Tall guarumo (*Cecropia*) trees lean over the rocky channel.

10

Birds Beside a Mountain Torrent

A short distance above the buildings of the experiment station in the Lancetilla Valley, a tall, slender tree with smooth, light gray bark stood amid a
tangled thicket that had sprung up on an abandoned banana plantation. A
remnant of the ancient forest, the tree towered so high above the upstart
vegetation crowding around its base that its lowest boughs were beyond
reach of the highest shoots of the new growth. Near the ends of many of
these boughs hung crowded clusters of long, gourd-shaped objects, the color
of weathered straw. One unfamiliar with tropical trees might have taken
these pendent bodies for gigantic fruits. But if he stood watching for a while,
he would see a big bird fly up and plunge rapidly into the top of one of these
seeming gourds, or perhaps emerge from one of them and fly with measured
wingbeats toward the dark forest that covered the surrounding hills. These
birds were largely rich chestnut in color, with black heads, necks, and
breasts, bright yellow outer tail feathers, and long, tapering bills. They were

Headpiece: The mountain torrent in the Honduran foothills where Royal Flycatchers
(*Onychorhynchus mexicanus*) nested.

Montezuma Oropéndolas, members of the troupial family, which also contains the grackles, cowbirds, meadowlarks, and American orioles. The clustered pouches were their nests, woven with skill and patience by those sharp, orange-tipped bills.

For more than a month, I had spent every available hour watching that tree, fascinated by the varied activities of the colony of yellow-tailed birds centering around those eighty swinging nests—nests somewhat like those of the Baltimore Oriole but much larger, their rounded bottoms hanging four or five feet below the slender twigs that upheld them. I had seen the oropéndolas rip long, pliant fibers from banana leaves and weave them into the pensile pouches, along with thin vines and strips of palm fronds, then carry in billful after billful of small fragments of dead leaves, until the bottom was covered by a yielding litter, which would prevent the two white eggs from rolling together and breaking when wind swayed the nests. I had waited while the female birds incubated, only guessing what was happening in those closely woven baskets so high and inaccessible; until finally the mothers' endless, tireless trips between the nests and the distant forest, from each of which they returned with fruit or some other morsel in their bills, told me that at last the hungry nestlings had escaped their shells. Soon the young birds' quavering, nasal cries proclaimed even more clearly their presence in the nests.

While the industrious female oropéndolas were engaged in these changing occupations, the males—bigger, heavier birds, the size of crows, and much less numerous than the females—strutted idly among the branches of the nest tree; or bowing profoundly, until their raised yellow tails stood vertically above their inverted heads and their lifted wings almost met over their backs, each in turn poured forth his curious liquid gurgle. Heard from afar, this was one of the most melodious and stirring of the forest sounds, but close at hand it was marred by screeching overtones, like the grating of a poorly handled violin bow.

I was eager to follow these amazing birds to the forest whence I had so often seen them returning, to watch them foraging in the crowns of the great trees. So one morning in early June, when I noticed them streaming between their nesting colony and a tree that rose above all its neighbors near the foot of the mountain on the eastern side of the valley—a landmark easy to follow until I plunged beneath the dense woodland canopy—I resolved to trail them there. I cut across a grassy field and through a plantation of young rubber trees, crossed on stepping-stones the broad, shallow channel of the Tela River, then took a path that led far up into the mountains.

After passing through a dense second-growth thicket, I reached the edge of the primeval forest, where I paused to listen to the monotonous *squeak, squeak, squeak* of a courtship assembly of Long-tailed Hermit Hummingbirds. Among the least musical of birds, these hermits performed with admirable persistence; I never passed that way without hearing their tuneless squeaking. Each perched so low, his brownish plumage blending with the brown fallen leaves and litter, that I could rarely see him, unless my sight was attracted by the pointed, white tip of his long tail, wagging rhythmically up and down while he called.

After listening for a while to these hummingbirds, I continued onward into the heavy forest. Soon I turned left from the trail and started down a steep slope of nearly bare soil between small Lancetilla Palms. The descent was so abrupt and slippery that I was often tempted to support myself by grasping their straight, slender stems, but the long, sharp, black spines that covered them sternly forbade this.

At the foot of the slope I came to a stream, beside which grew the tree I sought. It was shedding profusely a small, green fruit, about the size of a cherry, with a soft, nearly tasteless pulp and a single large seed. The oropéndolas were eating these fruits, but here in the forest they were as shy as they were bold and confident at their nest tree in the cleared valley. The stentorian *cack* of alarm of a vigilant male, some of whom regularly accompanied the females on foraging expeditions, sent off the whole party almost before I could focus my field glasses on them. In the same tree, I glimpsed a Rainbow-billed Toucan and heard the deep *coot coot* of some Blue-diademed Motmots, invisible among the high boughs.

I sat upon a mossy boulder, close beside the pellucid water of the mountain stream, to await the return of the oropéndolas. One feels the spell of such a spot before he can tell in what it consists. The elements that create the charm of any woodland scene, to which the spirit instantly responds, are too manifold to be grasped immediately by the conscious mind. We must rest long in such a spot, quietly receptive, before all the details of the picture, with all its varying lights and shades, impress themselves on consciousness, so that we may retain enough to reproduce at some future hour the scene that enchanted us. One who walks hurriedly through the forest does not know it; we must sit and lie and sleep in it before it becomes familiar. Although this is true of the lighter temperate-zone woods that we have roamed from childhood, it applies with much greater force to tropical rain forests, with their infinitely more varied life.

For a long while, I sat watching the clear stream tumble swiftly down its narrow bed, through a ravine covered by primeval forest whose giant trees interlaced their boughs high above the channel. Here it rushed over rock ledges in a spray of foamy whiteness, there dashed against great boulders verdant with delicate fern moss. The long aerial roots of epiphytes perching in the treetops dangled like slender cords above the torrent. A huge boulder, standing above the ordinary reach of the current, was almost hidden beneath a luxuriant overgrowth of mosses, ferns, begonias, and aroids of several kinds, the largest of which bore ample, heart-shaped leaves and stiff spikes of minute, green flowers. Above all this mass of foliage a scrambling Desmoncus palm reared the barbed tip of its newest frond, feeling for something higher to which it might cling and grow upward.

A stout bough projecting horizontally over the stream upheld a little garden of its own, with the rosettes of thick-leafed bromeliads growing above it, and slender, ferny shoots of an orchid hanging gracefully below. I was careful not to upset one of these bromeliads while passing beneath it, for the close-set leaves of each rosette had collected enough rainwater to drench me. The steep, rocky cliff across the stream supported great-leafed wild plantains, Panamá-hat plants, and many dwarf palms.

I waited long, but the oropéndolas failed to return and resume their feast on the green fruits of the tree that rose above me. A Greek of antiquity might have expected, had he tarried so long and quietly in so sequestered a glade, to see a naiad rise up dripping from the crystal pool at his feet, or a gentle-eyed dryad steal up behind the palm leaves to watch him shyly; a Goth, to surprise an elf emerging from the hollow of a great trunk, or a gnome peering out from a cleft in the cliff across the stream. Bred in a less credulous, or less imaginative, age than they, still I sought some sentient inhabitant of the dell, not a transient visitor like myself, but one that lived and breathed its exquisite tranquility during the long days when no human footstep intruded—perhaps some shy woodland bird, leading its life far from the prying gaze of man, attracted by the beauty or other advantages of the spot to build its nest and raise its young to the murmur and laughter of the hurrying water. A bird or shy furry animal adds the finishing touch to any wild woodland scene. Such a creature is needed to make it complete and give it the highest interest—significance as the abode of animate life. Without birds, the most beautiful forest glade is like a stage where no actors perform, or a house where the careless prattle of children is never heard.

The rushing current paused momentarily in a little, clear, rock-bound pool that spread before me. Here a brown Long-tailed Hermit plunged repeatedly, half submerging itself in the cool water. After each partial immersion, the hummingbird perched low in the neighboring undergrowth and twitched its long, white-tipped tail so rapidly from side to side that it became a blur of white and brown, almost as invisible as its wings in flight. Noticing me, it complained with a rapid twitter and darted off through the bushes, but soon it returned to repeat its dives. As far as I could learn, the hermit had no nest here; apparently, it was a transient like myself.

I waited longer, until at last I discovered the true genius of the dell, a brownish flycatcher of medium size, with a pale tawny-yellow breast and belly, a tawny-orange tail, and a conspicuous olive-colored crest that lay quite flat, projecting behind its crown. When the bird raised this crest slightly, I glimpsed bright red concealed beneath the dull outer feathers, but I did not suspect the full glory of what that crest concealed, for I did not yet know that the strange bird was a Royal Flycatcher. It hovered among the branches on the cliff-top, voicing at intervals a loud, piping whistle, somewhat like that of the Great-crested Flycatcher, but fuller and clearer, with a strange, melancholy quality as I heard it emanating from the dark shadows of the foliage, rising above the loud murmur of the torrent—a quality in keeping with the wildness of the setting and in turn enhanced by it. Soon I noticed this flycatcher's mate, similarly clad. Like nearly all the members of their family, they were active birds that darted swiftly from branch to branch, snatching up insects in their broad, flat bills.

Presently one of these flycatchers clung for a moment to a long, irregular mass of brown vegetable material that hung above the stream, attached to a slender, green vine. This had been in full view the whole time I sat on the boulder, seen yet unseen, like something devoid of interest. When the bird finally directed my attention to this brown mass, I took it for an accumulation of drift-weeds, such as one finds on branches above streams subject to flooding, stranded there by the last high water, although perhaps it was too high to have originated in this manner. But when one of the flycatchers flew up to it again, I began to suspect that the bird's purpose was not just to pluck a stray insect from its surface. Knowing that the nests of tropical flycatchers are of amazingly diverse forms, I surmised that this might be such a structure, although unlike any that I had seen.

Stepping carefully on slippery, moss-covered rocks, I crossed the stream and, balancing myself precariously on the out-jutting end of a ledge that overhung the water, pulled down the vine until I could reach the object that interested me. I found it a tangled mass, consisting of the long, thread-like, dry, staminate inflorescences of the small tree *Myriocarpa yzabalensis*, vegetable fibers, rootlets, a few dead leaves, and some tufts of green moss. My belief that I had found their nest grew stronger when the flycatchers fluttered above me, piping their protests, but the mass of dead vegetation was so flimsy and irregular that it baffled me; I found neither eggs nor any cavity that could hold them. If a nest, it was evidently incomplete. I withdrew to a more distant rock to wait for the birds to return and finish it.

The sweet, subdued song of a Gray's Thrush floated down from the trees above me and drew my attention to the songster, a plainly clad bird with an olive-brown back and cinnamon-buff breast, the size, shape, and a very close relative of an American Robin. Although these thrushes were abundant in the clearings and plantations of the valley, I was surprised to meet one here in the forest. My wonder increased when I saw the songster's mate fly into a slender, moss-laden tree close by the water with some muddy fibers in her bill, telling me plainly that she was building. Although the male thrush did not help, he escorted her on each visit to the nest, sang a few subdued notes while she deposited her material and shaped the structure, then followed when she flew off to seek more fibrous stuff and mud. Hers was the only nest of Gray's Thrush that I have ever seen in the forest. The cup of hardened mud was wholly covered with fern moss and lined inside with fine rootlets. How much more securely hidden it seemed, here among the mosses with which it blended so well, tucked almost out of sight above the moss-covered base of a great-leafed epiphyte, than atop a bunch of bananas or on a palm leaf in a clearing, where sometimes I found the nests of this thrush!

Another inhabitant of the dell was so obscure and quiet that I passed close beneath it without becoming aware of its presence. It was a large snake, glossy blue-black with a white belly, and it stretched out in a sinuous, perfectly immobile line along a slender horizontal branch that projected above the stream. The whole time that I watched, it moved not a muscle.

Late in the morning, hunger drove me back to the house in the clearing. The flycatchers had shown no further interest in the dangling mass of vegetation. In the afternoon I returned and sat again on the boulder, still hoping to see my new acquaintances resume work on their nest, if nest it was. An air of desertion hung over the stream; something essential was missing. The thrush no longer built her mossy cup while her mate sang; the flycatchers no longer darted back and forth above the stream, snatching up insects. Only the snake drowsed immobile on its perch above the water, exactly as I had left it five hours earlier.

Finally, tiring of a long, unrewarded vigil, I went again to examine the hanging mass of dead vegetation that so interested me. This time, pulling the structure lower for a more careful scrutiny, I found, near the middle of the slender, four-foot-long mass, a shallow niche or depression, open on one side, in which lay a single, reddish-brown egg. Its color blended so well with that of the materials amid which it lay that I had failed to notice it in the morning. From similar nests that I afterward found and from published records I learned that two is the usual number of eggs laid by Royal Flycatchers. Perhaps the second egg had been jarred out of the shallow niche when I carelessly permitted the supporting branch to spring up after I examined the nest in the morning. One egg had certainly been present then, as it hatched only five days later. In extenuation of my failure to find it, I might plead that the opening of the niche faced away from me as I stood precariously balanced on the ledge above the water—and it was excellently camouflaged. In my first contact with one of the most remarkable nests that I know, I had been thoroughly baffled.

The more I pondered over the Royal Flycatchers' nest, the more admirably contrived it appeared to me. No attempt had been made to conceal the structure. It hung over a woodland stream in full view, and it was very much larger than necessary, as though to make it more conspicuous. But conspicuous as what? Certainly not as a nest, but as a random tangle of dead vegetation, perhaps as a mass of brown drift-weed lodged there by a flood. Yet the eggs were not hidden in the middle of this mass, where nobody would expect to find them; they lay in a shallow niche, open to the gaze of any snake or monkey or predacious bird that might creep or leap or fly above them, if only the creature's vision were acute enough to distinguish them. Their cryptic coloration concealed them almost as effectively as though they were covered by an opaque screen.

Of the many eggs of American or "Tyrant" flycatchers that I have seen, that of the Royal Flycatcher is darkest. Of the numerous members of this great family of birds that build pensile nests, no other species, as far as I know, makes such a long, irregular structure. How closely nest and egg complement each other! Much as one departs from types widespread in the family, the other has not lagged behind it. I surmise that innumerable eggs were, during many generations, found and devoured by predators before this adaptation of egg to nest was perfected. Likewise, dull upper plumage makes the Royal Flycatcher less conspicuous as it sits in its nest, always facing inward, with its lighter-colored tail projecting into the outer air. The Royal Flycatcher seems to proclaim to all the multitude of enemies that lurk in the woodland and eat eggs or young birds: "Here is a tangle of drift-weed hanging from a dangling vine above the stream. Find my eggs or nestlings in it, if you can! Then climb to them, you who have no wings, if you are light and agile enough!"

To the thoughtful naturalist, nature offers two kinds of beauty: sensuous beauty—form and color and sound—which most of us recognize immediately; and the beauty of adaptation—the perfect adjustment of the organism to its habitat, of the organ to its function—which appeals to the thoughtful mind and is revealed only by long study and much pondering. The oropéndolas had led me to both.

Later, a hundred yards upstream, I found a second Royal Flycatcher's nest, quite similar to the first. It hung from a sapling on the rocky bank, on a branch that I could reach with a hooked stick and pull down until I could look into the shallow niche, where a blind and naked nestling, apparently hatched that very day, rested beside a single infertile egg.

Two days after my first visit to the mountain torrent, I had a delightful surprise. The morning was overcast and threatening. I sat on a log stranded on a mass of rock jutting into the middle of the stream, hoping to see the parent flycatchers come and feed the young in the second nest. After a few minutes, the rain came down, first as a light shower, but soon increasing to a heavy downpour. I sought shelter by crouching beneath an overhanging

ledge of rock, overgrown above with moss and ferns. My retreat proved dry
for only a few minutes; as water accumulated on top of the ledge, it collected
into little rivulets that trickled down the slanting lower side until they met
projecting points, from which they precipitated themselves in jets to the
ground.

I was trying vainly to adjust myself between the outer deluge and these
numerous streamlets falling from my roof, and becoming uncomfortably wet
and cold, when I noticed that the Royal Flycatchers had returned and
seemed to be enjoying what I tried to avoid. Far from attempting to stay dry,
they were flitting from twig to twig above the stream, spreading their wings
and tails to the shower, fluffing and preening their plumage with evident
satisfaction. While billing and arranging their feathers, they did not neglect
their dull olive crests, which lay quite flat and projected in a little knot
behind each bird's head, giving it an odd, hammerheaded aspect. The male
raised his crest and spread it fully, encircling his head with a splendid halo, a
scarlet aureole margined with spots of violet, steel blue, and velvety black,
compared with which the red or yellow crests of kingbirds and other large
flycatchers are but paltry adornments. I was amazed that such magnificence
could have lain, concealed and unsuspected, beneath the dull olive feathers
that most of the time covered the flycatcher's head. Then the female un-
folded her headdress, as widely spreading as her mate's and only slightly less
brilliant. These displays lasted less than a minute; the glorious aureoles were
folded down, and before my delighted eyes, the flycatchers were transformed again into plainly clad hammerheads, ready for the domestic cares
which at this season engrossed them. Cinderella had laid aside her gala dress
and resumed her station beside the hearth!

These superb crests are the Royal Flycatchers' regal splendors, their title
to inclusion among the nobility of birds. But on what occasions are they
displayed? What part do they play in the lives of their wearers? Beside the
mountain torrent, I saw a flycatcher spread its crest only once more, again
while the bird was preening in a shower. Collectors have recorded that when
a wounded Royal Flycatcher is held in the hand it expands its crest as it turns
its head from side to side in its death agony; but we are ignorant of the full
significance, in the bird's normal life, of this magnificent adornment.
Museums are filled with millions of birdskins gathered from far places, yet
few of the vital secrets of the birds who pass their days in the obscurity of the
tropical forests have been revealed to us. Slowly, very slowly, they will be
yielded up to those who have the zeal and devotion to watch the living birds
in their native setting.

I came almost daily to the mountain torrent, to enjoy the wild charm of the
scene and follow the progress of the nests that I had found. One morning,
when I brought a companion to help me take some photographs, a male

Prince Massena's Trogon perched upright on a bough above the stream and calmly watched us clamber over the slippery rocks. As big as a carrier pigeon and stockily built, he was clad in glittering green with a bright red belly—one of the unforgettable splendors of a region renowned for gorgeous birds. The following day, I noticed a hole, about a dozen feet up, in the trunk of a small tree growing beside the stream near the point where I first came to it. Picking up a stone, I tapped on the base of the tree. My pulse leapt up when the trogon darted out, not from the hole that I was watching, but from a large, black termitary attached to the other side of the trunk somewhat higher up. It was my first trogon's nest, in an unexpected situation.

The slippery trunk, encumbered with vines and epiphytes, repulsed my effort to climb it and see what the nest contained. In the afternoon, I returned with my assistant. We brought a machete and a supply of stout cord, and after we had cut two long poles and notched them, we tied short lengths of branches between them, and soon had a ladder that reached to the nest, twenty feet up. The trogon flew away while we worked. Eagerly I climbed up and stuck my hand into the blackness of the hole that opened in the side of the termitary, for even with a flashlight it was impossible to see much of the interior.

A short tunnel led slantingly upward and opened into the top of an ample, rounded chamber, on the bottom of which, lying only on some hard, black chips of the termitary, were three large eggs. Carefully I lifted them out, one by one, and found them palest blue, verging on white. A few soft-bodied, white worker termites and soldiers crawled over my hand, providing evidence that the thick-billed birds had carved their nest chamber in the heart of an occupied termitary, almost as hard as oak wood, and defended by a legion of soldiers that squirted a white, viscous substance from their syringe-like heads. However effective this gum may be in embarrassing the movements of ants and other insect enemies, it had not deterred the resolute trogons.

I replaced the eggs with affectionate care, for I was elated by the prospect of following, day by day, the progress of the nestlings that I hoped would hatch from them, until they sprouted feathers and flew off into the forest. Each time that I passed the tree, I tapped on its trunk to make the incubating trogon fly out, and thereby I learned about the division of labor between the two sexes. From eight or half past eight o'clock in the morning until about half past four in the afternoon, it was invariably the green male who emerged. Earlier or later in the day, it was his slate-colored, red-bellied mate, who passed the night on the eggs.

Usually the bird that I had chased from the cavity perched quietly on a branch not far off, slowly raising and lowering its tail in silence. The trogon, despite its spectacular darts to seize a berry or insect while poising on fluttering wings, is one of the more phlegmatic birds. One day at noon, however, the male flew to a bough above the torrent and called repeatedly *wuk wuk wuk wuk wuk*, the notes following each other rapidly with slightly ascending pitch. The call was not melodious. I thought it rather subdued, in keeping with the trogon's dignified, retiring comportment; but as I toiled up the mountainside, between the thorny palms, I learned that it carried far.

After I knew when the trogons replaced each other on the eggs, I decided to watch them do so. Late one afternoon, I hid behind a rock ledge at the streamside, upon which I had piled some leafy boughs through which I could peer. I had not long to wait; at nearly the expected hour, the female trogon dropped down from the treetops and rested on a branch above the torrent. Thence she flew to a perch directly in front of the doorway in the side of the termitary. The male stuck forth his head, and after a few moments flew quietly out and away. After another minute, the female trogon entered. The changeover had been effected without a note audible to me or any sign of greeting between the two partners.

The following morning, I was present at the final act of a tragedy. I hid myself early behind my ledge of rock, where, tormented by a cloud of insatiable mosquitoes, I waited quietly beside the rushing current for the male trogon to arrive and release his mate from her long night session on the eggs. The sweet, liquid voice of the Gray's Thrush, sounding even more ethereal here in the dimly lighted forest than in the cleared lands where I more often heard it, mingled with the murmur of the hurrying water and the wild piping of the Royal Flycatchers, who were catching insects across the stream. An hour dragged by, and still no trogon appeared, although it was now well past the time when I had learned to expect him.

Weary of feeding bloodthirsty hordes of mosquitoes without recompense, I was stealing forth from my hiding place when the long-awaited bird appeared, as though from nowhere, and alighted on his usual perch above the stream. Cautiously I slid back behind my ledge of rock, apparently unnoticed by the trogon, who remained sitting upright, moving his tail slowly up and down. Soon he flew to a second twig, then to a perch directly in front of the doorway in the termitary. When the female failed to reveal herself, he called in his lowest voice. To me, his notes were inaudible above the torrent's laughter, but I clearly saw the vibrations of his green throat. Still no response! Flying to the entrance, he poised momentarily on wing before it, then alighted and called again. Finally, he entered.

Seven minutes later, a trogon darted out of the termitary and flew off swiftly through the forest beyond the stream. In the dim light, its plumage looked so dull that I thought it was the female. Could two grown trogons find room in the chamber, and was this the hen? Advancing to the nest tree, I tapped on the trunk with a stone, but no bird darted out.

Still ignorant of the countless dangers that beset the nests of birds in the tropical forest, I had not taken the precaution to remove the big, black snake that rested above the torrent, nor to take away the cumbersome ladder between inspections of the termitary. Climbing up once more, I discovered on the topmost round some downy gray feathers, tipped with red. More feathers of the same kind lay on the broad surface of a heart-shaped aroid leaf just below this. I sensed disaster. If I had not seen the trogon fly out only a few minutes earlier, I would not have stuck my hand into the dark cavity without first inspecting it, for I had found snakes and wasps and fiercely stinging ants inside abandoned birds' nests often enough to make me cautious. I well understood the male trogon's hesitation to enter, when his mate did not respond to his summons to come forth, for the snake, or sharp-toothed mammal, or whatever else had raided the nest during the night, might still be lurking within.

I felt about on the bottom of the chamber in the termitary. My fingers encountered only a few fragments of eggshell and some seeds regurgitated by the incubating trogons, mixed with particles of the termitary that covered the floor of the cavity. It was beyond doubt the male trogon that I had seen steal away so quietly through the forest. Some animal had attacked his mate and her eggs during the long night shift, and a ravaged nest remained to greet him on his return in the morning. As I walked sadly back to the plantation, I should have given much to know what the male trogon thought or felt as he flew silently away from his desolated nest.

For weeks afterward, as I walked through the mountain forest, I tapped on the trunk of every tree with a termitary that I found, but never from another did a bird dart forth. Many years passed before I found another nest of the Massena Trogon. It was far to the south, in Costa Rica, in a massive, decaying stump in a pasture near the forest, but it had the same form as my first nest chamber in the termitary and like it contained three white eggs. Also like the first nest, it was pillaged by some unknown predator before these eggs hatched.

Meanwhile, the egg in my first Royal Flycatchers' nest had hatched. To follow the slow development of this nestling, and of the other in the nest higher up the stream, was consolation for the loss of my first trogons' nest. Another surprise awaited me, if that can be called a surprise which we watch as it slowly appears. In their juvenal plumage, the young of most flycatchers rather closely resemble their parents. But in this, again, the Royal Flycatchers asserted their individuality, for the upper parts of the newly feathered nestlings were barred with yellowish-buff and dusky, instead of being plain olive-brown, as on their parents. As I looked down into the niche where each young flycatcher rested, the thin network of its walls, with light passing through the dark, enmeshed fibers, created a mottled pattern of light and shade into which the barred nestling blended. It always sat facing the wall at the rear, where its bright eyes were least likely to betray its presence. Its stubby tail of lighter color was in the doorway. When one of the parents brought food, it clung beneath the entrance and passed the insect over the nestling's back. Finally, when three weeks old, the young flycatchers flew away, and their nests hung deserted above the rushing torrent.

I last visited this enchanting stream at the height of the rainy season in November, soon after the tremendous downpour of twenty-two inches in two days. With a thunderous roar, the current tumbled down its rocky bed; but the turbidity that had marked the crest of the flood had already disappeared, and the water, filtered through the forest mold, was again crystal clear. Since June, the aspect of the dark forest along the banks had changed little; still no bright flowers relieved the varied greens of the multiform foliage.

On the side of the termitary where the doorway of the trogons' nest had been, a roundish patch of dark brown contrasted with the dull black of the rest of the structure. I sought the ladder, in order to climb up and examine the changes that had occurred. But the warm dampness had long since rotted the cords that bound it together, and the flood had carried away the pieces of wood. With the midrib cut from a mighty palm frond, I prodded the black mass until a horde of white insects poured out. The termites had recovered what the birds had taken from them and repaired the hole in their nest. The tragedy of the trogons had been only an episode in the life of the termitary.

The Royal Flycatchers and Massena Trogons were the only birds whose nests I watched in the high forest at Lancetilla, but in the plantations and second-growth thickets, where nests of many kinds were much easier to find, I studied a number of birds, in addition to the oropéndolas. On the leafy litter that covered the ground beneath the oropéndolas' tree, I discovered two eggs of the Pauraque, and after they hatched, I watched a parent feed the downy nestlings in the deep twilight—an observation which, despite repeated attempts, I have never been able to repeat. In sandy banks along the lower reaches of the Tela River, where it flowed between bushy pastures and impenetrable thickets, I found burrows of Amazon Kingfishers and Turquoise-browed Motmots, each with four roundish, white eggs lying on the sand in an unlined chamber at the inner end, and I followed the development of the nestlings as they grew from ugly naked mites to lovely feathered creatures.

Amid the thickets, it was not difficult to discover nests of the Rufous-breasted Castlebuilder, huge edifices of interlaced sticks that seemed too large to have been built by retiring, sharp-billed birds hardly bigger than house wrens. In each was a closed nest chamber, lined with downy green leaves, that was approached through a long entrance hall carpeted with fragments of cast snake and lizard skin. Above the chamber was a thick thatch of coarse vegetable material that effectively shed the rain from the three or four white or pale blue eggs and the nestlings that hatched from them.

The least beautiful of the birds that I studied at Lancetilla, but certainly not the least interesting, were the Groove-billed Anis, lanky black birds with high-arched bills and beady black eyes set in bare black faces. Their nests of coarse sticks, lined with smooth green leaves that were constantly renewed, were built in fruit trees in the plantations, especially in thorny citrus trees. In some of these nests, three or four chalky white eggs were laid by a single female and attended by a single pair. Other nests contained from eight to twelve eggs, laid in a common heap by two or three females and incubated by the four or six parents sitting by turns, with a single male taking charge through the night. After the eggs hatched, all the parents of both sexes helped to feed the ugly, naked nestlings, which developed surprisingly fast. At a second-brood nest, a young ani of the first brood, only seventy-two days old, helped its parents to feed its younger siblings. Since in other books and articles I have told about all these fascinating birds, I need not repeat the details here.

In addition to the birds, I studied the anatomy of the massive underground bulb or rhizome of the banana plant, tracing the erratic courses of its conducting strands. With so many interesting things to engage me, I went harder than was wise in a climate that I found enervating; at one point, I broke out with huge boils, and I lost weight. In the evenings, I exchanged language lessons with Axel Pira, Jr., a Guatemalan boy who was serving an apprenticeship in horticulture at the experiment station, and who at the outset was as ignorant of English as I of Spanish. Each of us taught the other his native tongue. Foreign languages had been the great tribulation of my school days; through long and painful courses, I had learned to read a few, including French, German, and Latin, that I never dared to speak. But these lessons mutually given and taken were enjoyable. Using a textbook that gave English and Spanish words and sentences in parallel columns, within a month we were able to converse, haltingly, one day in his language and the next in mine. After we could exchange ideas more readily, my student-teacher gave me glowing accounts of his family estate in the Guatemalan highlands. When I needed a change of climate to restore my health, Axel arranged for me to visit his home, in the mountains a mile and a half above sea level. Early in November, I regretfully left Lancetilla and travelled by ship, rail, and automobile to the Guatemalan highlands.

Young male Montezuma Oropéndola (*Gymnostinops montezuma*) just after leaving the nest.

11

The Ohio River

I pass over my visit to the Guatemalan highlands, and to Barro Colorado Island where I went from Guatemala, as in after years I resided longer in both of these localities and shall tell about them in later chapters. After returning to the United States in the spring, I went to Cornell University to continue botanical research, but it did not go well. In the tropics I had found absorbing problems that kept me working at highest pitch and resulted in half a dozen substantial scientific papers and a number of shorter articles. Now my interests were changing; laboratory research had lost its zest; I could not revive the old enthusiasm. Older botanists advised me to specialize. Although I still loved plants, I did not find one aspect of their lives so much more absorbing than another that I could focus my interest sharply upon some narrow area of the whole great field of botany. While I felt so remote from my work, it was not difficult for my friend Winslow R. Hatch, then a graduate student in botany at Johns Hopkins, to persuade me

Headpiece: Young Green Kingfishers (*Chloroceryle americana*) ready to fly.

by letter to leave the laboratories and accompany him on a canoe trip down the Ohio River, which I did at the expiration of my fellowship from the National Research Council.

On August 15, 1931, we met at Harrisburg, Pennsylvania, and continued by rail to Pittsburgh. Here we took passage on the *Queen City*, a paddle-wheel steamer whose somewhat tarnished elegance reminded us of Mark Twain and the great days of river travel on the Mississippi and its tributaries. For three restful days, we wound down the muddy Ohio River, between the hills of West Virginia on our left and those of Ohio on our right. Disembarking at Huntington, West Virginia, we visited Dr. W. E. Neal, a former mayor of the city, to whom Win had a letter of introduction. He took us into his home while we bought a secondhand canoe and supplies for our voyage.

The start of our canoe trip was not encouraging. Leaving the boathouse of the Huntington Boat Club at four o'clock in the afternoon, we set forth down the river under a drizzle. After paddling about eight miles, we landed on the Ohio shore to look for a campsite among the willows that everywhere fringed it. The best we could find in the dusk was a patch of rough, sloping ground halfway up the steep bank, hardly wide enough for our pup tent. Unable to find dry wood for a campfire, we ate a meager supper of crackers and apples and crept into our tent early. Sleep rarely comes readily on one's first night on the ground; and here the usual difficulty of resting tranquilly was increased by the noises of freight trains rumbling along the tracks on both sides of the river, of electric cars in the nearby town of Kenova, and of barge-pushers with loudly splashing paddle wheels passing along the channel, stirring up waves that broke noisily on the bank. During the night the sky cleared and stars shone through the willows above us. Finally, my bones became reconciled to the asperity of the ground beneath them, and I dozed off.

Rain was falling steadily when we awoke next morning, but the cheerful whistles of the Carolina Wrens in the surrounding trees assured us that the new day was not as dreary as it seemed. While we sat in our tent eating a cold breakfast of cereal, bread, and apples, a Catbird peered in inquisitively but offered no encouraging song. All morning the rain continued, holding us in camp. Finally, it abated somewhat and we packed our damp equipment, to resume our voyage beneath the continuing drizzle.

An hour's paddling brought us to the second of the many locks through which we passed on this trip. The long Ohio River has been canalized, or rather converted into a series of lakes, by the construction of many dams along its course. Vessels were lowered, or raised, from one lake to another by a lock beside each dam. These locks were spacious enough to accommodate not only river steamers but the strings of big freight barges that were pushed up and down the stream.

The regulations provided that all vessels, no matter how small, must be passed through the locks at the end of every hour. But traffic along the river was light, and usually the lockkeepers did not make us wait. As soon as we arrived, we entered the lock, the great gates closed behind us, and thousands of gallons of water were spilled downstream as our little canoe, hardly noticeable in the huge lock chamber, sank steadily to the lower level. Then the gate in front of us opened and we paddled forth into the wide river with a fresh sense of importance. Only once in the course of our voyage did the lockkeeper suggest that we portage around his lock; but we remonstrated, calling his attention to our heavy cargo and the difficulty of portaging without a proper trail. He kept us waiting until the hour's end, then passed us through alone, as no other vessel had arrived. Beside each lock, the keepers dwelt in substantial houses in attractive surroundings, provided by the federal government. At the locks we filled our jug and canteens with water for drinking and cooking, as the river was far too dirty.

After passing through Lock 29, we continued downstream beneath intermittent drizzles, battling against the head wind that so often opposed us. We had counted upon the current to help us toward the Mississippi; but we had not expected the persistent westerly winds, which frequently cancelled the advantage that the current gave us in this canalized river, except when it was in flood. The dreary afternoon was made more dismal by clouds of dun smoke poured forth by forests of chimneys that rose along both shores, amid tall domes of Bessemer converters, in the great steel factories of Ironton in Ohio and Ashland in Kentucky.

After passing these grim sources of industrial strength, we went ashore for a belated lunch of more crackers and apples. At the water's edge, we found an unfamiliar plant and sat down with our constant companion, *Gray's New Manual of Botany*, to identify it. It proved to be the Water Willow, one of the few northern species of the large acanthus family. Then we paddled on again until other plants new to us called for another halt. When we had learned their names from our indispensable *Manual*, the afternoon was so far spent that we decided to camp at this point. In a small, nearly level, grassy opening amid cottonwoods, elms, maples, and Great Ragweed four or five yards high, we pitched our tent, covered the bottom with long grass from the shore, spread our rubber poncho and blankets over this, and made a fairly comfortable bed. Soon a cheery campfire was consuming the dry interior of a dead cottonwood that we split open, and we were roasting ears of corn.

Again the rain that started afresh during the night held us in our tent until late in the morning, although the bright songs of the Carolina Wrens seemed to mock our aversion to a soaking. Breakfast, putting our equipment in order, and more botanizing delayed our departure until nearly noon. Paddling downstream, we kept close to the soft green wall of low willows along the

Ohio shore, past Greenup in Kentucky and Haverhill in Ohio. Then we pushed across a wide, straight stretch of river, in the teeth of a head wind that sharpened into half a gale and raised whitecaps on the muddy waves, directing our course to Lock 30. The heavily laden canoe behaved splendidly on this first encounter with rough water, rising to every wave and rarely permitting any to splash across its bow.

We were passed through the lock without delay and continued downstream, keeping close to the bank to avoid the strong wind that swept up the open river. Soon a range of steep, flat-topped hills, several hundred feet high, rose close to the shore on the Kentucky side. Well wooded, intersected by deep, narrow valleys, they invited exploration, and we looked for a suitable campsite near their base. But behind a narrow fringe of riverside willows ran a railroad, with houses scattered along it. Hoping for something better, we paddled onward, until we passed beyond the hills and skirted a long, low peninsula, planted with corn. Directly ahead, the river curved sharply, and with the unquenchable optimism that persistently places the most desirable things just beyond the turn in our lives that hides them from us, we decided to paddle around the bend. Passing beneath the great steel railroad bridge at Sciotoville, we were greeted by a most discouraging prospect. The nearer shore was low and flat, with cornfields extending almost to the top of the bank, on which the usual fringe of willow trees was sparser and less inviting than it had been upstream. On the Ohio side, between the steep, scarred, riverward face of a flat-topped ridge and the barren shore, from Sciotoville as far downstream as our eyes could penetrate the pall of black smoke, the tall chimneys of a huge steel plant stretched for miles along the riverside. The pleasant campsite that we sought seemed more remote than ever; we were becoming discouraged and might have settled for any spot to pass the night if, after rounding the great northward loop in the Ohio, we had not, for once, been aided by a following breeze.

Just when we were in the region where the screen of pungent, sulphurous smoke from the steel furnaces hung thickest, and our regret for having left the inviting hills behind us was most poignant, we noticed a break in the willows on the opposite shore. It hardly looked promising, but it offered the best escape from our present predicament. As we paddled across the swift current, a Great Blue Heron, flying with slow wingbeats down the shore, turned to enter the narrow gap between the trees and continued inward until lost to view. Hopefully following the lead of the wide-winged bird, we threaded a maze of logs, sticks, and other flotsam that clogged the mouth of a narrow but deep and muddy channel.

As we pushed upward between the willows, sycamores, and Silver Maples that grew on the steep banks and interlaced their branches above the turbid water, Belted Kingfishers dived into the stream, breaking the opaque surface with a splash, then rising again to fly away at our approach. Little Green Herons abandoned their perches on the willows with a startled *kreek* and hurried upstream with craning necks. We passed beneath a railroad bridge and a highway bridge, rounded a curve, and found three boys gathering firewood by the shore. They told us that this stream was called "The Tiger," (Tygart's Creek was evidently its correct name), and they showed us a pleasant spot in an open, streamside grove where we could camp.

Fred, the oldest boy, led me to the house of a neighbor, Harry Hunt, where provisions were available. This generous farmer would accept no money for half a dozen roasting ears and a number of excellent apples; they were, he said, a reward for my honesty in asking for what many would have taken without permission. Meanwhile, Win had made camp; and as I returned to him in the dusk, a Black-crowned Night-Heron flew high overhead with a characteristic *quok*. As we were finishing supper, Fred returned with a gift of four cucumbers and an invitation to visit his home. These good people were doing their best to make two unshaven, disreputable-looking wanderers feel welcome.

When we awoke next morning, a chilling mist hung over the valley and a drizzle fell. Going to the streamside for my first wash and shave since leaving Huntington four days earlier, I was delighted with the abundant Tall Ironweed that displayed great spreading panicles of large, deep purple flowerheads on straight stems nine feet high. After breakfast, we paddled upstream for several miles, until we came to rapids against which our most strenuous efforts were of no avail. Then, tying our canoe in a quiet cove, we pushed through the dense riverside growth of Great Ragweed and found ourselves in a field of corn and pumpkins, which a woman in a sunbonnet was hoeing. Nearby stood her one-story log cabin, with whitewashed walls. We seemed to have entered a world remote in space and time from the great steel mills and bustling modern industry along the great river. In this field, we made the acquaintance of the Apple of Peru, an attractive if somewhat rank weed of the nightshade family. Along the field's border, Tall Bellflowers lifted splendid spikes of large, pale blue flowers.

Climbing a steep hillside, we came to an old apple orchard, beside which was an almost pure coppice growth of the Common Papaw, whose large leaves suggested tropical luxuriance, as well they might, as this small tree is the northernmost representative of the almost exclusively tropical custard apple family. Nearby, we found a magnificent Papaw Swallowtail Butterfly, which seemed to have drawn tropical splendor from the foliage on which it had fed while still a caterpillar. It clung amid low weeds for shelter from the strong breeze, giving us an opportunity to admire, without disturbing, its wide, pale bluish white wings boldly striped with black. A bright red dot adorned each hind wing, near the base of the long, slender "tail."

From a spur of this ridge, we enjoyed a wide view over the Ohio and the country through which it flowed. The surrounding hills, several hundred feet high, were all of nearly the same height, so that their crests formed a horizon almost as level as that of the ocean. These even-topped hills appeared to be the remnants of an ancient plain that had been uplifted, then dissected by the present river system. Tygart's Creek wound through a narrow, level valley, occupied by pastures and fields of corn, wheat that had already been harvested, and tobacco. Except for a few hillside or hilltop pastures, the enclosing ridges were wooded.

As we walked along a hilltop, we heard the little rattle of a Downy Woodpecker and the confidential *dee de de de* of Carolina Chickadees, intermingled with the small voices of a company of wood warblers. Although it was only August 22, the warblers were already migrating toward their winter homes. The Black-throated Green Warbler and the resplendent Blackburnian were beyond their breeding ranges; the American Redstart, American Parula, and Black-and-White Warbler were evidently travelling with them. Probably the Blue-gray Gnatcatchers that pursued scarcely visible insects with intricate aerial maneuvers were likewise on their way south, along with a few Red-eyed Vireos.

The creek's rapid current speeded our return to camp in the grove. We procured milk, apples, and corn from the Hunts, who would take money for only the first of these items. Later, Mrs. Hunt sent us a jar of delicious apple butter, of which she had made thirty-three gallons that day. Often, she told us, she put up two hundred or more gallons to use through the winter. As we were finishing supper, her son-in-law brought us a big bowl of vegetable soup, for which we could hardly find room. And in addition to all this kindness from strangers, we were invited to breakfast next morning, and fared well on hot biscuits liberally spread with the newly made apple butter. The meal over, Mr. Hunt showed us his farm, where he raised corn, wheat, hay, and cattle.

After passing the remainder of the morning studying the plants we had collected, we ate a late lunch, loaded our canoe, and paddled downstream to swim and bathe. We emerged from Tygart's Creek late in the afternoon to find the broad Ohio in flood. The opaque brown water flowed swiftly, carrying great quantities of logs, sticks, and brush. Riding the crest of the flood, we speeded past Portsmouth and the mouth of the Scioto River. At Lock 31, we found that the long dam, composed of many stout wooden sections or wickets about four feet wide, each hinged at the bottom, had been laid flat on the river bed to give the floodwater free passage. The water had risen to the concrete pavement beside the lock, the gates were open, and a strong current flowed through the lock chamber. The river had reached the eighteen-foot level, flooding all the machinery that operated the gates. Stopping only to replenish our supply of drinking water, we rode through the open lock without help from the keepers.

Some miles below this lock, we saw two young Little Blue Herons in pure white plumage standing on the shore, providing the finishing touch, the center of interest, of the wild scene formed by the wide river rushing between hilly, wooded shores. It was a pity that few such spectacular birds remained to adorn this waterway. Just at dusk, as we were rounding a bend in the river, keeping close to the convex shore, a great raft of barges suddenly loomed up ahead of us. The bargemen were also staying close to the shore, to avoid the swifter current amidstream. Tied three abreast and three deep, the huge, empty, steel barges reared their mass across our path like a solid wall. As they bore down on us, it seemed too late to cross their bows to the outer side, and we feared that the narrow passage between them and the shore would be churned to fury by the huge paddle wheel at the stern of the pusher. We passed some anxious moments until we had pulled up our canoe on a patch of grassy shore beside a cornfield, where we waited until the laboring steamboat had crept slowly past us.

By the light of a nearly full moon, we continued downstream, paddling easily and depending chiefly on the strong flood current to carry us along. Sitting in the bow, Win kept a close watch for the large logs and rafts of brushwood that were likewise being swept forward by the stream, and directed our course to avoid a dangerous collision with them. The wooded hills that bordered this beautiful stretch of river were transformed by the soft moonlight into distant ranges of lofty mountains; the scene was even grander than it would have been by day. Only here and there, at long intervals, a light shone from the window of a lonely farmhouse by the shore. When we passed the village of Vanceburg, Kentucky, soon after ten o'clock, it seemed already to have gone to sleep.

At the base of a high bluff on the opposite shore, we nosed our canoe inward under the willow trees, whose roots were inundated by the high water, and Win landed to seek a campsite. Passing through a thicket of Great Ragweed, he reached a grove of small Slippery Elm and Common Locust trees high on the bank. With hardly any undergrowth, this grove, upon which we stumbled in the moonlight, proved to be one of the most pleasant campsites of our whole trip. Abundant driftwood from the spring floods provided the fuel to cook our belated supper. Then, for the first time since we embarked, two conveniently spaced trees and favoring weather invited me to stretch my hammock, in which I was soon comfortably ensconced. Win preferred to sleep on the ground.

While I lay gazing out through a gap in the foliage upon the river bathed in moonlight and a high, rounded hill beyond it, a locomotive somewhere along the curving bank gave a prolonged peal of its particularly mellow whistle, which echoed from the hills along the crescent shore. Floating in from greater and greater distances, the reverberations dwindled away in one continuous sweep of sound, like the fading echoes of some rich organ melody. Grateful for this proof that man's mechanical contrivances some-times fit harmoniously into a natural setting and embellish it, I fell content-edly asleep. With the current's help, we had covered twenty-five miles in five hours.

Thereafter, whenever I found properly spaced trees on a clear evening, I slung my hammock between them. It was pleasant to lie in it and, until I fell asleep, gaze up through the trees that supported me, catching the stars' intermittent twinkle through the shifting interstices of foliage stirred by faint night breezes, listening to the interminable nocturnal disputations of the green Katydids in the treetops. Although doubtless all these male insects made the same sound by rubbing the thickened bases of their wing cases together, those closer to me persistently asserted *katy did*, while their more distant brethren declared just as emphatically *katy didn't*—and so the de-bate continued, until I tired of following it and slept.

When the wind blew, the swaying of the trees was transmitted to me through the taut hammock ropes, making me feel that I had severed contact with the firm earth and lived in closer communion with aerial powers, freer and more bird-like. Sometimes, when a breeze stirred the cottonwoods and their leaves danced merrily on flattened petioles, I mistook the sound for that of rain pattering on the foliage, and waited with drowsy irritation for the large drops to slip off the leaves and strike my forehead or extended hand, making me hurry to untie the hammock ropes and seek shelter in the pup tent, reluctantly deserting my airy couch for a less comfortable one on the hard ground, to keep my blankets dry. But presently it would dawn upon my awakening mind that what I heard was only the wind in the treetops, and I would remain contentedly in my hammock.

In the morning, I awoke to catch the sun's earliest beams striking through the boughs; to gaze across the river where they were dispelling the nocturnal mists; to hear the songs of the Cardinal and the Carolina Wren and the Red-eyed Vireo and almost feel that I belonged to the feathered nation, for my sleeping body was upheld by the same trees that supported them. Then to arise, and wash away all drowsiness by plunging into the river's clear water—but alas, too often the river was diluted mud or, what was worse, covered by a sickening, greasy scum, the unhealthy consequence of man's indifference to the purity of his waterways. This was the most unpleasant feature of our voyage. Years later, I travelled in a Peruvian gunboat for many hundreds of miles along the Río Marañon or Upper Amazon and its great tributaries, the Ucayali, the Huallaga, the Napo, and the Yavarí. These, too, were turbid streams, laden with silt from the high Andes; but they bore scarcely any trace of human pollution and their water supplied the ship's shower baths.

Next morning, from our camp opposite Vanceburg, we paddled across to the village to buy provisions, then cooked lunch and identified the new plants we found. In the afternoon, we continued down the swollen Ohio, between high, wooded hills, along the most picturesque stretch that we had yet seen. Our destination was the Ohio Brush Creek, whose name suggested that it would repay exploration. But its mouth was poorly indicated on our map; we were half a mile below it before, in response to our question, some men in a skiff put us straight, and we had laboriously to retrace our course against the strong current. Next morning, we ascended this stream until, even by wading and pulling our canoe through the shallow water, we could force it no higher. While a slight drizzle fell late in the morning, a large flock of Common Nighthawks, numbering between twenty-five and fifty birds, flew southward with a directness that suggested migration.

While we ate lunch on the bank, we were accosted by three boys, sons of the farmer on whose land we were trespassing. They had never seen a canoe, so we gave them a ride in ours. In a neighboring stubble field where wheat had grown, a wide tangle of vines of the Man-of-the-Earth adorned the ground with a profusion of great white flowers. Never having seen the root of this morning glory, which attains huge size, we borrowed a mattock from one of the farmer's boys to unearth it. The thick root stood upright in the ground, with its crown, from which the slender twining vines sprang, just deep enough to escape the plowshare. A foot below the surface, it was six inches thick. Here it divided into two massive branches that promised to descend to considerable depths. Since we were not digging in our own field, we thought it proper to stop here. Morning glories of the genus *Ipomoea* sometimes become more massive than the slender vines that we commonly notice; in the drier parts of northern Central America, they grow to be small trees with large white blossoms.

In the evening, we bought peaches from the farmer's wife and started downstream. As we passed a ford, all the boys in the neighborhood seemed to have gathered to watch the passage of the first canoe they had seen. After a second night in our camp along the Brush Creek, we resumed our voyage down the Ohio. As we coasted along Manchester Island early in the afternoon, we noticed thunderclouds densely banked above the river ahead, so black and menacing that we landed on the island. We had reason to be thankful for our prudence, for in a few minutes the storm burst in falling sheets of water and a fury of wind that lashed the river into choppy waves which might have swamped our loaded canoe. The storm was too violent to continue long; in half an hour the worst was over, and we resumed our voyage in a drizzle.

Next day, we stopped at the quiet little town of Ripley, Ohio, to buy provisions and call for mail. We had returned to the canoe and were about to push off, when some men loitering at the foot of the main street hailed us.

Returning, I found the postmistress, who had walked several blocks to tell us that we had mailed a card without an address. I accompanied her back to the post office to rectify the oversight. Where could one find a more considerate postal service?

The floodwater continued to help us on our way, despite opposing westerly winds, until we had passed Cincinnati and the mouth of the Miami River and had Indiana on our right hand. On a mud flat beside an island, we found a large flock of Killdeers foraging and bathing in the shallow water. At Warsaw, Kentucky, where we stopped for more provisions, we seemed to have somehow reached Europe: one road-marker pointed to Sparta and another to Ghent, while across the river in Indiana was a county called Switzerland. The dome of a church had a Russian or Polish bulge. The flood was now subsiding, and as we approached Lock 39 we found a maneuver boat setting up the heavy wooden wickets to re-form the dam that had been laid flat when the river was high. At the far side of the channel, where the dam was incomplete, we might have passed downward, but the water poured through with such a rush that we thought it safer to wait until the men returned to the lock and lowered us through it. Then for an hour we paddled down a long, straight stretch of river, placid in the evening calm, with picturesque hills rising in front of us. A gravelly beach on the Kentucky shore provided a good camp site.

At the mouth of the Kentucky River, we landed at Carrollton to call for mail and replenish our provisions. Learning that we were on a canoe trip, the proprietor of the drugstore told us of the deep, impressive gorges through which this southern tributary of the Ohio flows in its upper reaches, in the "Switzerland of America," as he called it. He praised his river so highly that we were persuaded to turn aside from the Ohio and explore it. Soon after noon, we started up the Kentucky, paddling along a broad, deep, almost currentless channel, through water the color of strong coffee heavily diluted with milk.

By the middle of the afternoon, we reached the first lock, which was operated by hand instead of by electricity as were the locks along the Ohio River. According to the government regulations, the lockmaster could request aid from the crew of passing vessels, so I lent a hand to turn the big, two-handled cranks that swung the gates in and out, and also helped to open and close the wickets in these gates. We were required to fill a "manifest" of crew and cargo, a formality that never detained us on the larger river. The friendly lockmaster told us that this river was first made navigable by dams and locks ninety years earlier. Seventeen of these were built between its mouth and Beattyville, the head of navigation, two hundred and sixty miles upstream. The impressive scenery began about eighty miles above Carrollton; and he assured us that if we ascended this far, we would not turn back until we reached Beattyville and saw all that was to be seen.

Unfortunately, we never reached the spectacular upper course of the Kentucky. For a day and a half we followed the winding, willow-fringed channel, between steep, flat-topped, rocky ridges, whose long curves paralleled those of the river, as along much of the Ohio. These ridges were covered with light second-growth woods or poor, scrubby pastures. The lower ground along the river was planted with corn and tobacco. The latter was now being harvested by whole families of men, women, and children, whose homes hardly suggested prosperity. Finding that by steady paddling against the slight current we made only about twenty miles a day, we concluded that to reach the mountains would entail too protracted a side excursion. To enjoy the scenic upper Kentucky River, one should put his boat into the water at the head of navigation and travel with the current. Within hearing of the overflow from Lock 2, we turned back, late in the afternoon of the second day.

The rain that had long been threatening now came down in a deluge and continued until my teeth began to chatter and cold chills coursed along my spine. All our baggage got wet. My companion would not be satisfied with any campsite except a certain "grassy knoll" that he had noticed on our upward voyage, but this enticing spot continued until nightfall to elude us. Finally, departing from our usual practice, we stopped at a farmhouse whose lights we saw shining through the rainy night, and obtained permission to sleep in a shed.

Next morning, while tent, blankets, mosquito nets, changes of clothes, and odds and ends were spread along the shore to dry in the sunshine, I envied the chickadees who foraged among the locust trees on the bank. They, too, had weathered the deluge of the preceding evening and were now enjoying the sun's earliest rays, none the worse for their experience. Not theirs to waste hours of precious sunshine recovering from the effects of a drenching that they had doubtless already forgotten. Having acquired no wettable property, they had given no hostages to the weather.

After everything had been spread out in the sunshine, we noticed a fine yacht speeding upstream, throwing out swelling bow waves that threatened to break over our drying equipment and undo all our work. Hurriedly, we moved our things farther from the water's edge. To the passersby, this bit of shore must have resembled the bank of a Jamaican stream on a wash day. As the luxurious vessel swept proudly past us, we reflected that if, instead of "wasting" our time on vagabond excursions, we applied ourselves diligently to some lucrative occupation for the next few decades, we, too, might travel the inland waterways in the grand style. But even while recovering from the effects of our recent soaking, we did not envy the people in the yacht as much as we envied the chickadees. It seemed better to be enjoying this free life before the years softened our hardihood and tamed our spirit of adventure.

At our next camp, while Win cut a forked stick from a sycamore tree, two young Song Sparrows fluttered into the water, then turned to swim toward the shore a foot distant. Then he noticed their nest, eight feet up in the sycamore. We picked the two fugitives from the water and replaced them in the nest, where the third member of the brood had remained. All were well feathered and almost ready to leave. This nest, found on September 3, was the last nest of any bird that we noticed that year. Below the first lock, we spied a lone Pied-billed Grebe swimming in the river, and for a quarter of an hour we pursued it with the canoe to make it dive. It remained under water for from ten to fifteen seconds, and always came up in an unexpected spot.

Returning to the Ohio after an absence of four days, we turned our course downward along a beautiful stretch of the river. Our persistent adversary, the head wind, blew so strongly upstream, creating a heavy ground swell in addition to curling whitecaps, that it seemed safest to paddle close to the shore, at the price of following all its windings, instead of shortening the distance by crossing the broad channel from cape to cape, as was our custom. At sunset the wind died away, leaving the river covered with long, smooth waves that were pleasant to ride over. That night we camped comfortably in a grove at the mouth of Clifty Creek.

Next day we enjoyed the deep gorges, high waterfalls, winding woodland trails, and ferny limestone outcrops of Clifty Falls State Park in Indiana. Our progress through the park was slow, for we carried our *Gray's Manual*, and whenever we found a plant new to us—and there were many—we sat down then and there to identify it. One of the most beautiful of the waterfalls was formed by a rivulet that flowed through a deep, shaded ravine, between upland meadows. Suddenly it plunged over the edge of a flat, overhanging ledge of hard rock into a great, rounded, almost domed amphitheatre walled by the underlying soft limestone. At the head of the cascade, a mere trickle of water spread out into a thin fabric of beaded drops and slender threads, which atoned by its tenuous delicacy for what the fall lacked of the thunder and roar of plunging torrents that create the majesty of a waterfall. In the profound gorge below the fall, the steep soil slopes on either side were covered with the most luxuriant growth of jewel weeds that I have seen. I could not decide which was prettier, the large, clear yellow blossoms of the Pale Touch-me-not, hanging singly like oddly shaped fairy bells with maroon dots in their throats, or the more deeply colored but smaller flowers of the less abundant Spotted Touch-me-not, which often hung in pairs, each handsomely dotted with deep orange on an orange-yellow ground. We could not resist touching green pod after green pod, to make its turgid valves curl up so suddenly that they shot out the jewel-like seeds.

After two days at Clifty Falls, we resumed our voyage down the Ohio. At Westport, Kentucky, I went ashore to buy food. As I was returning to the landing along a path through a cornfield, I noticed a girl in pink pajamas gathering roasting ears. Believing her to be the farmer's daughter, or at least a member of his household, I asked whether she would sell a few ears. She replied that the corn was mostly too hard for roasting, but if I could find a few ears that were suitable, I might help myself to them. I searched until I found half a dozen that were not too old, then started back toward the river with them. Meanwhile, the girl had vanished.

When I reached the shore where Win was waiting in the canoe and related my little adventure to him, he questioned me so minutely about the girl's appearance that I was annoyed. Then he revealed that, while he waited there, a girl such as I had described came to the landing in a skiff, furtively entered the cornfield, and soon afterward emerged with an armful of corn. She confided to my companion that she had suffered a great scare, for at first she had taken me for the farmer who owned the field and felt sure she had been caught pilfering. Then she re-embarked in her boat and continued upstream. By her coolness of mind and ready tongue, she had diverted suspicion from herself and led me to take what otherwise I would not have touched. One thief had made another!

I entered the canoe, and we paddled out to drift downward in midstream while we lunched, as we often did. The flood had now completely subsided, and the sluggish current did not take us far. While we floated idly, sounds of loud talking and boisterous laughter, mingled with the notes of a phonograph, were borne to us across a wide expanse of water from a large gasoline yacht tied up beside a boathouse. Just as we finished our meal and resumed paddling, the yacht began to move, and soon we noticed that it was coming straight toward us. As it approached nearer and nearer, without revealing any tendency to swerve, we made some uncomplimentary remarks about yachtsmen who inconsiderately toss smaller craft in their rough wake, when they have a wide channel in which to pass. But when a few rods behind us, they cut their motor and glided past with diminishing speed, calling to ask whether we would like to be towed toward Louisville. So far we had travelled wholly under our own power, with whatever help natural forces happened to give us, but we felt that a ride behind a powerful launch would not be unwelcome. Paddling up astern, we threw our painter to a man in the skiff that the launch towed and were tied behind it.

We remained sitting in our canoe, until invitations to come aboard the yacht became so pressing that it seemed churlish to remain aloof. Clambering over the rail of the yacht *Whileaway*, barefoot and unshaven, we felt like a couple of pirates boarding a prize. We introduced ourselves, and in turn were introduced to the four men and women of this jolly party, which included a former mayor of Louisville. I found it difficult not to smile when they pressed us to be seated, and Win could not without certain special arrangements, as he had developed a boil in a basic situation; and of course he would not divulge why he persisted in standing. When the yacht stopped at a picnic ground on the Indiana shore, we wished to continue under our own power; but Captain Alderson persuaded us to remain in tow and have supper on the yacht. With Win's experienced help, an excellent meal was prepared in the galley and served on deck. Then followed dancing on deck to the music of a phonograph, while the dishes were washed.

As we continued down the river, a large passenger vessel, passing upstream on our port side, stirred up waves that made our canoe rock violently, until the towline snapped and she fell behind, adrift in the night. The yacht's engine was promptly slowed, but we passed some anxious moments until her powerful searchlight spotted the runaway canoe, floating free and upright. Soon the truant vessel was captured and tied to the yacht with a stouter rope. As we approached Louisville, uncertain where we should camp, the captain suggested that we sleep in his vessel while she lay at the landing pier. After instructing us how to lock up the yacht when we left in the morning, and presenting us with a signed copy of *The Ohio River Pilot Guide* that proved very helpful, the captain departed with his guests and our sincere thanks. Next morning, we breakfasted on our own provisions in the yacht's galley, and I brought my journal up to date, before we locked up the *Whileaway* and paddled down the river past Louisville.

We had some difficulty finding the canal that leads around the Falls of the Ohio. Finally entering it, we paddled about two miles through quiet water to the lock at its lower end. Here we were kept waiting about an hour until another vessel arrived, a pusher with a barge at its bow. With it we entered the spacious lock chamber, six hundred feet long by a hundred and ten feet wide, which still seemed empty. The huge, electrically driven gates swung closed behind us with powerful, organ-like vibrations swelling from their rusty hinges; then the escape valves were opened. Quickly we dropped thirty feet; with more deep groans, not unmusical with all their volume and power, the lower gates swung open, and we emerged into the swift current of the Ohio below the Falls. For many miles above and below this point, the land on both sides of the river was level and well cultivated, in contrast to the hilly country through which we had passed upstream.

On our second night after passing the Falls, we camped on a rocky shore a short distance below Mauckport, Indiana. Along this stretch of the river, the White-eyed Vireo was more abundant among the fringing willows and cottonwoods than we had found it higher up. I often awoke to listen contentedly to his quaint medley of whistles and short phrases, somewhat like the Catbird's melange, but without the latter's harsh mews. Carolina Wrens still sang, but Cardinals were falling silent as September advanced. Behind our camping site, in a forest managed by the State Department of Forestry, we found one of the most impressive stands of temperate-zone hardwoods that I ever saw, rivalling tropical forests in magnificence. For a mile, we walked admiringly between the stately trunks, massive and straight, of the Beech, Sugar Maple, White and Red Oaks, and Shag-bark Hickory. Few shrubs grew in the deep shade of these noble trees, and the herbaceous ground cover was sparse. Here for the first time we saw, in its native habitat, the Kentucky Coffee Tree, whose large leaves, twice pinnately compound, had a distinctly tropical aspect.

Beyond the forest, we passed through a barren woodland pasture to a hilltop, from which we descended across a weedy stubble field and a dry streambed to the county road. The day was oppressively hot; the sun, beating down on the dusty road, made a task of walking, but all discomforts vanished when we came in view of a splendid display of large Passion Flowers, on vines that clambered over a barbed-wire fence and roadside weeds. Each three-inch-wide blossom, bearing a fringed crown in its center, was white, tinged and banded with purple and a shade between magenta and maroon. We sampled the aril-covered seeds that filled the large, egg-shaped fruits, known as maypops, and found them pleasantly acid; but they were scarcely ripe. In a neighboring peach orchard, many more of these lovely flowers were borne on vines that spread over the sandy soil. Like the Kentucky Coffee Tree, they carried my thoughts back to the tropics, where many species of passion flower grow wild and a few are cultivated for their edible fruits.

As we continued down the Ohio toward Rockport, Indiana, we noticed two Caspian Terns resting in shallow water on a bar. With what stately grace they rose into the air at our approach! At the Rockport post office, we found a letter informing us that by September 28 we should be at Johns Hopkins, where I was to teach botany for one semester during Professor Johnson's temporary absence, and Win was to be my laboratory assistant. It was already the fifteenth, and it would have been contrary to the spirit of our leisurely voyage to race down to the Mississippi, still two hundred and twenty six miles away, in the eight or nine days that we could remain on the Ohio. Besides, although our trip had been on the whole enjoyable, I was tired of trying to keep clean in such a dirty river. We decided to sell our canoe, if we could, and go home by rail. Finding no buyer in Rockport, we paddled ten miles downstream to Owensboro, where there was a boat club. Here we met an elderly member of the club who expressed his distrust of canoes, which he had never entered, and intimated that it would be difficult to sell such an unstable vessel—which had never capsized on our long voyage! Finally, Win said that if I would relinquish my half-share, he would bear the considerable expense of shipping the canoe to his family's summer cottage in Massachusetts, and to this I agreed. We decided to return to Rockport, where the railroad station was conveniently near the river.

We spent the following morning in Owensboro, visiting the barber and making ourselves more presentable for the homeward trip. Starting up-stream in the afternoon, we made our last camp in a pleasant grove of cottonwoods on the Indiana shore. After breakfast next morning, we crossed to the Kentucky side for a final excursion inland. Landing in a grove of willows and cottonwoods, we passed through a small plot of tobacco in front of a farmhouse, and were forcing our way through a weedy fence-row, when my eye fell upon an inconspicuous plant new to me, a Broomrape. A thick, yellowish stem three or four inches high, with leaves reduced to tiny scales, bore a dense spike of rather large, purple, two-lipped flowers. Digging up one of these parasites, we found its bulbous base attached to a thin root of the Great Ragweed, from which it drew all its nourishment; it had neither chlorophyll nor roots of its own.

While we sat hidden among the tall weeds, examining and discussing this interesting plant, the farmer's daughter passed by and, suspecting our inten-tions, called her father, who promptly arrived and demanded to know what we were doing there. Win, who might have won distinction in the diplomatic service if he had not become a professor and educationist, soon soothed the farmer's ruffled temper. When we showed him the plants we had gathered and told him about our trip, he thawed, explained that all the stealing along the river had made him distrustful of strangers, and excused himself for his abrupt questioning. We assured him that, under the circumstances, it was quite proper, and had given no offense. Presently he was telling us all he knew about the Broomrape, how it lived at the expense of other plants and sometimes, especially in low, moist ground, grew so thickly on the roots of his tobacco plants that it killed them. He directed us to places where we could find more specimens. All along the river, we found the farmers friendly, eager to tell us what they knew and to show us their farms, and occasionally, as has been told, treating us most generously. We wondered whether it was just by chance that nearly all the hospitality we received came from the southern side of the river. To the north, the people were cordial but not so open-handed.

We stood talking with this farmer while he pulled excess "suckers" from the tops of his tobacco plants. He told us that four or five acres of this crop was all that one man could handle, and would keep him fairly busy from the time he prepared the seedbed in February until he plucked the leaves from the dried stems in November. To what trouble people put themselves, or their neighbors, to satisfy their craving for needless, even harmful, luxuries! For war and their vices, men squander wealth with less hesitation than for anything else.

On September 19, exactly one month after we put our canoe into the river at Huntington, we crated it in the baggage room of the railroad station at Rockport, with the advice and assistance of the stationmaster and telegraph operator, who good-naturedly tolerated our hammering, sawing, and bustling about in their domain. That afternoon, which was swelteringly hot, we left for Lincoln City in a combination passenger and baggage coach at the end of a string of freight cars. With two changes of trains, we reached Louisville, which on the following evening we left for Baltimore on a faster and more luxurious train. In thirty days on the Ohio, we had travelled four hundred and fifty miles by canoe, not counting our side excursions up some of its tributaries. We had found and identified seventy-five species of plants new to us, although, as was to be expected, we found nothing new to science in this country already well combed by naturalists—that remained for later travels in fresher fields. And, in those days at the height of the Great Depression, our living expenses during our month on the river came to slightly less than eleven dollars for each of us.

From what we saw of this rich valley, now so thoroughly tamed and subjugated to man's uses, it was hard to imagine how it had appeared to pioneer naturalists, to Wilson and Audubon, Say and Rafinesque, or to still earlier explorers who saw it in all its pristine splendor, when majestic forests, of which we saw only a few small remnants, lined the shores, Wild Turkeys strutted in the woodland, and tremendous flocks of Passenger Pigeons darkened the sky. It was sad to think that all this vast midland of North America was "developed" without much consideration for values not given by corn and wheat, cattle and tobacco. In later years, I visited valleys having a more varied flora, richer and more colorful bird life. These, too, are threatened with the same ruthless exploitation that overtook the Ohio and its tributaries. They can be saved only if a too materialistic and too rapidly increasing humanity undergoes a change of aims and values so swift and radical that it could be compared only to a mass religious conversion. Although I saw only a valley too recklessly shorn of its natural wealth, I am glad that I made this leisurely voyage through the heart of my native land, and experienced some of the kindness of its people, before finally leaving it for fresher fields.

Nestling Great-tailed Grackles, eighteen days old, on a coconut frond on Alsacia Plantation, Guatemala.

Nestling Brown Jays, twenty-three days old.

12

A Border Farm

When I returned to the United States after my first visit to Central America, I looked forward eagerly to learning from books more about the habits of the birds of many kinds that I had seen there. As I searched the ornithological literature, I was disappointed. I learned that many thousands of birds had been collected in tropical America, that they had all been named and classified, minutely described to the last spot of color, and measured to the last millimeter. Indeed, this business of measurement and classification had been carried so far that, when an average difference of a few millimeters in the lengths of individuals of the same species from different parts of its range could be discovered, or when a slight variation in color could be detected when two specimens were laid side by side, these so slightly differing birds had received distinguishing names. Yet in the field, the two races so designated could not always be distinguished by appearance, voice, or habits.

When I looked for what had been recorded about the habits, the mode of

Headpiece: The Río Morjá, with Alsacia house among coconut palms on the hilltop. Three kinds of kingfishers, Turquoise-browed Motmots, and Rough-winged Swallows nested in burrows in the farther bank, at the edge of the banana plantation.

life, of these birds collected in such great numbers and so minutely catalogued, I found the situation quite different: the notes on habits were as scant as the necrologies of specimens were voluminous. Frank M. Chapman and Josselyn Van Tyne on Barro Colorado Island and William Beebe and his associates in British Guiana had published careful studies of the life histories of a few species. In Brazil, Carl Euler and Emilio A. Goeldi had provided less detailed accounts of the nesting of a number of widespread tropical birds. Bird collectors had recorded many incidental observations on habits and described a number of nests and eggs, not always correctly identified. But the total amount of information was slight, covering only a very small proportion of the birds of tropical America. The bare appearance of the nests and eggs of the majority of Neotropical birds remained unknown to science.

The subject had begun to fascinate me in no ordinary degree. My intimate studies of the lives of a few species of tropical American birds, and fleeting glimpses of numerous others, sharpened my desire to know more about them. I thought of the thousands of kinds of lovely feathered creatures, leading their beautiful, well-ordered lives among the forests on the plains and mountains of that wild, sparsely inhabited region that I was beginning to know—leading their lives obscurely, in ways never yet watched or recorded by man. It seemed to me, as a scientist and lover of beauty, that the most worthy cause in which I could engage, the highest endeavor to which I could dedicate my own peculiar endowment, temperament, and training, was to uncover the secrets of the lives of the tropical American birds, and to make them known to those of my fellow men who are so fortunate as to be interested in these matters. It seemed, too, that by striving to understand these birds as living, breathing creatures, I could help to bring to fruition and completion all this tedious labor of description and classification of dead specimens, which, although held to be fundamental to scientific knowledge, is in itself dry and uninspiring; that I could, in some measure, justify the sacrifice to science of so many thousands of bird lives.

An even wider consideration impelled me to undertake these studies. From childhood, I have been troubled by what might be called the problem of unrealized and lost values. Those poignant verses of Gray,

> Full many a flower is born to blush unseen,
> And waste its sweetness on the desert air,

are to my mind among the most melancholy in our language. The thought of such neglected flowers always distresses me. Their loveliness ought to be contemplated, their fragrance inhaled, by some appreciative being. Similarly, a bird's beauty ought to be enjoyed, its song heard with delight, all its habits observed with interest, not only by its feathered companions, but by ourselves, who perhaps can appreciate them more deeply.

The world in which beauty, goodness, or any other value, is wasted seems incomplete, lacking something necessary for its perfection. And who or what should appreciate all the lovely and amiable things that this world contains if not ourselves, who, of all animals, seem most highly endowed with aesthetic feeling and understanding? One might contend that our most important role on this planet, our *raison d'être*, is to complete or fulfill the world process by grateful, cherishing enjoyment of everything good and lovely that it has produced. I believed that by studying the way of life of some of the splendid feathered creatures that live obscurely in tropical forests and thickets, and sharing my discoveries with others, I could in small measure advance this great, rewarding task of bringing to fruition all the values that the natural world contains potentially and offers to us for realization.

To find financial support for such an undertaking was not easy, especially in the dark years immediately following the great financial debacle of 1929. Had I intended to collect birds in Central America, I have little doubt that I could have sold their "skins," even then, at a fair price. Had I wished to bury myself in a museum and pore over the dry and lifeless bird specimens already amassed there, it is probable that I could, with time, have found some institution to pay me a modest salary for doing so—although my special training in another branch of biology might have made it more difficult to obtain such employment. But I was already convinced of what subsequent years have amply confirmed, that few new species of birds remained to be discovered in Central America or, indeed, in any part of the world, and that the advancement of our understanding of bird life must be sought in a different direction. From these considerations, and because of my deep-rooted aversion to destroying life of any kind, I decided that I would not attempt to finance my studies of tropical birds by collecting them.

Fortunately, I was still sufficiently young and imprudent to be capable of a bold step. Although I had no income, I decided to return to Central America and continue to study the birds, devoting my small savings to the enterprise, and trusting to fortune that, somehow, I would be able to replenish my capital before it was exhausted. My father encouraged me with the dry remark that, of the many shiftless people he had known, not one had died of starvation! My decision was the more readily made because, when Professor Johnson returned to Hopkins to take over the botany course that I taught while he was on leave of absence, I would be unemployed, and any definite course of action was preferable to idle waiting. But these considerations, I am now convinced, hardly influenced the step I took. Sounding from afar, coursing with the swiftness of electric pulses over the convexity of the earth, the calls of the tropical birds fell upon my inward ear during those short, bleak winter days—and they were calls that I could not resist. Moreover, had I not already concluded that to learn about the lives of these birds was the highest endeavor of which I was capable?

My young friend, Axel Pira, who was employed as timekeeper on a United Fruit Company plantation in the Motagua Valley of Guatemala, arranged for me to stay at "Alsacia," a neighboring independent banana plantation situated opposite Los Amates, where I arrived in mid-February, 1932. I vividly remember my first moonlit journey to Alsacia. It was late in the afternoon when I set forth from the railroad station at Los Amates with Eugene Pellman, the overseer of the plantation, in one of the little red motorcars then commonly used in the banana-producing districts. After speeding for two miles up the main line of the railroad that joins Puerto Barrios with the Guatemalan capital, we crossed the Río Motagua by a high bridge. Then, reversing the car, we ran for miles down the right side of the river, chiefly through thickets of tangled second growth. It was now the cool of evening; insects were flying freely; and many struck stingingly against our faces, some even entering our squinting eyes, as we pounded along the rails in the small vehicle open on all sides.

When we reached the Río Jubuco, it was nearly dark. The moon and stars were shining brightly before we finished transferring the baggage and provisions from the motorcar to an open flatcar, drawn by a mule over the light tramline through the banana groves. It was pleasant to recline on the baggage during that leisurely three-mile ride and gaze up at the broad banana leaves that arched over the tracks, their mobile vanes spread out horizontally in the cool night air. The trotting mule drew us through an enchanted chasm, fretted with broad alternating bands of deep shade and bright moonlight, between tall, columnar trunks glistening where stray moonbeams struck their glossy surfaces.

As we emerged suddenly from the banana grove, we faced an elevation which, seen in the dubious light, might have been anything from a lofty hill to a low mountain. Beyond, far higher ranges loomed up in the shimmering moonlight. At the foot of the slope, where the tramline ended, we found horses awaiting us. Mounting, we followed a winding road up the hillside, while the baggage followed more slowly by oxcart. Supper was ready for us in the long, gray plantation house on the crest. Although the altitude was well under a thousand feet, it was cool enough to sleep under a light blanket, here at the foot of lofty mountains, this February night, and on many that followed. How different from the hot, breathless nights of April and May!

We jokingly called the plantation "La República de Alsacia," because it lay on the then unsettled boundary between Guatemala and Honduras, and owed allegiance to neither republic, although I was told that, as a matter of policy, it paid taxes to both. Perhaps more properly, it was a crown possession, owned by a king, who lived in El Salvador and rarely visited it, and his viceroy, the Finnish overseer with whom I lived. The unsettled nationality of the farm resulted from a long-standing controversy between the Hondurans, who claimed all the lower Motagua Valley, and the Guatemalans, who insisted upon a boundary along the divide between this river and the Río Chamélecon. In the daily presses of the two sister republics, the vindication of their respective claims consumed enough printers' ink to blacken the waters of these considerable streams. In a hundred garrisons in both countries, soldiers were drilling, to be ready to prove by force of arms the justice of their nation's demand, should either side, by an ill-advised move, give the other an excuse for attacking.

I had blundered into the midst of one of those innumerable, interminable boundary disputes which, ever since they won independence, have continued to excite the former Spanish possessions in America, and in aggregate have caused much bloodshed. Happily, in this particular instance, the controversy was settled peacefully by arbitration in Washington, early in the following year, when it was agreed that the summits of the Sierras of Merendón and Espiritu Santo should be the boundary between the two republics, and that Guatemala should retain its outlet to the Caribbean Sea. Thus Alsacia became definitely part of Guatemala.

As an independent territory between the two countries, the plantation became a haven of refuge for fugitives from the law of either, for neither would permit representatives of the other to enter and capture them. The overseer numbered several murderers among his employees. The first man that I encountered, on the morning after my arrival, bore a long scar on his cheek, stretching from an ear nearly to the corner of his mouth; but such facial decorations were a not uncommon sight along the banana coast, where machetes were sometimes used to attack things other than the rank tropical vegetation. In a region where strong liquor flowed freely and almost every man earned his daily bread with a tool that was also a formidable weapon, where many strapped on a revolver as they left their houses as automatically as they reached for their hats, a man could too easily become a murderer without premeditation; and I did not find my neighbors on this plantation more dangerous than men elsewhere in this unruly region.

On pay days, Don Antonio, the Honduran *comandante* at the neighboring hamlet of Lancetillal, arrived with a few soldiers to keep the peace. In order not to prejudice the question of the farm's sovereignty, it was understood that the *comandante*'s presence was a personal favor to the overseer who requested it, rather than an official act. Be that as it may, the monthly pay day always went off with a minimum of drunken celebration and random shots, and, at least while I was there, no casualties. On some banana plantations, pay day was a time of bloodshed, especially in intervals of prosperity when there was no lack of money to buy *aguardiente*. During my residence at Alsacia, I heard of only one machete fight, an ugly affray which I was glad that I did not witness. The overseer urged me to carry a revolver whenever I left the house, as he did, but with my equipment for studying and photographing birds I had enough things hanging around me without this piece of hardware.

The comfortable plantation house stood on the upturned end of a sharp spur projecting from the rugged Sierra de Espiritu Santo into the valley of the Río Motagua. The dwelling was several hundred feet above the level floor of the valley; the land sloped away on every hand, but especially to the front and sides, permitting a spacious vista in every direction. In front, the broad, flat valley stretched afar to the base of the precipitous Sierra de las Minas in the north. Through the green sea of banana plants that covered the level land the muddy current of the great stream wound its shifting way. Somewhere amid that great expanse of verdure, hidden from our view, stood the famous Old Mayan monoliths of Quiriguá, which for well over a thousand years had guarded amid the forests the secrets of the ornate symbols that covered them. Where could they be more impressive, or speak more meaningfully to an understanding spirit, than in their silent forest setting?

Toward the rising sun, the house looked up the narrow valley of a tributary stream, the Río Morjá, that flowed through banana plantations between steep, pastured hills. At the head of this valley towered Cerro Azul, whose steep, wooded slopes swept up to a summit more than a mile high. Behind the house were foothills, which from the distance appeared to be mantled by primeval forest, although actually their more accessible parts bore only older second growth, as I afterward learned to my regret. The sun set over a great expanse of tangled *guamil* or recent second growth, interrupted by scattered fields of maize and small banana groves.

The lower Motagua Valley contained vast expanses of this forbidding *guamil*. For seventy miles inland from Puerto Barrios, the railroad passed through these riotous, impenetrable thickets, interrupted by banana plantations, small clearings, and an occasional broad, grassy savanna with scattered pine trees, a surprising anomaly in humid tropical lowlands. Proceeding upriver, the *guamil* ended only where the arid zone began, for this type of vegetation is the product of much rainfall.

Guamil represents the first disorderly efforts of exuberant tropical vegetation to recapture land that has been wrested from it by man, flood, or landslide. In this region, it occupied chiefly banana plantations deserted because of disease and the abandoned milpas of the natives. The coarse grasses and weeds, the first plants to spring up after cultivation has been relaxed, are before long choked out by the shrubs and tangled vines that overshadow them and intercept their light. The struggle for a place in the sun is intense; every bush, almost every branch, is weighted, nearly smothered, by vigorous climbers and creepers, which cannot wait until the upstanding vegetation has provided supports for them, but start life by trailing among the low herbs.

Although there is a superabundance of vegetation in this second growth, there are few plants; at least, everything is so intertwined and entangled with everything else that it is difficult to delimit any single plant. So great is the competition, so eager is every growth to overcome and bear down its neighbors, that few shapely specimens are formed; there is little to delight the botanist's eye. For the bird-watcher, these impenetrable tangles are hardly more fertile; tied and tripped by the ubiquitous vines, grappled by thorny branches and blistered by stinging nettles, he cannot move without vigorously wielding his machete and making enough noise to frighten the boldest of birds away. But if trails, along which he can walk quietly, intersect these tangled thickets, he will find them the home of such noteworthy birds as the Spotted-breasted Wren, the Rufous-breasted Castlebuilder, the Barred Antshrike, and the Little Hermit Hummingbird.

Finally, rapidly growing trees, especially the guarumo or *Cecropia* and species of the leguminous *Inga*, manage to push their heads above the tangle of vines and bushes and begin to overshadow them. The guarumo is preeminently fitted to colonize abandoned land. Its thick, hollow shoots, almost always inhabited by a special kind of ant, are so impelled by the vital necessity to rise above the surrounding vegetation that they grow many yards high before they branch; they concentrate all their resources upon upward growth. In their first impetuous skyward spurt, they neglect even to strengthen their own bases. Later in life, when they develop into tall, coarsely branched trees, they seem unable to do so, but support themselves by prop roots that grow obliquely downward all around the trunk, sometimes from points several feet above the ground. The base of the trunk remains too slender to uphold the tree. The cecropias and the ingas are the precursors of the forest; but many years must pass before the high forest reestablishes itself and forms a canopy dense enough to starve out the vine tangles by cutting off their light.

As the dry season advanced, I understood the reason for the great amount of second growth in this valley. It was the old story of the shifting agriculture of tropical lowlands. To start a milpa, the small farmer fells a patch of forest or older second growth. After the smaller trunks and branches have dried, he sets fire to the impassable, tangled mass of fallen plants. When the flames have died away, he sows his maize, without previous cultivation, in holes opened by an iron-shod pole in the bare ground, between the charred remains of the trunks and thicker branches that resisted the conflagration. In this hole he drops five or six grains, perhaps presses down the earth above them with his foot, then advances a step or two and repeats the operation.

The maize plants are usually not thinned; cleaning is limited to chopping off the weeds with a machete, with the result that, after a year, the ground becomes so congested with the roots of grasses and other weeds that the farmer finds it less laborious, and more profitable, to make a new clearing than to dig up the land already used and prepare it for another crop. Accordingly, fresh inroads are continually made upon the forest and the older second growth, where the more troublesome weeds have succumbed to light-starvation beneath the canopy of woody plants and vines. Yet the result of all this sacrifice of the woodland to axe and machete is not a pleasing rural landscape of checkered fields neatly cultivated, but a wilderness of tangled thickets which are rarely permitted to rest for the many years that they would require to revert to climax forest.

The dry season is the time to prepare for the milpa. At intervals through the day, we heard from the nearer hillsides the thud of some mighty tree crashing down to make way for corn plants. In the distance, the air was hazy with the smoke from burning brush. The amount of burning that continued from February until May was amazing. Hardly a day passed without several extensive brush fires becoming visible somewhere in the wide sweep of hill and plain that we overlooked from our lofty perch on the hilltop: the pall of smoke by day, the flames by night. Sometimes at night we watched an irregular line of fire stretching thousands of yards over a distant mountainside beyond the broad valley of the Río Motagua. One night in May, a whole hillside, separated from us only by the narrow valley of the Río Morjá, on which for weeks the felling of trees had been in progress, was enveloped in crackling flames. In the darkness, it made a stirring but distressing spectacle when one thought of the destruction of the vegetation, of the nests and young of birds and many small, harmless animals, which it involved. The following morning, the whole long slope was a black and lifeless ruin, where from the thicker stumps and logs columns of smoke continued to rise for several days while vultures soared above, seeking half-consumed victims of the conflagration. After a few weeks, green shoots of maize pushed above the charred soil, but hardly in a way that a thrifty farmer of the Corn Belt would approve.

The second-growth thickets resulting from this shifting agriculture are the home of the Plain Chachalacas—large, long-tailed, brownish, arboreal birds, related to the curassows and guans. Although nearly as long as a hen pheasant, they are lighter; with ease and grace, they walk along high branches hardly more than an inch thick. Their full morning chorus in March is unforgettable. Although their voices are harsh and unmelodious, their performance is so vigorous and spirited, such a wonderful expression of wild nature, that one overlooks their harshness in the exhilaration of listening to them. How the stentorian calls reverberate back and forth over the brushland, like successive echoes arising and dying gradually away in hilly country! Now a bird in the tree above me takes it up, and I feel the full power of the voice that pours out the loud *cha-cha-lac, cha-cha-lac*. Other birds in the nearest group join in, setting up a din of vigorous calls. Now the enthusiasm of the near flock wanes; but off in the bush, a quarter of a mile away, other birds are shouting, in voices subdued but not softened by the distance. As this outburst dies away, other groups in still other directions take up the challenge, which now sounds faint and far off.

Soon the clangor surges back again, coming nearer and nearer, until the six or eight chachalacas directly overhead resume their raucous calls, almost deafening at such close range, and drown those of all their more distant rivals. It is a great chorus of rounds, in which probably hundreds or even thousands of chachalacas over miles and miles of territory participate; for when all the nearer fowls are silent, I hear in the distance voices so faint that they are extinguished by the chirping of small birds. I seem to be in the center of it all, but were I to stand near those distant birds whose calls fall so weakly on my ears, I would doubtless receive the same impression of being close to the central group.

For half an hour, the great chorus surges back and forth over the valley, now rising to an ear-splitting volume of sound, now diminished by distance to a whisper, but never silent. It is not melody that I listen to, but something grander, the harmony of great spaces and the joy of splendid free birds. I notice that some voices in the flock of chachalacas nearest me are weaker and much higher in pitch than others; probably these are the females, since the strong voices seem to lead and these to follow. In the intervals between their outbursts of loud calling, the birds converse in contented, throaty clucks and cackles, and chase each other mildly back and forth among the branches. Some perch on the topmost boughs of the tallest, epiphyte-laden, old trees, survivors of the former forest that tower above the recent growth. Their slender, long-tailed, brown figures stand sharp and clear against the morning sky. As time passes, their responses grow shorter; eventually the calls cease and are seldom renewed on the same day. In northern Venezuela, I heard similar, but less intense, choruses by the Rufous-vented Chachalacas; but the Gray-headed Chachalacas, among which I have lived for many years in southern Costa Rica, have never given a comparable performance.

When I arrived at Alsacia in February, banana leaves were covered with exceedingly fine, gray pumice, which had settled like fine roadside dust upon them after the recent eruption of the Volcán de Fuego, a hundred and twenty miles to the west. Two carloads of fruit had been ruined, I was told, because the pumice-covered leaves used for packing them had abraded the skin, spoiling the appearance and market value of the bananas. On a neighboring plantation, clothes hung up to dry were so soiled by falling volcanic dust that it was necessary to wash them again. The fine powder stuck so tenaciously to the foliage of the banana plants that repeated heavy rains failed to wash it all away; it was quite noticeable on the older leaves after three or four months.

The house on the hilltop was shaded by tall coconut palms, whose great, stiff fronds, which rustled and creaked in the wind, were the nightly shelter of a numerous company of Great-tailed Grackles. I was surprised to find these birds around an isolated dwelling, for I had come to associate them with towns, not only the seaports of northern Central America, but also the cities and villages of the interior, up to an altitude of at least seven thousand feet in the Guatemalan highlands. So familiar are these grackles to Guatemalans that, in a country where most birds are nameless, or have only family names such as *carpintero* for all the woodpeckers and *chorcha* for many different orioles, the male and female are distinguished by separate designations. The big, long-tailed, yellow-eyed male grackles, elegantly attired in glossy, iridescent black, are known as *clarineros* or trumpeters, while the smaller, brownish females are called *sanates*. As the sun sets, one can often watch these versatile foragers streaming in from the surrounding fields to roost among the shade trees of a village or on palms in the central plaza of a small town.

The grackles that established their headquarters in the coconut palms at Alsacia were amusing, but not always restful, neighbors. Every morning for four months, I awoke to the sound of their voices. In the dim light of early dawn, the clarineros repeated over and over, in a calm, subdued voice, a long-drawn note between a screech and a whistle, which sounded pleasant and contented, and reminded me of someone running up the whole musical scale on a stringed instrument with a single deft stroke. As the day brightened, the commotion among the coconut fronds increased, with much excited calling by the males and clucking by the females, until finally they all flew down from their high roost to seek breakfast. Some went directly to a pink-flowered shrub of the melastome family that grew abundantly on the grassy slopes below the house, to gather the small, sweetish, black berries. Others settled on the road, the cowpen, or the lawn, where they walked about searching for small, creeping things.

One morning, I watched four sanates cleaning a gaunt old cow, who stood alone in the pen. One bird stood on her back and pecked at vermin among her hair. After a slight show of resistance, the first sanate permitted another to alight beside her and share the feast. Two more moved about on the bare ground between the quadruped's feet, at intervals jumping up to pluck something from her flanks or belly. They climbed on her legs and tail, performing the same beneficial operation, while the cow stood patiently still, as though familiar with, and grateful for, this cleansing service.

Other grackles flew from their roosts directly down into the valley, and as the morning wore on the rest followed, singly or in small flocks. Here they foraged on the moist shore of the river or in the shallows, or searched among the piles of driftwood and washed-out banana plants stranded in the shoals.

On hot afternoons, they delighted to bathe in the shallow water at the stream's edge, with vigorous shaking of wings and tail that raised showers of crystal drops to sparkle in the sunshine. One afternoon, I watched a sanate stand so close to a bathing clarinero that she was sprinkled by the spray that he sent up, as other birds take advantage of the shower from a suburban lawn-sprinkler. After the bath, all flew up to the willow or guarumo trees on the bank, vigorously shook the water from their plumage, then carefully preened it with their slender bills.

Later in the afternoon, the clarineros often joined the cowbirds turning over small stones on the bare flood plain, as will be described in more detail in the following chapter. Then, as the sun declined and the air cooled, they flew up the hill in small groups, often cackling like a flock of Purple Grackles, to congregate again in the coconut palms. On the way, some would pause in the pink-flowered *Conostegia* bushes on the hillside for a dessert of berries, before going to roost. From their arrival until darkness fell, our hilltop presented the liveliest scene. The varied calls and squeaks of the clarineros mingled with the constant chatter of the more numerous sanates. They seemed to have great difficulty finding satisfactory roosts, and flew from one frond to another perhaps a dozen times before they finally settled down to rest. The strong breeze that often blew up from the valley at sunset, tossing the great palm leaves, helped to keep the birds stirred up. Caught by the wind, the clarineros' long tails flagged back and forth while they perched on the palm fronds, causing them evident inconvenience. I loved to watch their graceful maneuvers in the breeze, when they hovered, soared, and poised with dangling legs above the crowns of the palms, as gulls disport above a windward shore.

On some unusually breezy evenings, the clarineros engaged in spectacular encounters, meeting face to face and rising high above the treetops, until the wind took hold of them and twisted them around, making them forget their opponents and devote all their attention to prevent being blown away. The clarineros seemed to enter these sparring bouts in a playful rather than an angry spirit, the more to enjoy the wind by their vigorous exercise in it, as boys engage in sham battles in the water. The sun hung well above the western hills when the grackles began to gather in the palms; the last red glow was fading from the sky when finally they had ensconced themselves out of sight among the inner fronds, and their last sleepy notes gave way to the awaking calls of the Pauraque. But, especially at the outset of the nesting season, the clarineros slept lightly, often waking in the night to shatter the monotonous chirping of insects with their shrill calls.

The range and power of the clarineros' voices won admiration; they seemed to lack only a set song. At one extreme, they uttered a pretty, little, tinkling note; at the other, calls so loud and penetrating that they were best heard at a distance. If one may say that a bird with such a varied repertoire had one most characteristic utterance, it was the long-drawn, rising call, intermediate between a squeak and a whistle, that often started off their day. They also voiced a resonant *tlick tlick tlick*, and a vigorous, rollicking *tlick-a-lick tlick-a-lick*, that seemed the outpouring of rare good spirits. A rolling or yodelling call, very vigorous, contrasted with a lazy, screeching note, like the slow swinging of a gate on rusty hinges. From time to time, they invented a new verse; when one pleased them, they repeated it again and again. One clarinero fell in love with an engaging phrase that sounded like *wheet-tóck*, and uttered it continually for a week or more, until at last, like a popular song, it lost its appeal and was rarely heard. Another seemed to be enchanted by a bugle-like call, *ta-dee ta-dee ta-dee*, which was really very martial and stirring, especially at a distance, as it had great carrying power. After delivering a call, the clarineros often perched with their necks and long, sharp, black bills directed straight upward, a pose that displayed to best advantage the sleek glossiness of their purple necks. Sometimes two of these handsome birds, resting side by side on a coconut frond, pointed upward simultaneously and held this stance for the better part of a minute, looking quite self-conscious.

At the end of February, the sanates started to build their nests, sometimes in orange or lime trees, or even in low, thornless bushes; but the majority chose the coconut palms where they roosted. Here their preferred site was in the very center of the crown, between the two youngest expanded fronds, which stood upright face to face, and, between their broad green surfaces, provided a cozy nook in which to support the structures. But this most coveted position would not have appealed to the prudent, for these young fronds were still growing upward at different rates and bending outward toward different sides. Moreover, the thin, ribbon-like divisions of the fronds were not strong enough to support the bulky open cups, heavily plastered with mud or cow dung. Despite our efforts to tie them up, most of the nests built between the inmost fronds tilted strongly and spilled out the eggs or nestlings.

The successful nests were firmly set, against the trunk, on the broad bases of the older fronds. Here the sanates laid two, or more often three, beautiful, glossy, blue eggs, so variously marked with blotches and intricate scrawls of black and brown that it seemed possible for each bird to recognize her own, if all the eggs in the colony had been mixed together. As she had built alone, so the sanate incubated her eggs and reared her nestlings without ever receiving help from a clarinero. These birds never paired off.

At first, I was happy to have such lively, entertaining birds living so close around me, where I could conveniently study their habits as they had never been studied before. But as the weeks passed, I wished them elsewhere.

Most of the nests of other kinds of birds that I tried to study on the hilltop were prematurely destroyed, and I strongly suspected that these grackles were the chief culprits. They so effectively kept other large birds at a distance that I ruled out other avian predators, and I once surprised a clarinero in the act of pillaging the oven-shaped nest of a tiny Yellow-throated Euphonia. On July 6, after a restless night, the great majority of the grackles inexplicably deserted the coconut palms on the hilltop for an unknown destination, leaving only a few sanates who still had young in the nest, one whose belated eggs were just hatching, and two faithful clarineros. The early mornings were strangely silent after their departure.

Another noisily conspicuous bird at Alsacia was the White-tipped Brown Jay. Depending upon appearance alone, one familiar only with the blue or gray jays of temperate North America would probably not detect any family resemblance in these crow-sized jays, but they have only to open their mouths to reveal their affinity; in voice and mannerisms, they are unmistakably jays. From eastern Mexico down the Caribbean side of Central America to western Panamá, they roam in family groups through shady pastures, banana plantations, and open second-growth woods, and in Costa Rica they ascend to eight thousand feet on the great volcanoes. They never enter heavy forest.

In March and April, loud, complaining cries, sounding afar over the bushy hillsides and through the banana groves, led me to bulky open nests built of coarse sticks and lined with loosely matted, fibrous roots. They were situated in vine-draped trees or in the crowns of banana plants, and in each sat a female Brown Jay, warming two or three mottled eggs. The loud cries, which betrayed the nest's position to me, also brought attendants with food for the incubating bird, who sat continuously for hours. I soon noticed great individual variation in the bills of these noisy birds; some bills were almost wholly yellow and some were black, but the majority were pied with these two colors in the most varied patterns, enabling me to recognize individuals with greater ease than if I had placed colored bands on their legs. Continued watching convinced me that the jays with the blackest bills were breeding adults, while those with most yellow were usually non-breeding yearlings, although a few breeding jays had more yellow than black on their bills. The hunger cries of the incubating female brought not only her mate, but also one or more of these yearlings, with food for her. They kept her so well supplied that she left her eggs only briefly, at long intervals. While she was absent, her mate stood like a sentinel, guarding the nest.

After the downless nestlings hatched, the yearlings helped the parents to feed and guard them. At one nest, seven grown birds—the two parents plus five helpers—kept three nestlings abundantly supplied with food. A second nest had three helpers; and two other nests had, respectively, two and one.

Later, in Costa Rica, I also found helpers at Brown Jays' nests, one of which was high in the mountains. Sometimes an attendant bringing food would pass it to another for delivery to the nestlings; a helper might entrust it to a parent or another helper, and a parent might pass it to a helper or the other parent. Thus the closest cooperation prevailed among all the attendants of a nest.

This was the most exciting, and probably the most important, of the discoveries that I made at Alsacia. At that time, it was known that in a few exceptional species of northern birds, especially among bluebirds, swallows, and gallinules or moorhens, the young of an early brood might help their parents to feed their younger siblings of a later brood in the same season. But, as far as I know, no case of yearling helpers had yet been discovered—certainly none had been studied and reported in detail. Indeed, such helpers are not likely to occur among the migratory birds of the northern lands where practically all the careful studies of nesting birds had then been made; for migration breaks up families, and, moreover, the high mortality that it often entails makes it imperative for birds to nest at an early age and raise large broods. But as, in recent decades, the life histories of permanently resident tropical and subtropical birds in the Americas, Africa, and Australia have become better known, more and more instances of yearling helpers have come to light. It is becoming evident that delayed breeding, with the non-breeders assisting at the nests of their elders, plays an important role in the economy and population dynamics of birds constantly resident in the milder climates.

At nests of two kinds of woodpeckers I learned what has since been verified in many other species, that the male, rather than the female, incubates the eggs and broods the young by night. Even by day, he may work harder carving the nest cavity and attending the young than his mate does. I found two burrows of Rufous-tailed Jacamars, long-billed, gemlike birds, so glittering and sprightly that they remind one of overgrown hummingbirds; after years of intimacy, I still find them one of the most exciting of all birds to meet. I continued to watch Rufous-breasted Castlebuilders make and attend their great mansions of interlaced sticks, which I had begun to study at Lancetilla two years earlier. I watched a pair of Citreoline Trogons carve their nest chamber in a hard, black termitary, still swarming with its little, white makers. This was prosaically situated on a fence post near a tramline, rather than beside a picturesque mountain torrent, like the Massena Trogons' nest at Lancetilla; but the male and female followed much the same schedule of incubation that the forest-dwelling trogons had done. These and many other birds kept me busy from morning to night while I dwelt in the house on the hilltop.

13

A River on the Plain

We gravitate to rivers like water from the surrounding hills. Their attraction does not depend wholly on the greater abundance and conspicuousness of birds and other animals along their course; even in a land devoid of animate life, we should be drawn to them, perhaps more strongly than where fields and forests are enlivened by moving creatures. Probably the secret of their attraction is that, to a high degree, they exemplify that which we too seldom realize in our own lives, the felicitous union of permanence and change.

We yearn for permanence, yet we welcome adventure and change. We wish our souls, and even the bodies that support them, to endure forever, along with those dear to us and everything we cherish. Yet spirits as restless as ours would be bored to tears in Eternity, where time does not intrude and nothing changes or perishes but everything remains unalterably the same. We crave novelty, movement, and growth; we welcome pleasant surprises, all of which belong to a world in constant flux, where things grow and decay; they

Headpiece: A nestling Ringed Kingfisher (*Ceryle torquata*), 29 days old.

are difficult to reconcile with permanence. More than most things beneath the sun, flowing streams realize this happy combination. They are never the same for two consecutive moments, a fact that Heraclitus of old recognized when he declared that you cannot step twice into the same river. Yet even if the water at the ford is always different, the river itself remains the same while the creatures that live along its course are born, grow old, and perish, as Tennyson recognized when he made the brook proclaim

> For men may come and men may go,
> But I go on for ever.

This paradoxical union of permanence and change, of sameness and difference, is not the least of the river's attractions.

The Río Morjá, which wound through the banana plantations that stretched like a sea of verdure at the foot of the elevation on which the house at Alsacia stood, drew me as strongly as the mountain torrent at Lancetilla. I passed many pleasant, instructive hours with the birds that lived along it. Doubtless, back in the forested hills where it was born, the young Morjá dashed down a rocky channel as impetuously as the torrent above which the Royal Flycatchers nested; but before it reached the broad, flat valley where I knew it, its character changed completely. Here, in the drier months early in the year, it flowed clear and smooth over a broad channel that in many places I could cross by wading through water that did not reach my waist.

The concave bank of the curving channel was a vertical wall of deep, rich, alluvial soil that supported tall banana plants in long, unbroken ranks. Their great, wind-frayed leaves arched over the limpid water. On the convex side, bare expanses of sand or shingle sloped gently up to a dense stand of tall Wild Canes, which were invading the growing sand flats by means of long runners that sent up young leafy shoots. That the river was not always as gentle as I found it in February and March was attested by the freshly cut banks on the concave side, where banana plants had tumbled into the water, along with the light rails and pressed-iron ties of a washed-out tramline. As this bank receded, the exposed sand and shingle on the opposite side of the channel advanced toward it.

Stretched at ease in the rich sunshine on an exposed sandbar beside the Río Morjá, in the dry season when the water was low, I sometimes imagined myself transposed to some familiar stream twenty degrees farther north, at a later month. Clear water flowing over a stony channel speaks the same language in every climate, and sparkles with equal brilliance in the sun's rays. The willow trees along these banks look much the same as those that border some tributary of the Susquehanna or the Ohio; their light sprays of pale green foliage contrast strongly with the background of dark tropical verdure that clothes the distant mountains. The Spotted Sandpiper, still with the immaculate breast of its winter dress, and the Louisiana Waterthrush who tilter over the shore are the same who a few months later will hunt along northern watercourses. Although of wholly different lineages, these two birds of similar habitats and modes of foraging have acquired similar color patterns and mannerisms to a degree that excites my wonder. I cannot predict whether the Green Heron, who with infinite patience stands on the farther bank poised to spear some unwary fish, will build its nest in the thicket beyond the bend in the river, or beside some other stream two thousand miles nearer the North Pole. Nor can I tell whether the Rough-winged Swallows who course above the water will raise their young in an abandoned kingfishers' burrow in the neighboring bank or in some hole or cranny in a far northern land.

The rattle of a wintering Belted Kingfisher falls in with my mood—I almost expect to see her enter one of the burrows in the bank, and later bring forth her brood. The Dowitcher and the Lesser Yellowlegs, coming up from nearer the equator in April or May, will fit immediately into the scene. Yellow Warblers and Magnolia Warblers glean insects among the willow boughs, or dart out to seize some passing insect in the air; the Catbird's familiar *mew* sounds in the thicket behind me. The painter who introduced into this northern scene the Turquoise-browed Motmot emerging from its tunnel, or the Rainbow-billed Toucan flying overhead, might be accused of unpardonable ignorance of the geographic distribution of organisms.

But to preserve the illusion of a transposition in space and time, I must continue to bask in the warm sunshine in a mood that is not too critical. Yonder, where the current is undermining a steep bank, a cecropia tree intermingles its stiff, large-leafed boughs with the feathery spray of the willows. Beyond the fringe of young willows on the stony floodplain are thickets of wild plantains and shell-flowers with huge, unmistakably tropical leaves. Behind these, banana plantations stretch away toward the foothills. At intervals, distinctly tropical birds raise their melodious or raucous voices.

In such a setting live five of the six species of kingfishers that inhabit the Western Hemisphere. The Belted Kingfisher is a winter visitor that one may not meet every year. The Ringed, the Amazon, the Green, and the Pygmy Kingfishers are residents, a splendid assemblage ranging in size from that of a crow to that of a sparrow. Each excels in its own peculiar manner: the Ringed and the Pygmy are most handsome in plumage; the Green is most graceful in form; the Amazon is most pleasant in voice. Farther south, in Panamá and Costa Rica, the elegant Green-and-rufous Kingfisher ranges up from South America; and during the northern winter, when the migratory Belted Kingfisher is present, all six of the American kingfishers may be found along the same waterway. The presence in a single locality of all the representatives of an avian family inhabiting a hemisphere is rather rare; but aquatic habitats are more uniform over vast areas than are terrestrial habitats, and aquatic species of birds have correspondingly wide ranges.

Of the four kingfishers resident in northern Central America, the rarest is the Pygmy. It is, as W. H. Hudson said of the European Kingfisher, a truly gemlike bird; but it is a more fiery gem, as befits the more torrid climate in which it lives. Its back is deep metallic green and its breast deep orange. In this little bird the Ringed Kingfisher's anvil *klang* is attenuated to a faint tick; when excited, it crowds these ticks into a rapid, low rattle, hardly more than a buzz. From a low perch, it plunges headlong into water so shallow that its larger relatives would hardly risk the dive. Since birds as a rule are wary in proportion to their size, it permits the closest approach by man of all the kingfishers. It is said to nest in termitaries in mangrove swamps.

Sometimes, as I sat on the bank putting on my shoes after wading or swimming in the refreshing water that brought mountain coolness to the warm lowlands, I watched the little silvery-scaled minnows gleam and flash in the current. As each at intervals turned on its side, it reflected a silvery beam, which vanished as soon as the fish righted itself, to become almost invisible against the background of the sandy bottom with which it blended so well. Now here, now there, from a score of points, but only for an instant in each, came the bright flash from as many different minnows, just as the fire-flies' sparks gleam momentarily in a hundred places above a meadow on a warm night, but rarely twice in the same spot. I wondered whether these silvery glints were the betraying signals for which hungry kingfishers waited as they perched motionless on a bough overhanging the stream, or hung above it between two misty circles of rapidly beating wings. If so, they must indeed be quick to capture their meal, for the revealing gleam of silver is more evanescent than the firefly's spark.

Although, unlike some of their cousins in the tropics of the Eastern Hemi-sphere, the American kingfishers are specialists who devote their lives to perfecting the art of fishing, they do not find it an easy way to earn a living. They seem to miss more fish than they catch; it is only because they will not be discouraged by failures that they finally obtain a meal. Often the kingfisher waits long and patiently for a minnow to appear in a suitable position beneath himself, only to have it dart away as he drops toward it; he planes off before striking the water, and with a loud rattle returns to his perch for another attempt. Often, too, he plunges beneath the surface, only to reappear with an empty bill, shake the sparkling drops from his plumage, and fly to a branch where he resumes his watch. Once I saw an adult male Amazon Kingfisher fail to catch anything on four successive dives into a stream that abounded in minnows. How much more difficult is this method of foraging than that of a flycatcher, who operates in his own element and can twist and turn in pursuit of an elusive insect, seldom missing his prey! What patience the young kingfisher must need to become proficient in his art!

In the light, sandy loam of vertical banks beside the Río Morjá, five kinds of birds nested early in the year, while the current was low. Four dug their own burrows in the yielding earth; the fifth, the Rough-winged Swallow, depended upon tunnels that the others had made. The Ringed and Amazon Kingfishers and the Turquoise-browed Motmot excavated their burrows in bare, exposed banks, where they were easy to find. The Green Kingfisher's burrow was usually situated behind a curtain of vines or other vegetation that screened the bank, or else among exposed roots, where it was almost certain to escape detection unless I saw a kingfisher enter or leave.

As long as all these burrows were tenanted by the birds who dug them, they were traversed by two parallel ruts or grooves, made by the occupants' short legs as they shuffled in and out. When, after the departure of the original tenants, Rough-winged Swallows took possession of a tunnel, numerous, fine, irregular toe-scratches replaced the clear-cut, parallel furrows in the entrance. Although usually I had to wait long to see a kingfisher or motmot enter or leave its burrow, the swallows darted in and out so frequently that, even without the revealing footprints, it was not difficult to learn which tunnels they occupied.

Along the Morjá and other streams, as well as in roadside banks, I watched several kinds of kingfishers and motmots excavate their tunnels. All these related birds work in the same manner, which is likewise that of the burrow-nesting species of jacamars, members of a different avian order. They loosen the earth with their bills, then kick it backward with their feet. As one of these birds enters a tunnel at which it is working, two parallel, intermittent jets of earth spurt out behind it. The jets follow the worker inward: the backward kicking that produces them is doubtless continued until the bird reaches the head of the excavation, so that each time it returns to its task, the loosened earth is gradually shifted outward. Rarely, as I saw in the Great Rufous Motmot, a small lump of earth is carried out in the bird's bill. In all kingfishers, motmots, and jacamars, as far as I know, the two sexes alternate at the task of digging, and, as a rule, neither will continue to work unless its mate is resting nearby—which is quite different from the way of woodpeckers, either sex of which will continue to carve the nest cavity in a trunk or branch while its mate is beyond sight. In Amazon Kingfishers, Turquoise-browed Motmots, and Rufous-tailed Jacamars, I have seen the male feed his mate while, or shortly before, they prepared their nest.

The burrows of the big Ringed Kingfisher, six inches wide and four inches high at the mouth, were easy to recognize by their greater diameter. In February, I found three under construction in freshly cut banks beneath the banana plants. Much as I would have liked to open them before the eggs were laid, in order to learn how long they took to hatch, I desisted from doing so from fear of causing desertion. After I was sure that incubation was well advanced, I prepared two of these burrows for study. First, I measured their length by inserting a long, flexible vine, slowly, so as not to injure a parent, if one happened to be within, or break an egg. Then, estimating as well as I could where the inner end of the curving tunnel would be, I dug down into the soft earth, hoping to connect with the rear end of the chamber where the eggs lay. Here I made an opening large enough to admit my hand. After each inspection of the brood chamber, I closed my private entrance with a stone, then filled the pit I had dug with earth and covered it with litter, to hide all traces of my work from hungry animals and prying men. I used this method of preparing nests for study with a variety of burrow-makers, who never deserted if I waited until they had started to incubate, although they were likely to do so if I meddled with their nest before they had eggs.

The two burrows of the Ringed Kingfisher that I prepared in this way were between seven and eight feet long. The third burrow, which I measured but did not open, was only six feet and seven inches long. The enlargement at the inner end of the first two burrows was twenty-three inches below the surface of the ground. When I broke into this brood chamber, the buzzing of green flies, making a peculiar hollow resonance in the long subterranean tube, assured me that incubation had begun, for the kingfishers, while sitting on their eggs, regurgitate the indigestible bones and scales of the fish they have eaten and these attract the carrion flies. Lying flat on the ground, I could scarcely reach the four white eggs that rested on the bare, sandy floor of each of these chambers; kingfishers, like motmots, provide no softer bed for them. After measuring the eggs, I carefully replaced them. Even before I closed the aperture that I had made at the back of the first burrow, the male, who had fled when I started to dig above him, returned to it; but when he noticed the unaccustomed light at the rear, he retreated with considerably more agitation than he had shown when first disturbed.

I was curious to know how the male and female Ringed Kingfishers arranged their turns on the eggs, but winning this information proved to be far more difficult than I had anticipated. It is easy enough to tell the sexes apart when they perch overhead: both are crow-sized, crested birds, with strong black bills, slaty blue upper plumage, and a broad white collar around their necks. The underparts of the male are uniformly rich chestnut; the female has a slate-blue chest, separated by a narrow white band from the chestnut of her abdomen. But the curvature of the long burrow made it impossible to see the birds, by looking in the front with a flashlight, as they sat on their eggs. On leaving, they usually shot out and flew low above the water, to vanish around a bend in the river without giving me a good view of their undersides. Sometimes, but not always, I could distinguish their sex as they perched on a banana leaf before entering. And they entered and left the burrow so infrequently that I had to wait long in order to see them at all.

I would probably have abandoned this problem in despair, if I had not already a growing conviction, which has since been amply confirmed, that incubating birds do not come and go at random but follow a more or less definite schedule. Day after day, I passed long hours sitting on a folding camp stool, half hidden among the Wild Canes that were invading newly formed land on the convex shore of the river.

Before me is a wide expanse of sandy, then stony, floodplain that slopes gently down to the swiftly flowing, limpid stream. On the farther side, in a vertical bank as high as my head, is the oval hole, the doorway to the kingfishers' burrow, on which my eyes are fastened. The hardest thing on which to concentrate, I believe, is nothing; and a hole is a no-thing, a vacuity or absence of substance. To assure myself that no kingfisher had left or entered the nest in the brief intervals when, despite myself, my attention wandered to some of the life and movement around me, I set a slender twig, the size of a match stick, upright in the mouth of the tunnel, where it will be pushed over by any bird going in or out. So long as it remains upright, I am confident that there has been no movement at the kingfishers' burrow.

As for the kingfishers themselves, except briefly twice a day, they are out of the picture. One is in the darkness of the burrow, warming the eggs. Its mate is away upstream, perched upon some overhanging bough, watching, with a patience I am learning to emulate, for some unwary fish, or just drowsing in the midday heat. Although it is for them that I wait, they do nothing to enliven the long hours of my vigil. Other creatures keep me company; at the risk of missing what I most want to see, I cast sidelong glances at them. In the early morning, wintering Orchard Orioles sing sweetly in the clump of young canes where they have roosted; Rainbow-billed Toucans straggle in undulatory flight over the banana plantation and congregate to sing with frog-like voices in some tall trees upstream; Brown Jays fly raucously about; a pair of Blue Grosbeaks alights on the seemingly sterile sand to pick up minute seeds; beautifully variegated lizards scuttle about their obscure affairs over the warm, bare floodplain; the voices of many birds fill the air.

In the early afternoon, all is still. The sun has poured his ardent rays upon the naked sands and stones until they burn my bare feet. I am apt to drowse in the heat and glare, and throw off my clothes to steal a dip in the cooling stream, with one eye on the hole in the bank. Emerging, I dry quickly in the warm air, pull on my thin garments, and resume my station. As the uneventful hours slip by, I am likely to fall into a reverie once more and miss the kingfisher's swift exit that I have waited so long to see.

As the sun sinks in the west, the bird world comes to life again. Flocks of cowbirds, grackles, and blackbirds alight on the shingly shore to search for food; toucans gather in the treetops to sing their vespers; Gray's Thrushes delight me with the season's first, brief, musical phrases; Little Blue Herons stalk their prey in the shallows upstream; Amazon and Green Kingfishers fly past, sounding their rattles, or pause to plunge for fish; but the Ringed Kingfishers remain absent from the lively scene. As daylight fades, I wade back across the river, set up my little sentinel twig in the burrow's mouth if it has been pushed over, dress again, and follow the tramline through the darkening banana plantation toward the house on the hilltop, where lights are already shining.

After I had watched for ten days, alternately elated by a significant observation and dejected when a kingfisher shot out of the burrow and flew away without permitting me to distinguish its sex, their incubation pattern became clear to me. They operated on a forty-eight-hour cycle, rather than a twenty-four-hour cycle like every other inland or freshwater bird that I know. With most small land birds, one day at the nest is much like the next, and the same individual, usually the female but in some families the male, covers the eggs every night. With the Ringed Kingfisher, each sex took charge of the eggs on alternate nights. They replaced each other only once a day, usually between half past seven and nine o'clock in the morning, sometimes as late as ten o'clock. If on Monday morning the male relieved the female, on Tuesday she replaced him, and so on through more than three weeks of incubation. During its spell of duty of approximately twenty-four hours, each partner took a single recess, suddenly darting out of the burrow at some time between one and four o'clock in the afternoon. The eggs were then left unattended until, after an absence of from thirty minutes to an hour, the same bird silently returned to them, after which it remained continuously in the burrow until relieved the following morning, sixteen or seventeen hours later.

To me, this seemed a rigorous schedule; I was inclined to pity birds obliged by the custom of their race to spend such long hours sitting inactive in the dark earth. But, one day, I saw a kingfisher behave in a way that changed my mind. While I was watching a nest of the Green Kingfisher, a female Ringed Kingfisher flew into a balsa tree beside the creek with a fish, half as long as herself, dangling crosswise in her bill. For more than two and a half hours, by my watch, she held it so, shifting her position only from one branch to another of the same tree. Finally, when I was ready to leave, she started to beat the dead fish against a branch, evidently to prepare it for swallowing. After this exhibition of stolidity, I no longer felt sorry for the kingfishers. Long intervals of inactive sitting are wholly in keeping with their character.

These kingfishers were sticklers for formality. One morning, sitting among the young canes across the river from the burrow, I witnessed an amusing episode. Instead of waiting for his mate to relieve him, as he usually did, the male inexplicably emerged from the tunnel and flew upstream at half past seven. Five minutes later, the female came downstream and perched in the cecropia tree where she often rested before she entered the burrow. Soon the male reappeared and alighted near his mate, who greeted him with a rapid, low rattle that seemed to be a scolding. After a minute, he returned upstream. Now I fully expected to see the female enter the burrow, as it was her day to incubate; but evidently such an unwelcomed entry would be beneath the dignity of a well-bred kingfisher matron. She delayed another minute, as though considering what to do, then flew off after the delinquent.

Soon they returned, but separately, and perched on banana leaves above the burrow, where they seemed to engage in an argument. Finally, after flying back and forth in front of the burrow several times, the male reentered—rather sheepishly, I surmised. After waiting long enough for him to compose himself decently for her reception, his mate followed him into the nest. In about one minute—less time than usually elapsed between the entry of one partner and the exit of the other—the male shot out and turned upstream, klecking loudly, a free bird at last!

The Ringed Kingfishers' neighbors, the Amazon and Green Kingfishers, followed less rigorous schedules of incubation, the male and female replacing each other several times a day. The male Amazon Kingfisher took the larger share of daytime incubation, sometimes sitting all afternoon, until his mate relieved him at sunset. Now I understood why I sometimes found him fishing in the dusk. After sitting all afternoon, he evidently needed several minnows to satisfy his hunger.

On March 28, the female Ringed Kingfisher arrived in the neighborhood of the burrow at sunrise, an unusually early hour for her appearance. When she was last inside, her eggs had not hatched; but on the preceding afternoon I had found two of them pipped, and doubtless she was aware of what was happening. She flew excitedly around and clashed briefly with the neighboring pair of Ringed Kingfishers, an event I had never before witnessed. Finally, at five minutes after seven, she entered the burrow, nearly half an hour earlier than on any other morning when I watched. When I opened the burrow a little later, I found two nestlings. Going then to the burrow downstream, I found three newly hatched kingfishers. Soon there were four in each burrow.

With pink, transparent skin utterly devoid of down, and dark knobs on the head where their eyes would be, the nestlings were decidedly not things of beauty. Curiously, the lower mandible of each young kingfisher projected well beyond the tip of the upper mandible, giving it a markedly prognathous aspect. They could already stand upright, supporting themselves on their bulging abdomens and their feet, which had callous pads on the heel joints, to protect them from abrasion on their nursery's sandy floor. Indeed, they could already walk in a tottering fashion when only a few hours old; when the light appeared through my private entrance at the back of the burrow, some of them retreated forward into the tunnel beyond my reach.

Kingfishers are exceptionally devoted parents. One Ringed Kingfisher remained in its burrow while I probed the length of the tunnel with a vine, dug down to make an opening at the rear, removed the eggs for measurement, replaced them, fitted a stone in the aperture I had made, then packed down the soil above it, all of which occupied well over an hour. When I stuck my flashlight into the rear of the neighboring burrow, the incubating bird dealt it a resounding blow with its great beak, then retreated backward into the entrance tunnel. Jumping down the bank, I looked in the front and saw its quivering form blocking the passage, so near the mouth that I reached in and gently touched its finely white-barred tail, which was turned outward.

Two days after the eggs hatched, this burrow was damaged by an unknown human intruder, who left a wide opening at the rear, where my private doorway had been, and with a machete enlarged the front entrance until it gaped like an alligator's opened jaws. The four nestlings had escaped, probably by retreating to the middle of the tunnel, where they were beyond reach from either end, but one was pale, cold, and unable to stand; eventually three succumbed. After I closed off the back of the burrow as well as I could, the parents continued to feed and brood the surviving young in the exposed and gaping burrow.

Even the little Green Kingfisher sticks devotedly to its nest in the face of danger. A female with newly hatched nestlings remained brooding them while I dug a hole at the back of the chamber, reached into the darkness, and lifted her out, struggling weakly in my grasp and trying to bite with her slender, sharp, black bill. When, after examining the nestlings, I replaced her upon them, she stayed brooding while I closed the burrow. Yet no kingfisher—indeed, no bird of any kind that nests in a burrow—has ever tried to lure me from its eggs or young by "feigning injury," as ground-nesting birds so frequently do, and even tree-nesters may do. Such distraction displays would probably remain unnoticed by predatory animals entering the mouth of a burrow or digging it out from above.

Stuffed with increasingly large whole minnows as it grew older, the surviving Ringed Kingfisher nestling in the gaping burrow grew rapidly. When ten days old, its eyes were open and it answered its parents' rattles with a high-pitched, almost trilling utterance. Two weeks after it hatched, its upper mandible caught up to the lower in length. Now it began to defend itself energetically, biting hard with its great mandibles whenever I removed it from the burrow to examine or photograph it. A few days later, it squealed and fought furiously.

Although this nestling was one of the most vociferous that I have ever handled, its struggles and cries were mild compared with those of the single young Ringed Kingfisher who survived in the burrow downstream. When, after two weeks of neglect, I opened this burrow for inspection, the young bird, now fully feathered, fled down the tunnel and jumped into the river, where it spread its wings, turned upstream, and flapped its way slowly against the current. When it encountered obstacles of stranded brush, it hooked its bill over them and scrambled across. Wading in the muddy shallows, I pursued until a fallen banana plant blocked its wayward flight and I overtook it. With deafening screams and fierce attempts to bite, it struggled violently in my hands for ten agonizing minutes. Finally, it quieted down and perched on my arm until its plumage dried, after which I replaced it in its burrow. Of all the nestlings I have known, only the Red-billed Tropicbirds on Swan Key were equally pugnacious and noisy.

Each of the burrows that I prepared for study produced a single fledgling, who left when from thirty-five to thirty-seven days old. These Ringed Kingfishers remained in the nest about a week longer than the smaller Amazon Kingfishers, and ten days longer than the still smaller Green Kingfishers. I did not see them go, but I had the good fortune to witness the departure of a fledgling from the third burrow, to which I had given little attention. Early on a morning in mid-May, the young kingfisher sat in the mouth of its tunnel with only its head and breast visible, repeating a loud rattle. As it bobbed its head up and down in typical kingfisher fashion, the level rays of the rising sun fell upon its broad collar, which gleamed with snowy whiteness. Soon it launched forth, turned downstream, and rose to alight in a willow tree. On its very first flight, this young Ringed Kingfisher, who had never been able to spread its feathered wings in the confined space of the burrow, covered two hundred feet on an ascending course that rose thirty feet.

While I watched the three kinds of kingfishers, I saw much of the Turquoise-browed Motmots, who also dug burrows in the sandy banks of the Río Morjá. When I studied these motmots at Lancetilla two years earlier, I wondered how they kept their lovely plumage so fresh and clean, and preserved their seemingly fragile racquet-tails in such good condition, while digging a burrow and raising a family underground. One pair of motmots along the Morjá, starting a burrow only ten inches below the top of a river bank that sloped downward on the landward side, soon found themselves digging out into the light and air. Abandoning this misdirected effort, they dug another nearby, in which the female laid her eggs when it was still shorter than any other Turquoise-browed Motmots' burrow that I have seen. Only forty inches long, it was so straight that I could look in at the front and see the lovely birds warming their eggs, as I could do at no other burrow.

Only at night, while they slept, could I be sure that my visit did not make them alter their position in the burrow. So, taking my flashlight, I walked down the hill and approached the river through a silent grove of tall banana plants, whose glossy trunks glinted in the beam of light. Emerging on the bank, I frightened a Boat-billed Heron, who rose into the air and flew downstream with a weird *quok quok quok quok oo-wa-ee*. On the sandbar on the farther shore, a raccoon eating at the water's edge looked into the flashlight's beam with two brilliant orbs, then turned and walked deliberately away. Jumping down the bank, I cautiously approached the motmots' burrow and directed the beam into it.

What could that be, sleeping at the inner end, where I expected to find a motmot? Had some grayish furry animal stolen into the burrow, devoured the motmot and its eggs, then curled up to rest and digest its meal in the bird's place? But no, a patch of turquoise on the creature's head and blue on its sides could only belong to the motmot; it must surely be the parent bird, safe and sleeping peacefully on its eggs, but so transformed in the yellow light that I did not at first recognize it. The bird's back, and all the parts that by day are soft green, appeared yellowish gray in the rays from the little electric bulb. The soft, loose plumage was fluffed out and resembled fur. But the blue on the wings, and especially the turquoise on the brow, shone with so bright and radiant a luster that I switched off the light, half expecting to find them self-luminous. But all remained dark in the burrow.

The motmot was facing inward. Its tail extended forward into the entrance tunnel, without flexure. The black and blue racquets on the end were so far behind the bird's body that they appeared not to be connected with it; they looked like detached pieces of feather caught in the ceiling. I returned several times and always found the motmot in the same position, with its tail running out into the entrance tunnel, where alone it found ample room without becoming bent. This, then, was the secret of how these birds kept their tails in good condition while they incubated their eggs and brooded their nestlings. Years later, in Costa Rica, I found Broad-billed Motmots sitting in their burrow with the same orientation.

While the kingfishers and motmots dug their burrows and raised their families, Rough-winged Swallows traced irregular courses above the river and its shores, or rested, chirping harshly, on high, exposed twigs. The most widespread and probably the most abundant swallow on the American continents, nesting from southern Canada to northern Argentina, the grayish brown Rough-wings were much more numerous along the Río Morjá and neighboring rivers than were the prettier, steel blue and white Mangrove Swallows, which generally flew lower above the water. In February, while kingfishers were still excavating their tunnels in the bank, the Rough-wings took an interest in them, alighting in their entrances and sometimes venturing inward a foot or so but hesitating to explore their dark depths. They would have to wait many weeks before these burrows were available for their occupancy.

Ever watchful, the swallows moved into the tunnels as soon as the original proprietors moved out. Within five days after four young Amazon Kingfishers flew from a burrow in the bank, a pair of swallows had started to build in it. They did not, like kingfishers and motmots, lay their eggs in contact with the earth but provided an insulating layer of dry vegetable materials, which would help to keep them at a higher temperature. Although the original occupants preferred to sit at the inner end of a long, curving tunnel, where they had no view and were invisible from the front, the swallows placed their nests farther forward, where the incubating bird could look out upon at least a narrow circle of the river and the opposite shore. The entrance tube was quite wide enough for the thick, shallow cup of grass, dry leaves, rootlets, and other scraps of vegetation in which each female laid four or five immaculate white eggs, chiefly in late April, when young kingfishers and motmots were already flying.

Whereas the contemplative original tenants of these burrows incubated continuously for long hours, the mercurial swallows, with their less substantial food and swifter digestion, were capable of no such displays of patience. When I started to study their mode of incubation, I set a twig in the mouth of the tunnel to assure myself that no bird had entered or left while my attention wavered, as I had done with the kingfishers; but the sentinel was pushed over so frequently that I grew tired of wading across the river to reset it, and relied on direct observation alone. I had to keep my eyes glued constantly on the hole in the bank, for an instant's inattention might cause me to miss the swallow's swift inward or outward dart.

A swallow seldom remained with her eggs for more than fifteen consecutive minutes, and her longest diurnal sessions rarely exceeded half an hour. During her outings, which were almost as long as her sessions, her eggs were left unattended; for, again unlike the kingfishers and motmots, the male took no share in incubation. In order to confirm this, I caught a swallow in another burrow and marked her wings conspicuously with white paint before releasing her. To my chagrin, this nest was then abandoned. Had her mate been in the habit of helping to incubate, he might have continued to attend the eggs, thereby restoring her confidence, since I did nothing to frighten him.

After sixteen days of intermittent incubation, the eggs in other burrows hatched; and now both parents darted in and out, taking tiny insects to the nestlings. Since their nursery opened upon a broad expanse of swiftly flowing water, where a weakly fluttering departure, such as many fledglings make, might result in drowning, the young swallows, like their predecessors the kingfishers and motmots, remained in their nest until they could fly well. Although well feathered when two weeks old, they delayed their departure for another week, when they were able not only to fly across the forty-foot channel but to continue for an equal distance above the opposite shore.

As a young swallow left its nest, flying strongly, several adults that had been circling above the river rushed after it. One flew above and apparently in contact with it, gradually forcing it down into the vegetation on the farther shore. I have witnessed such "shielding flight" in several other kinds of birds. The parents seem to direct the strongly flying but still inexpert fledgling down into sheltering vegetation where it will be safe from the attack of a passing hawk; or perhaps they behave in this manner to divert a raptor's pursuit from the fledgling to themselves, who are better able to escape it.

When, after slithering along in the soft, oozy mud at the foot of a high river bank, I looked into a certain swallow's burrow, I was amazed to see a fully feathered fledgling lying prone and trembling at the feet of a big toad, beneath a mouth that seemed wide enough to engulf it whole. Whether the amphibian had designs on the young swallow, I could not tell. As I prodded the intruder with a stick, the fledgling slipped past it, to join its nest mates deeper in the tunnel. I removed the toad; but a few days later, after the last of the swallows had flown, it or a similar batrachian occupied the otherwise vacant burrow, which it could have reached only by climbing up or down an almost vertical bank. In a deserted motmots' burrow, I found another toad. These amphibians, then, were the last inmates of the tunnels that, in the dry season, the kingfishers and motmots dug in the loamy banks of the Río Morjá.

In addition to the residents of the burrows, the stream was frequented by many birds who came only to forage. As the sun declined in the west, flocks of dusky birds settled on stretches of low-lying shore that were covered with small, water-worn stones. Among them were Giant Cowbirds, Red-eyed Cowbirds, Melodious Blackbirds, and Great-tailed Grackles, especially the glossy, long-tailed males of the latter. All these black members of the troupial family busied themselves lifting or turning over small stones, using the same method. Inserting their bill under the near edge of a stone, they raised it by pushing forward with the upper mandible, while the lower mandible was dropped and the bill held slightly open. If nothing edible was revealed beneath the pebble, it was dropped; if some small creature, or perhaps a seed, was hidden beneath, the stone was turned over or pushed aside, and the object was eaten.

The chief of these stone-turners were the Giant Cowbirds, who slip their eggs into the long, woven pouches of oropéndolas and caciques, then neglect them. Their strong bills were well adapted for moving small stones. While the female cowbirds hunted over the beach with great diligence, the bigger but less numerous males strutted pompously among them, appearing too haughty to earn a living by such a humble occupation. Walking with a sort of goose-step, a male would stiffly approach a busy female and plant himself squarely in front of her. If she retreated, he might forget his dignity and hop ungracefully after her. Facing the object of his pursuit, he drew himself up until he towered above her, seeming thrice her height. Arching his neck, he lowered his head until his bill rested among the outfluffed plumage of his breast, while the erected feathers of his cape surrounded his head as an iridescent, purplish black ruff, in the midst of which his red eyes gleamed like rubies. Sometimes, as though still further to accentuate his height and impress or overawe the indifferent female, he bobbed up and down in front of her by flexing his legs with plumage still puffed out. But in these evening gatherings, food interested the hen more than did the pompous suitor; a peck and a sharp note were all the recompense of his efforts. If he persisted in annoying the female, she fled, and with sonorous wingbeats he pursued her across the river and out of sight.

As though to emphasize the old adage that "birds of a feather flock together," still another dusky bird was sometimes present in the evening. While the four dark relatives of the orioles turned stones on the shore, a few Muscovy Ducks often dabbled in the neighboring shallow water, or perched high in trees along the shore. With their black plumage glossed with green, pure white specula in their wings, and red caruncles above their eyes, the drakes were handsome birds. It was difficult to believe that these splendid free creatures were of the same species as the Muscovies that waddled obesely in rural dooryards, their soiled plumage variously marked with white or else wholly blackish, all their charm gone. How domestication dims the brightness and dulls the intelligence of animals!

As the sun dropped down behind the banana plantation, the Giant Cowbirds flew noisily downstream, to a roost that I never found. The Great-tailed Grackles went in the opposite direction, with labored wingbeats rising to the hilltop, where they slept in the coconut palms in front of the house. The Red-eyed Cowbirds and Melodious Blackbirds did not go so far, but retired into the dense stand of young canes that had sprung up on newly formed land, just behind the barren flats where they had been turning stones. Sometimes, in the waning light, I sat on the sand beside the cane-brake to enjoy the blackbirds' clear, soothing whistles that emerged from its depths. As the birds grew drowsy, their notes became lower and more widely spaced, until in the deepening dusk they fell silent. Then I waded across the river and followed the grackles up to the house on the hilltop.

Although they did not join the stone-turning party on the bare shore, Orchard Orioles, which wintered abundantly in the lower Motagua Valley, roosted with their less colorful relatives in the cane-brake. While in Central America, they sing more than any other migrant, not only after their early arrival in July or August, but for a month or more before their spring departure for the North. Indeed, in early April, before Gray's Thrushes came into full song, the Orchard Orioles produced more melody than any other bird along the Río Morjá—just as, in April and early May, migrating Swainson's or Olive-backed Thrushes collectively sing more than any other bird in the forests of southern Central America. As though eager to display their newly acquired vocal power, the black-throated, yellowish, yearling, male Orchard Orioles sang as much as, if not more than, the mature males in chestnut and black. When they awoke in the cane-brake at dawn, their chorus of cheery voices was delightful to hear. Orchard Orioles lingered at Alsacia until April 21, when I saw a single female. Their absence from the region was brief, for on July 20 I met them higher in the Motagua Valley.

As the year advanced, the weather deteriorated. February and March had been dry, bright, and not disagreeably warm. May became oppressively hot and sultry. The atmosphere, polluted by smoke from innumerable fires set to clear land for planting maize, and charged with the gathering vapors of the wet season, lost the fine transparency of the early months of the *verano* or dry season. In late May, the rains began. The downpours of June swelled and muddied the once limpid current of the Río Morjá, which rose to cover the bare expanses of sand and gravel on the farther shore, where I had passed so many pleasant, rewarding hours watching the birds, and it threatened to inundate or wash out the burrows in the opposite bank. An Amazon Kingfisher and a Rough-winged Swallow, who had lost earlier broods, and the motmot who had dug a second burrow, were the only birds who laid in these burrows after May 1; and after the middle of the month, I noticed no further attempts to raise families in them. Back in the hills, too, most of the broods that I had been watching had flown from their nests by mid-June.

For four months I had been extremely active, watching the birds most or all of every day, developing photographs and writing up notes far into the night. Amid heavy tropical vegetation, where nests are so elusive, I regarded every one that I found as a never-to-be-repeated opportunity to gain knowledge, a challenge to learn all that I could about it. Now, when the weather became so oppressive, I paid for going too hard in a climate that I found hard to bear, as had happened at Lancetilla two years earlier. Overcome with feverish lassitude, I dosed myself with quinine until my ears rang; every high-pitched sound fell upon them with preternatural sharpness; and the shrill calls of the male Great-tailed Grackles cut into my ears like a knife. But this self-prescribed treatment failed to improve my condition, so I resolved to seek a change of climate for my health's sake.

Although Alsacia Plantation had proved to be a splendid locality for my work, and never elsewhere have I learned more about birds in an equal interval, I did not depart it as regretfully as I have left many another place where I have studied nature. At first, I was so delighted with the broad prospects that spread on every side from the house perched on the hilltop, that I overlooked certain shortcomings of the immediate surroundings. As the months passed, I discovered, as John Burroughs did when he abandoned "Riverby" for "Slabsides," that scenery on the grand scale, however inspiring, cannot in the long run compensate for lack of intimacy—the unfrequented byways along which one loves to stroll; the quiet, sequestered spots that become dearer to us the longer we loiter in them. Alsacia, a plantation developed for monetary returns, provided few such spots within easy reach of the dwelling. The nearer terrain was occupied by steep, pastured hillsides, covered with tall bunch-grass, infested in the dry season by ticks and redbugs; with bushy, vine-entangled second growth, penetrable only at the price of vigorous exercise with the machete; and the bananas, noble plants indeed, but as monotonous in the endless sameness of a great plantation as a field of wheat would be to a race of Lilliputians. A pleasant stream that flowed down a rocky bed in the hills behind the house had been spoiled when the tall second-growth trees on the surrounding slopes were cut and burned to make a milpa. And now, with the rising water, I could no longer wade across the Río Morjá to the exposed sandy beaches of happy memory—indeed, they were no longer exposed.

I left Alsacia early on a sunny morning after a dark and rainy night. A pearly mist filled the atmosphere, softening the outlines of the distant mountains and tempering the rays of the sun. It was a morning almost spring-like, the mild radiance of the new day lending a charm to the humblest weed. But for the overseer of the plantation and me, it narrowly escaped becoming tragic.

When we reached the main line of the railroad in the little motorcar, it was necessary to telephone to the dispatcher in Zacapa for an order to proceed down the tracks to the station. We were told to await the passing of a southbound freight train, but the dispatcher was not certain whether it had yet reached our spur. When, after waiting a reasonable interval, the train did not appear, we concluded that it had already passed and proceeded down the railroad. As we rounded a curve bordered by banana plants that shortened the outlook, a locomotive at the head of a long string of freight cars was bearing down on us! The engineer jammed on his air brakes; but before the heavy train could come to rest, we had stopped the car, pushed it into reverse, and were scuttling backward in front of the still-advancing locomotive, with only yards separating it from us. It was a narrow escape, which fortunately resulted in no accident more serious than the cancellation of Pellman's license to operate his car on the main line. Thenceforth, he was obliged to hand over the controls to his motor-boy.

A hedge of thorny cacti in the dry Motagua Valley.

Volcanoes Fuego (left) and Acatenango rise above a rushing stream of Guatemala's Pacific slope.

14

The Dry Tropics

High plateaus and chains of lofty mountains running in every direction give Guatemala a marvelous diversity of climate and scenery. It supports humid lowland forests composed of a bewildering variety of tropical trees and vines, and high mountain forests dominated by a single species of conifer of northern type. Regions of great rainfall alternate abruptly with arid valleys overgrown with cacti and thorny scrub. Its natural vegetation runs the entire gamut from coastal mangrove swamp to alpine meadow; its crops range from banana, cacao, and sugarcane to wheat, oats, and potatoes. In my travels about the country, I have been exhausted by heat and benumbed by cold; I have felt burning thirst and lived through days so humid that it was hardly necessary to drink. Here, in an area approximately equal to that of the state of New York, is a diversity of climate, scenery, and natural products that rivals, on a small scale, that of the whole United States. I believe it scarcely possible for one who knows only a region in the temperate zone where

Headpiece: A dry watercourse winds like an avenue through thorny scrub in the dry Motagua Valley.

climatic conditions are essentially uniform over great areas, as in Europe or eastern North America, to form a true conception of the abruptly diversified character of a mountainous tropical land.

Nevertheless, northerners crave sweeping generalizations about the tropics. The returning traveller is expected to confirm widespread misconceptions by vivid accounts of insupportable heat, teeming, venomous reptiles, and deadly impenetrable "jungles"; if he attempts to disabuse his hearers, in conversation or in print, they yawn and turn away. When anyone asks how I managed to survive tropical heat, I point to the indisputable facts that I have spent years in the tropics and am still alive; but when someone asks how I endured tropical cold, I know that I have an intelligent and interested listener. One need only be familiar with a mountainous tropical country, such as Guatemala or southern Mexico or Colombia or Ecuador, to be acutely aware of the futility of trying to generalize about the earth's broad central zone, of which each of these countries is a small fraction. We have already in our elementary lessons in geography been taught the one generalization that it is quite safe to make: that everywhere within this zone the sun passes directly overhead at noon twice in the course of its annual swing through the heavens.

The two-hundred-mile railway journey from Puerto Barrios to Guatemala City gives the thoughtful traveller a revealing, although far from comprehensive, picture of the country's natural zones. To one interested in climate, vegetation, and man, and their mutual interactions, this ride, which formerly took all day, is a fascinating experience; one does not wish it shorter or more hurried. It was along this line that I first penetrated to the interior of Central America, and I found sights to hold my attention every mile of the way. Since then I have crossed the backbone of the American continents, in their tropical portions, along a number of other routes; although the scenery was sometimes grander, no other presented such strong contrasts in vegetation and climates in so short a distance. Costa Rica, along the railroads joining the two coasts, is too uniformly humid to afford comparable contrasts. The Ecuadorian Andes, as I saw them in crossing from Guayaquil to the eastern foothills in the Pastaza Valley, are too thoroughly denuded of their original vegetation in all their central and higher parts, too amazingly empty, to present a satisfying picture of natural conditions. On that altitudinal journey of twelve thousand feet, I found the Andean scenery magnificent, but depressing in its vast emptiness, so different from the richly varied landscapes in the lower mountains of Central America.

Despite our narrow escape on leaving Alsacia, we reached Los Amates in time for the passenger train, which took me deeper into the country. The railroad reaches the highlands chiefly by way of the Motagua Valley. On my way to Alsacia earlier in the year, I had traversed most of the humid lower part of this valley, with its vanishing forest of great, epiphyte-burdened trees; its lush, large-leafed herbage; its great banana plantations; its miles of rank, undisciplined second-growth thickets; its willow trees seeming so out of place beside the brakes of giant canes along the river banks; its surprising savannas where palm and pine grow side by side on open, grassy plains. Indeed, save for the presence of pine trees, this was much the same type of country as I had earlier known on the Caribbean coast of Honduras and Panamá.

As I proceeded inland, the picture rapidly changed. The pine-covered mountains, which for miles below Los Amates I had seen in the distance, now pushed in close to the river. The valley was becoming more arid, and the banana plantations, which a little lower had spread over the wide plains, were here reduced to narrow strips along the stream. After a few more miles, they disappeared. The vegetation rapidly became lighter; the woods were more open and contained smaller trees. Epiphytes and tangles of vines still burdened them, but not in the same profusion as nearer the coast. The foliage, consisting of smaller leaves in lighter sprays, was in large measure responsible for the altered aspect of the country. The type of agriculture had changed with the natural vegetation; cornfields, instead of being log-cluttered temporary clearings, were clean and tilled by primitive ox-drawn plows, and appeared to be permanent. Here, too, I began to notice roads, which in the lower valley were almost unknown; railroads and light tramlines served the plantations, and travel among the foothills was over narrow, muddy mule trails.

The railroad, deserting the main stream, followed up the valley of a tributary, then cut through low, pine-clad hills and swung back again to the Motagua at Gualán. Noon was now approaching, and when the train stopped at the station, it was surrounded by women and girls who wore mantillas or substitute bath towels over their heads but often no shoes on their feet. Their smooth, dark skin betrayed their largely Indian ancestry. On their heads they bore baskets or trays filled with tamales, rolls of wheaten bread, boiled eggs, tortillas of maize wrapped in white cloths to keep them warm, oranges and other fruits, all of which they displayed under the carriage windows to tempt hungry passengers. Currency of the United States, which nearer the coast passed freely in all denominations, was here likely to be rejected. The woman who sold tortillas or hard-boiled eggs held it a sacred obligation to provide the buyer with the essential coarse salt, even if she had to run along beside the departing train to pass it up to him.

Gualán, a small town situated amid acacia trees upon a low, arid hill overlooking the river, had an aspect very different from all the settlements nearer the coast and was much more picturesque. It was the first typically Guatemalan town to greet us as we travelled inland from Puerto Barrios.

With adobe walls usually plastered white and roofs of red tiles, the little houses had an air of permanence, as though they belonged to the land and would remain. This impression was, no doubt, to be attributed to the native clay, lime, and logs of which they were built. In the lower valley, the better buildings were made of lumber that was often imported and roofed with corrugated sheet-iron which at that time invariably came from abroad. The humbler dwellings were ephemeral, palm-thatched shacks, with walls of palm leaves, or flattened trunks of bamboo, or poles stuck upright, with wide chinks between them.

Above Gualán, the railroad was pushed close to the river bank by steep hills covered with low, thorny scrub and cacti. On the opposite side of the valley, high, pine-clad mountain ridges rose into the sky. Wherever the Motagua's swift current—here no longer deep and calm but turbulent and broken—washed close beneath the tracks, on looking down I noticed all sorts of old railroad iron. These scattered car wheels, locomotive boilers, girders, and the like raised disquieting visions of trains that had jumped the rails and hurtled into the river, perhaps killing or maiming most of the passengers. The apprehension aroused by these repeated suggestions of disaster was intensified when our train was held up by a freight train derailed on a high trestle, until I began to feel that it would have been more prudent to have remained on the coast, despite its fancied terrors. On a later trip, however, a railroad official made me more comfortable by assuring me that these rusting irons were not relics of trains that had plunged into the river, but outworn equipment that had been dumped there to retard the erosion of the banks. Despite these efforts to curb it, in the wet season the river often undermined the tracks, while landslides from the steep cuts at times buried them under tons of earth and rocks. To keep open this thin line of communication through a sparsely settled country meant constant struggle against shifting earth and water. Occasionally, during the rainy season, through-train service was interrupted for days or even weeks at a time.

Approaching Zacapa, the railroad again temporarily parted company with the river and cut across a semi-desert plain to the town, where a half-hour pause permitted passengers to eat a hot lunch in the station restaurant. Zacapa, halfway between Puerto Barrios and Guatemala City, was also the junction of the main line with the then newly completed branch leading to El Salvador. This is the single railroad linking together neighboring countries in Central America; the railways of all the others are isolated systems, which at most extend through coastal plantations on the other side of an international boundary. Soon loud tolling of the bell warned diners that the train was about to leave. Slow eaters gulped down last mouthfuls and rushed back to the coaches.

The midday heat on this dry plain was so intense that I was not sorry to be moving again. Passing between lean pastures enclosed by cactus hedges, both more picturesque and more efficient than the barbed-wire fences in the lower valley, the railroad gradually converged with its old companion, the river. Here all nature bristled with forbidding thorns. The trees were low, flat-topped and thorny, those of the acacia type predominating. The cactus family was well represented by tall, stiffly branched organ cacti; lower trees of the prickly pear group, with more open crowns of cruelly armed, flat-jointed branches; low, barrel-like melon cacti; and orange-flowered Pereskias that looked more like ordinary, branching, broad-leafed, thorny trees. The cattle grazing on the scanty herbage of these barren plains, green only in the wet season, looked thin and underfed. Fruit trees and perennial crops like bananas and cassava grew only on low-lying ground near the river, but in the rainy months swiftly maturing maize yielded harvests on some of the higher valley land and even on the slopes of the enclosing hills.

Across the river, the great rampart of the Sierra de las Minas, which paralleled the stream for many miles, still rose into the clouds. Here and there a bright cascade tumbled down the steep, pine-shaded slopes, in picturesque contrast to the arid valley below. This towering barrier is responsible for the valley's dryness, for it intercepts the moisture-laden trade winds from the Caribbean. Wherever in Guatemala, at whatever altitude, a narrow valley is enclosed by lofty ridges that lie athwart the prevailing winds, that valley is arid. Sometimes clouds pour over the crest of the ridge for days, only to dissolve as they are warmed by their descent into the lower atmosphere, and not a drop of rain falls in the valley; but if one crosses to the northern or eastern side of the range, he will find rain.

The low, rounded hills near the tracks were, indeed, seamed by deep watercourses, strewn with boulders that attested to the force of the current that sometimes rushed through them, but they were now as dry and parched as the surrounding slopes. A low, massive, white-walled church, standing amid clustered huts across the river, reminded me that the influence of Old Spain had extended into this region. In the lower valley, I noticed no such monuments of the colonial period.

At the village of El Rancho, one hundred and thirty-six miles from the sea by the railroad, the line had risen only nine hundred feet. Here I left the train and tarried for most of July, in a little hotel fronting the tracks. After the forests and densely tangled thickets of the lower valley, this arid country gave me an exhilarating sense of freedom. What a pleasant change, to be able to walk freely in any direction that fancy dictated, with no hindrances save those forbidding cactus hedges—without having to fight my way with a machete through stubbornly obstructing vegetation, whenever I left the beaten track!

But these hills, so bare and open, seemed also, by contrast with what I had just left, to be almost as barren of life as the polar regions. Yet on longer acquaintance, I learned that this semi-desert was by no means devoid of life and interest. There was charm in following sandy- or stony-bottomed dry watercourses, bordered by low, thorny trees, which wound like shady avenues between rounded, rocky hills. Diminutive White-lored Gnatcatchers, much resembling their northern cousins, flitted through the sparse foliage, uttering the same thin, nasal mew. The long-drawn, lazy, undulatory cooing of the White-winged Dove, the only pigeon known to me that really sings, seemed to express the very spirit of this parched valley, where man and nature live so unhurriedly, drowsing beneath a scorching sun. Later, when I heard this same dove sing on frosty mornings in the highlands, it invariably carried my thoughts back to this hot, dusty valley.

Nor were other birds lacking. When I arrived late in June, Tropical Mockingbirds were singing lively medleys on the bushy hillsides. Flocks of blue, White-throated Magpie-Jays, with clusters of long, recurved, black plumes waving over their crowns, straggled through the thorny trees, uttering harsh calls mingled with clear, ringing notes. These big jays interested me greatly. The incubating female cried out for food, much in the manner of the White-tipped Brown Jays that I had just left, and so led me to her bulky nest, situated among high, thorny boughs. Here she was fed not only by her mate but also—as I saw later on the Pacific slope of Guatemala—by helpers, who together kept her so well supplied with food that she left her eggs only briefly, at long intervals, to preen rather than to forage.

Motmots of two kinds, both the Russet-crowned and the Turquoise-browed, were conspicuously abundant; sandy cliffs and the scarped terraces of the hillsides were penetrated by multitudes of their nesting burrows. Big Rufous-naped Wrens, members of the cactus wren group, built between the prickly joints of the opuntias the deep, downy pockets that served them as nests and dormitories. Black-throated Orioles, yellow and black birds with a clear, whistled song, hung their gourd-shaped, woven pouches from the tips of the upper branches of the low trees. All these birds flitted among the thorny boughs and alighted on perches bristling with the sharpest spines without appearing to pay the least attention to these formidable projections, yet they seemed never to suffer the slightest inconvenience from them. This was to me a continuing source of wonder.

The fauna of this semi-desert differed from that of the rainy lower valley as greatly as the flora. Most of the few species of birds of the humid zone that extended so far inland were confined to groves and moist thickets along the river. The Central American lowlands support two groups of birds that have few species in common. One group occupies the humid tropics and spreads over the Caribbean littoral from end to end, but on the Pacific coast is well represented only south of the Gulf of Nicoya. The second group, which thrives in the more arid tropics, occupies the Pacific lowlands from the Gulf of Nicoya northward but on the Caribbean side is restricted to mountain-walled arid valleys such as this. Here, in the middle Motagua Valley, I began to become familiar with the birds of the drier regions.

Almost every day, while I dwelt at El Rancho, somebody asked me "How many birds have you killed?" or "How many doves did you shoot?" Since I never carried a firearm, I suspected that these simple swains regarded my binoculars as an apparatus for concentrating the emanations of an evil eye into rays deadly to birds. I was sorry to discover that these uneducated people were aware of no possible relationship between man and free animals save that of hunter and hunted. In other parts of the world, less primitive, I have met people to whom the only intelligible transaction between man and man is to gain money or some other form of wealth. I do not know which kind of spiritual blindness is more to be pitied.

The ruined churches and convents of Antigua, Guatemala's capital in colonial times.

The highway leading westward from Antigua. The ornate fountain in the wall to the right imparts antique charm.

15

A Fantastic Journey

From El Rancho, I went by motor truck to Cobán, capital of rainy Alta
Verapaz, a department renowned for its coffee, its Indians who staunchly
preserved their native tongue in preference to Spanish, and its varied bird
life. I found the little town attractive. It was situated on a long, low, narrow
ridge, so that, walking down the side streets that led out from the central
plaza, I looked across green valleys to pine-clad ridges haunted by beautiful
Bushy-crested Jays with golden eyes set in black heads. Such a town, unless
insufferably dirty, always appeals to me. The best hotel left much to be
desired; disorderly and unclean, it had a weedy patio littered with empty
cans and no facilities for bathing. (I know that Cobán has now acquired a
better hostelry.) I was told that the massive old church had been built early
in the sixteenth century by Fray Bartolomeo de las Casas, indefatigable
champion of America's oppressed Indians. I noticed many neat, substantial
houses, some of which were the homes of prosperous North American and

Headpiece: A farmhouse near Zacapa in the arid middle reach of the Motagua Valley.

German coffee planters. Most of the larger mercantile establishments were owned by Germans; indeed, the Alta Verapaz was sometimes called "Little Germany."

The northward journey from El Rancho to Cobán had taken a day and a half. On the return trip, it took us nearly four days to cover about sixty miles by motor; that painfully slow ride remains in memory as the longest sequence of mishaps that I have ever experienced.

We set forth from Cobán at three o'clock in the afternoon. Don Antonio, the genial Galician who had come up with me from the railroad to sell brushes of North American manufacture to the housewives of Cobán, was the only other passenger in the truck. We made a brave start, for the highway as far as the Indian village of Tactic was excellent. The last few miles had been newly repaired, and the completion of the work was being celebrated that very afternoon with appropriate dedicatory ceremonies. We rode beneath verdant arches of intertwined boughs, between which the roadside was lined with leafy woodland branches set upright in the ground. These were chiefly of pine and the starry-leafed Sweet Gum, the latter a noble, lofty tree common in northern Guatemala at altitudes of from four to six thousand feet and hardly to be distinguished from the *Liquidambar* of the southern United States. At Tactic we found the Indians dancing to the music of the marimba, a sweet-toned xylophone played by three men together. They were dancing when we went early to bed; they still danced to the strains of the marimba when we arose at four o'clock next morning to resume our journey. Indian celebrations usually continue through the night.

Between Cobán and the Motagua Valley the road crossed three major ridges between five and six thousand feet high and a number that are lower. The rough, unpaved road wound up precipitous slopes with grades so steep that we travelled many miles in lowest gear, and with turns so sharp that they could be rounded only by backing the car, sometimes more than once—a delicate and dangerous operation. The driver was assisted by a boy whose duties included placing a block behind a rear wheel whenever the car stopped on a steep slope, especially at the hazardous curves where it was necessary to go into reverse; in helping to turn the front wheels by pulling on a tire; in filling the radiator, which was continually boiling over, at the streams we forded; in putting chains on the tires and removing them several times a day, as the road changed from muddy to rocky; in loading and tying on the cargo. The young driver's assistant had plenty to do and was almost indispensable at the curves, where frequently the road sloped outward, so that if the car skidded nothing would save it from hurtling down hundreds of feet to the stream at the foot of the steep, bare declivity. I usually found some excuse for dismounting at these perilous turns, for I felt safer with my feet on the ground.

From the head of the winding valley of the Río Cobán, we climbed without mishap to the summit of the first high ridge, which supported a light growth of low, broad-leafed trees draped with long, gray streamers of "Spanish Moss" or *Tillandsia*, and heavily burdened with other epiphytes. Thence we descended into an elevated valley where pine trees grew scattered over grassy slopes devoted to grazing. At the bottom of this valley, we forded a rocky stream and began a second ascent. When halfway up, our troubles began. Heavy rains had fallen; the car sank deeply into the mire; and we passengers joined the driver and his helper in filling ruts with stones brought from a distance, then jacking up the buried wheels and placing solid material beneath them. Advancing thus laboriously, we rounded the next curve, only to find that one side of the road had washed away. Luckily, we discovered some pine logs lying nearby and with them formed a makeshift bridge, then covered its approaches with brush and logs. This was a mistake, for the loose material was pushed ahead by the wheels and piled up in front, making their advance still more difficult.

After we had labored at highway-making most of the afternoon, extending our search for loose stones far up and down the next brook that flowed across the road, we were joined by two men who had been driving a two-ton load of coffee from Cobán to the railroad at El Rancho. Their truck had thrust a rear wheel through the decayed logs of a small bridge over a mere trickle of water, and they had been waiting here for two days. Ours was the first vehicle to arrive from either direction in this interval. With our new reinforcements, we finally succeeded in maneuvering our car to a firmer part of the road, when the sun was about to set.

We reloaded our truck, then walked ahead to unload the hundred-and-fifty pound sacks of coffee from the larger truck, in order to extricate it from the position where it had been resting so long. Then, by the flickering light of a fragrant pinewood torch, we jacked up the wheel that had broken through the bridge and stuck timbers beneath it. When all was ready, we watched breathlessly while the motor started. All our work was lost; the logs separated and the wheel fell again. Then, rain falling, we toiled feverishly to replace the coffee in the truck, for a wetting would spoil it. In the morning, all those heavy sacks would have to be unloaded again!

Gerardo, our driver's helper, had been sent back to a solitary farmhouse to seek something for our supper, and now he returned with a woman and a small boy, bringing tortillas, brown beans, an omelette, a mixture of chayote and egg that I found very savory, and strong black coffee already sweetened. For travelers stranded on a remote mountainside, this was a feast; but poor Don Antonio, unable to accustom himself to native food, could not enjoy it like the rest of us. After the woman and boy had carried away the empty dishes in a basket, the four of us settled down on the seats in the truck to pass the night as best we could.

From the distance came the melancholy wail of a coyote, a soul-stirring ululation that, heard for the first time in this solitude, brought moisture into my eyes. Don Luís, the owner and driver of the truck, then told us that once his car, stalled on this mountain, had been surrounded by a pack of coyotes, who did not permit its lone occupant to sleep that night. I fervently wished that we would be similarly visited, for it would have been much more diverting to cast the beam of a flashlight around a circle of glowing eyes than to sit shivering in sleepless misery through much of that wet, chilly night. When finally I dozed, the car was shaken by an earthquake tremor, the only one that I have ever felt in an automobile.

It was noon the next day before we had maneuvered both cars over the rotten bridge. Then we crossed another high ridge, covered with grass and scattered agaves, with an open pine wood on its crest. At the foot of this mountain, we forded a wide stream and started across the plain to Salamá. The level valley was evidently very arid in the dry season; but now the broad roadway, defined only by deep ruts, was an almost impenetrable morass into which our wheels sank up to the hubs. To avoid getting stuck, it was imperative to move fast; we careered wildly onward, skidding and swaying from side to side, jolting and jarring, sending out to right and left wide-spreading showers of muddy water and watery mud. At the conclusion of this mad ride, which would have been a worthy subject for a ballad by the author of "John Gilpin," we crossed a river by a covered bridge and entered the town of Salamá, capital of the department of Baja Verapaz. Here a policeman recorded the license number of the car and the names of all the occupants. During the years when I travelled in Guatemala, this was the usual procedure when one entered or left every town or larger village; on a long journey, it was sometimes necessary to give one's name and destination to the police five or six times in a day.

We slept that night on mattresses of the fibrous, brown leaf sheaths of palm trees—probably the palmetto, which is abundant in the arid valleys of this region—cut into small rectangles and loosely stitched together. These mats were laid over ropes strung crisscross over a wooden bed frame, a combination none too soft. But the hardest bed could not have kept us awake after our exertions and the preceding nearly sleepless night. We were aroused at five o'clock in the morning by the clear notes of a bugle and the tattoo of a drum from the barracks across the street from our little hostelry. Soon a company of ragged soldiers filed past our window. After a meager breakfast, we set out early and unsuspecting upon a day of mishaps.

Without adventure, we crossed the plain to San Gerónimo, and with much grinding of gears reached the beautiful forest of Sweet Gum and pine on the summit of the ridge to the south. Here a rustic shrine and some rude crosses stood at the wayside. Throughout Spanish America, the points where travellers have been killed are marked with crosses, but we did not learn the details of the tragedies that had occurred here. Our own private griefs quite filled our minds, for now we discovered that a bolt had been jarred out of a front spring, letting the truck's body rest upon the axle. But this annoying accident was, for me at least, soon compensated by a splendid view of the upper half of the shapely cone of El Volcán de Agua, standing out clearly fifty miles away.

At the foot of the mountain that we had just crossed, we were delayed nearly an hour by two passengers who started to prepare for their journey only after asking whether we had room for them. A few miles farther on, a valve was somehow torn from one of the rear tires. Since we carried no spare tire, the casing was stuffed with burlap and blankets and placed on a front wheel, while the good tire from the front was shifted to the rear. After this, we continued slowly onward with a rocking-horse motion. When we stopped at a roadside farmhouse for lunch, Don Luís discovered that a package containing six pairs of shoes made in Cobán had been rocked out of the car. Poor, overworked Gerardo, who was responsible for the lading, was sent back to retrieve them. He appeared at El Rancho two days later—without the shoes!

We galloped slowly into Morazán, a dirty, unattractive village among arid hills, where a telegram was sent ahead requesting that a spare tire and gasoline be sent to meet us. Here we tried in vain to buy oranges or water-coconuts to assuage our thirst, for we mistrusted the local water. Finally, we succeeded in purchasing a flavorless pineapple. The padre, we were told, visited the church thrice a year, to marry, baptize the accumulated infants, perform the burial rites for people who may have been interred months earlier, and hear the parishioners confess their sins. Morazán was redeemed for me by a pair of Black-throated Orioles, resplendent in golden plumage, who were feeding nestlings in a long, woven pouch suspended from the tip of a coconut-palm frond in the central plaza. Great-tailed Grackles were also nesting here, as they nest and roost in nearly every town and village in Guatemala, from sea level up to the high western plateau.

As we rumbled slowly into the next hamlet, it appeared that the whole population had turned out into the road to stop us. A sick man, they said, wished to be carried to El Rancho, where he could take the train to the hospital in Zacapa. Don Luís at first refused him transportation, because the truck was already quite full with the passengers it had picked up during the day, and, moreover, it was bumping so much that it was not fit to carry one in his condition. Presently, the ailing man himself appeared from among the crowd, a fortnight's stubble on his wasted face, his open shirt revealing a lean and hairy chest, and began his plea by reminding us that "There is a God above, and we are all his children."

Glancing around at the mongrel crowd, the haggard women and dirty children, the dull, apathetic faces, I resented this reminder of our common brotherhood and thought that the supplicant had made a stupid beginning. But when presently the invalid dropped into the roadway, writhing, groaning piteously, and muttering *ai ai ai* in an outburst of agony too intense to be feigned, our common humanity asserted itself. I offered to give him my seat if his neighbors would find a horse to carry me to El Rancho. Don Antonio also relinquished his place on the front seat. After some discussion, we resumed our journey with the two of us standing on the running-boards of the now heavily overloaded truck.

We had not gone many miles more, through hills covered with low, scrubby vegetation, when, ascending a grade, the motor sputtered, gasped, and died—out of gasoline. I feared that our invalid would expire, too, for he jumped into the road, where he dropped down and grovelled in the dust, uttering his agonized *ai ai* and heart-rending groans. He had borne bravely the jolting of the car, which must have been a gruelling ordeal to one in his condition, but now he completely lost control over himself, begged us to shoot him, and, this plea failing to move us, looked around for a *barranco* into which he could leap and end his agony. Full as Guatemala is of profound chasms and ravines, none happened to be at hand in this spot.

Not finding it pleasant to watch the contortions and hear the groans of this poor, pain-wracked creature, for whom we could do nothing save give him the deflated inner tube to ease his contact with the road, I suggested to my Galician companion that we walk ahead. We were still about ten miles from our destination. In the dusk, after we had gone some miles, we met another truck bringing succor to our own. Presently we were overtaken by the returning car, in which we soon completed our journey. A good supper, a shower, then a bed with real springs, helped us to forget the hardships of the past three days. Our sick passenger survived at least long enough to board the train the following morning, after which he passed beyond our ken.

Such was travel by road into the Alta Verapaz in 1932. After so long an interval, I could hardly feel sure that I had made the fantastic journey here recorded, did I not find all the details thereof minutely set down in my journal, written upon its conclusion. Already great trimotored airplanes were carrying passengers, cargo, and mail between Guatemala City and Cobán in forty-five minutes. Yet for thrills and adventure, travel by air is prosaic compared with the passage of the El Rancho–Cobán road, as it then was in the rainy season.

A month after arriving in El Rancho, I continued feverish. Although it was painful to drop the studies of dry-country birds that I had begun there, I resolved to go up to the capital for a medical examination. At the railroad station, while Indian girls tempted the passengers with water-coconuts, zapotes, mangoes, and the green-skinned oranges of the region, the locomotive tender filled with water, for there was hard work ahead. The end of its journey was only sixty miles distant but more than four thousand feet higher. The railroad now at last took leave of the river and started to climb into the hills. Twisting and doubling up the steep slopes, it rapidly gained altitude, crossing deep ravines on high trestles, as the engine, throbbing with pent-up energy, pulled ever higher into the mountains.

On another of my journeys up to Guatemala City, a derailed freight train blocked a bridge, forcing the passengers from the upward-bound train to transfer to the downward-bound train and *vice versa*. As each of the two trains retraced its journey of the morning with a locomotive coupled to what had been its rear end, I had the novel experience of riding at night immediately behind the engine. It was weird to see these deep gorges half-illuminated by the intermittent flashes of burning oil that accompanied each exhaust of the cylinders. Beheld in this manner, each *barranco* seemed an unfathomable abyss, penetrating far into the central darkness of our planet, much more mysterious and awe-inspiring than it ever appeared by day, when the bottom was visible.

Soon we passed Progreso, a small town with much the aspect of Gualán. Then the railroad wound for miles through low, rounded mountains, following the narrow valley of some shallow, impetuous mountain stream, now upward, now downward, as would best advance its progress into the high interior of the country. The mango trees along these watercourses, with great, dome-like crowns of dark foliage, contrasted sharply with the thin, scrubby native vegetation on the higher slopes. A timeless atmosphere pervaded these rocky, sterile hills; the railroad seemed an upstart intruder that had been wholly ignored by them, for the advent of the steel line had evidently done little to modify the character of the country or the life of its people. The scattered inhabitants dwelt in small, white-walled, red-tiled houses, some of which had large, domed baking ovens under a shed in their yards, or else in still more modest homes with thatched roofs. On the stony slopes, they cultivated their fields of maize much as their ancestors must have done while Guatemala was still ruled by Spain, or even as the Indians did before the white invaders burst upon them.

As I traveled seaward through these hills three years earlier, the train suddenly stopped far from a station, and a railroad official, with whom I had been talking, arose and stood silently in the aisle, with bowed head, for about a minute. When the train moved on again and he resumed his seat, he explained that, at that precise moment, Miner C. Keith, the builder of this railroad as of several others in Latin America, was being buried on Long Island, New York, and all the trains in Guatemala and El Salvador stopped simultaneously to honor him.

As we continued upward into the highlands on this later journey, the air acquired a bracing coolness, and passengers put on garments that they had laid aside in the warm lowlands. Tropical forms of vegetation were left below us; the mountains were thinly forested with pines and oaks. I first noticed the latter at an altitude of about twenty-five hundred feet. Beautiful vistas over distant mountains, purpling in the soft light of late afternoon, spread before us as the train passed along a steep open slope, to be blotted out as it rumbled into a deep cut. Finally, as evening fell over the earth, we speeded through a long, open cut and emerged on the rim of the broad plateau where Guatemala City lies, nearly a mile above sea level, and the low, white walls of the capital gleamed a welcome across the plain. To the west, the great volcanoes Agua and Acatenango stood silhouetted against a blaze of sunset in the evening sky.

Founded after the partial destruction by earthquake, in 1773, of the former capital, now called Antigua, and largely rebuilt after its own disastrous earthquake in 1917, Guatemala City was almost entirely modern. Although it contained good hotels, stores, and comfortable homes, there was then little of architectural or historic interest to be seen in this, the largest city of Central America. To a North American, the streets, with low walls rising directly from the sidewalks and no grass or verdure in sight, appeared plain and severe, as though the inhabitants did not appreciate green things. But when I climbed to the roof of one of the few three-story buildings and peeped down into patios smiling with shrubs and flowers, I was convinced that the inhabitants were not devoid of love for plants; but the houses, in the Spanish style, turned their best face inward rather than to the street. In the suburbs I saw broad, tree-shaded parkways lined with modern villas, with no dearth of living vegetation and blossoms. At a middle altitude in a fairly dry zone, Guatemala City enjoys about the best of tropical climates. But the surrounding plains, with few trees to cover their nakedness, looked bleak and windswept, especially in the dry season when they were dusty and brown.

Scarcely anything in the city fascinated me so much as the great outdoor relief map of Guatemala in the Parque Hipodromo. I lingered on the concrete towers that one climbs to look over the wide-spreading map in high relief, and came away with a vivid impression of the amazingly rugged surface of the country, of the altitude and great extent of the highlands, of the irregularly jumbled mountain chains. Beginning in the northwest, and cutting like a great trench through the heart of the republic, was the long, nearly straight valley of the Río Motagua, along which I had approached the capital. Standing on the southern or Pacific side of the map, I looked across the broad coastal plain, channeled by short, swift streams, and rising gently to a rampart of mountains with an average elevation of well over a mile above sea level. Along this mountain rim, the edge of the central highlands, tower the great volcanic cones, whose southern slopes sweep up from the coastal plain to summits two miles or more in the air. The loftiest of these, Tajumulco, near the western end of this line of volcanic sentinels, reaches an elevation of nearly fourteen thousand feet and is the highest peak in Central America. But others, Tacaná to the west, on the border of Mexico, then to the east Santa María, Zunil, San Pedro, Atitlán, Fuego, Acatenango, and Agua are impressively high. Behind or to the north of the file of volcanoes stretches the plateau, which lies mostly at six or seven thousand feet above the sea, but is ridged by higher mountains, and everywhere furrowed by profound *barrancos* or canyons.

Trunks thickly clothed in moss, in second-growth forest near the summit of the Sierra de Tecpán, Guatemala, at 10,000 feet.

16

Cypress Forests and Hummingbirds

To those who dwell in the temperate zones and know the tropics only through the media of books and pictures, their mention almost always calls up visions of palm-fringed shores, of lush, luxuriant forests teeming with gorgeous butterflies and birds, of hot days and balmy nights. It was of such tropical regions that I had read as a boy; it was such that I first desired to see. But after I had become somewhat familiar with the Central American low-lands, I became increasingly inquisitive about the high interior of the great isthmus, upon whose outlying mountains I had so often gazed while dwelling in the coastal lands. I knew them only from hearsay and books, and a brief visit to Guatemala City and Antigua between steamship calls at Puerto Barrios, on my roundabout journey from Almirante to Tela in June, 1929.

My first visit to the still higher country of western Guatemala came toward the end of the following year, when I spent three weeks on the Pira estate on the Sierra de Tecpán, to regain strength that I had too wantonly dissipated in

Headpiece: Cypress trees (*Cupressus lusitanic*) near the summit of the Sierra de Tecpán.

a lowland environment that stimulated my mind at the same time that it enervated my body. Again, after leaving the Motagua Valley in late July of 1932, I visited the hospitable Pira home, where during twelve days in the bracing highland air I recovered from the debility that I could not shake off in the hot country. Before returning to the United States in mid-August, I arranged to spend the following year on this fascinating mountain, tutoring young Axel while I studied its natural history. I wished to follow the changes of the seasons, here more pronounced than at lower altitudes, and to learn how they affected the vegetation, the animals, the agriculture, and the people. And, of course, I desired to discover all that I could about the habits of the birds, which at this elevation were almost all different from those that I knew in the lowlands.

I sailed from New York to Puerto Barrios in late December, 1932. Certain governmental regulations detained me for several days in Guatemala City, and I could not leave for Tecpán until early January of the following year. Except in and around the larger cities, Guatemala had then no paved roads. For sixty miles our *carro de linea*—a touring car in which each passenger paid for his seat—was trailed by the dense cloud of gray dust stirred up by its passage. The dry season was then two months old, and the desiccated volcanic soil had become light and powdery. The driver and passengers held handkerchiefs before their noses. Horsemen that we passed along the road looked suspiciously like bandits, for they had handkerchiefs tied over their faces below the eyes, and some carried revolvers on their hips. But their masks were worn for no purpose less legitimate than that of filtering some of the dust from the air they breathed, and the carrying of firearms for self-protection was still widely practiced in the country.

The highway, which on steep mountainsides was a shelf cut deeply into the white pumice that covers this region, led over a high divide, then steeply downward into the historic old town now called Antigua (the Ancient). This, of course, was not always its name; in its heyday, when it was the seat of colonial government for all Central America, it boasted the grandiloquent title of *La Muy Noble y Leal Ciudad de Santiago de los Caballeros de Guatemala* (The Very Noble and Loyal City of St. James of the Gentlemen of Guatemala). After its partial destruction by earthquake in 1773 and subsequent abandonment as the capital, it became a sleepy provincial town living on memories of a more glorious past.

The site of Antigua is superb. It nestles in a narrow valley dominated by the nearly perfect volcanic cone of Agua and the towering sister volcanoes Acatenango and smoking Fuego—at once an inspiration and a menace. On the other sides, it is enclosed by high, verdant ridges. Its architectural remains suggest its wealth and power in bygone days. The imposing palace of the colonial governors, fronted by two long tiers of massive arches, is still well preserved, but most of the other ancient structures have suffered heavily from the ravages of time and earthquakes. One who delights in antiquities might roam for days among the many ecclesiastical ruins, for hardly a block in the town is without some solid testimony to the former great influence of the Catholic Church in Guatemala—an influence which, hardly less than earthquakes, caused the civil government to seek a new site. At the time of my visit, Indians squatted among the crumbling masonry of ruined convents in wretched hovels, and cattle grazed in their grassy courtyards.

The highway leading forth from Antigua was bordered by white-plastered walls that enclosed flourishing coffee plantations, where Grevillea trees from Australia shaded the glossy-leaved coffee bushes. A roadside fountain set in one of the walls lent an antique charm to the scene. Winding up a narrow valley, above a rushing stream, the highway rose gradually to the plateau of Chimaltenango. Our cloud of dust trailed us across broad, almost treeless plains, covered with the stubble of maize and wheat, brown and barren after months of drought, frost, and wind. In the distance, the plains tilted up to meet the high, wooded ridges that dominated the horizon. Growing along the roadside and in the hedgerows, Maguey plants with compact clusters of huge, sharp-pointed, fleshy leaves imparted an exotic character to the landscape. The highland towns through which we passed—Chimaltenango, Zaragoza, Patzum—with their low, crowded houses of white adobe, their narrow, dusty, cobbled streets, were not attractive enough to invite the dust-tortured traveller to pause.

At noon we reached Tecpán, the end of my journey by public car. Much like the others through which we had passed, this town is situated on a broad, mountain-rimmed plateau at an altitude of seven thousand feet. It rests at the foot of a long and lofty ridge, which rises steeply to the north and stretches, gradually descending, for miles toward the northeast; while on the west a mighty spur sweeps like a protecting arm about the town. This apparently simple ridge is the face of an extensive mountain complex, an intricate system of sharp ridges and deep valleys covering many square miles. For a year I explored these valleys and ridges, and by no means exhausted their secrets.

In Tecpán, I was met by some of the younger Piras in their family car. After a short run across the plain, we turned left and began to climb steeply. Here and there along the roadside banks grew *Wigandia kunthii*, a large, coarsely branched shrub with broad, stiff leaves and spreading panicles of rich purple flowers. A modest, prostrate herb with finely divided foliage and a profusion of delicate pinkish blossoms, for which I could learn no name more poetic than *Loeselia glandulosa,* did its best to brighten the roadside, despite the clouds of dust that rolled over it from every passing vehicle. As the highway zigzagged higher and higher, these pretty plants were left below, but the bushy second growth near the mountaintop was adorned with great masses of bright yellow flowers of various shrubby or arborescent composites in profuse bloom. Not far below the summit of the mountain, the private road to "Santa Elena" branched off to the right. Along this we wound for nearly a mile, through stands of young Cypress trees and flowery thickets, to the two-story dwelling, which stood in a spacious, Cypress-rimmed natural amphitheater in the mountainside.

On the broad stairway leading up to the front door, I was welcomed by my host, Don Axel Pira, a cultured Swede who had lived in Guatemala for a third of a century. He led me into the spacious living room, where now in the early afternoon a cheerful log fire burned in the ample corner hearth, which projected into the room and radiated heat from two sides. This fireplace was used almost every day throughout the year. The house, situated at an altitude of ninety-seven hundred feet, was the highest permanent dwelling of a white family that I had seen. At this height, nights were invariably cold. Although the outside air would warm up in the middle of sunny days, its heat did not last long enough to penetrate the thick walls and make the inside of the house comfortable without a fire.

In the evenings, the hearth was the center of interest; we would all sit within range of its generous warmth, while we read or talked over the events of the day, and the women of the family sewed. Or we would comment upon the latest news in *El Imparcial*, the daily from the capital, and propose remedies for the social and economic ills of a tortured world as earnestly as though we could do something to alleviate them, as is the pleasant but ineffectual habit of men who discuss public affairs, whether in a crowded metropolis or on a lonely and remote mountaintop.

In a bookcase against the opposite wall, I was glad to see the volumes on birds from Brehm's *Tierleben*. My host was interested in natural history and, having in his youth passed several years in Germany, read German as easily as he read Spanish and his native Swedish.

Near the house stood the sawmill. Too near the mountaintop for enough water power to turn the machinery, it was operated by steam generated in a wood-burning engine. The logs, chiefly Cypress, were dragged from the surrounding forests attached to two-wheeled trucks drawn by four or five yokes of straining oxen. They yielded a superior lumber, light, close-grained, soft, and easily worked, but very resistant to the attacks of termites and to decay when exposed to moisture, as was convincingly attested by fallen trunks in the forest, which had remained sound while seeds had germinated upon them and grown into trees of fair size. The products of this sawmill were shipped to Guatemala City and other distant points by motor truck or oxcart, for this part of the country was wholly devoid of railroads. Indians who dwelt in huts about the sawmill did nearly all the work in the forest and the mill.

The combination of altitude and humidity at Santa Elena was unfavorable for agriculture. Beside the house was a kitchen garden, which, in the frost-less months of favorable years, yielded such hardy vegetables as carrots, spinach, and radishes; but these plants grew with exasperating slowness and led a precarious existence. In the exceptionally wet year 1933, this garden was a total failure. Neither corn nor wheat was grown on the Sierra de Tecpán much above nine thousand feet; although farther west in Guatemala, as on the southern side of the Sierra Cuchumatanes, a somewhat different climate permits the cultivation of wheat, potatoes, and other vegetables as high as ten thousand feet and even more. Here oak rather than Cypress appears to have been the dominant tree in the vanished forests, while on the Sierra de Tecpán I noticed few oak trees above nine thousand feet. On the slopes of Mt. Chimborazo, near the equator, barley, potatoes, Quinoa, Oca, Windsor beans, and Melloco are grown successfully as high as twelve thousand feet. Beneath the clearer skies of central and southern Peru, culti-vation extends as high as thirteen thousand feet.

I loved to wander through the Cypress forests, especially on the ridges farther from the sawmill, where they grew in almost pure stand, mixed with pines and a few broad-leafed trees such as the towering cornel (*Cornus disciflora*), and had not yet felt the destroying axe. In subsequent wander-ings through the Guatemalan highlands, I saw no stands of Cypress that approached these in magnificence. The great, fluted, columnar trunks, sometimes seven feet in diameter at shoulder height, stretched grandly upward nearly a hundred feet to the lowest branches. Their topmost twigs rose at least fifty feet higher. In the heaviest stands on the backs of the ridges, the undergrowth was too sparse to conceal the full majesty of the mighty boles, or to distract the eye that followed them upward to the lofty canopy. And the spirit, led by the eye, was exalted, until I was overcome by a feeling between reverence and exultation that vanquished the fatigue resulting from a long climb through the thin atmosphere.

More tolerant of altitudinal extremes than many plants, this tree that flourishes on high mountaintops grows well when planted at much lower altitudes. In Costa Rica, where no conifer of northern type is native, this Cypress is frequently planted and does well even as low as two thousand feet.

The undergrowth in the Cypress forests on the Sierra de Tecpán was largely of bamboo, which on steep slopes, where the vertical separation of the crowns permitted the influx of more light, formed dense thickets through which it was difficult to pass. Here on the cloud-bathed mountaintops, a thick garment of moss clothed the limbs of every kind of tree, and formed huge swellings, as thick as pumpkins, on the lofty boughs of the Cypresses. This envelope of mosses was fertile ground for a variety of epiphytic plants, some of which were very beautiful. The most magnificent of these was *Fuchsia splendens*, a shrub with long, cable-like roots that crept over the moss-covered branches, and woody stems several yards high, from which dangled a profusion of big, bell-shaped flowers, with a deep red calyx and four green petals. Another attractive epiphyte was a False Solomon's-seal (*Smilacina salvini*), which anyone familiar with the white-flowered herb of northern woods would recognize as such. The rootstock of this air plant was a series of bulbous swellings, more than an inch thick, embedded in the mosses that covered the trunks and branches of trees. From the newest joint of this rhizome sprang a leafy shoot a foot or more long, terminated by a nodding raceme of small pink flowers. I never found this Solomon's-seal growing in the ground.

Birds were not as numerous in these dark, humid Cypress forests on the crest of the Sierra as at lower altitudes, but some of the kinds that I met here were rarely or never found among the oak forests on lower slopes. Golden-crowned Kinglets hunted in lisping flocks among the sombre foliage of the Cypress trees, which they seldom cared to abandon. Sometimes I would see a Mountain Trogon, resplendent in metallic green, bright red, and white, or I would glimpse a pair of Emerald Toucanets, great-billed birds that seemed out of place in the same woods with kinglets, Brown Creepers, Red-shafted Flickers, and Hairy Woodpeckers.

The rarest and most fantastic of the feathered inhabitants of these heights was the Horned Guan or faisán. One morning, while walking along the narrow back of a ridge covered with heavy, broad-leafed forest, I was startled by an explosively loud, guttural outcry. As I proceeded cautiously down the ridge, the croaking calls suddenly ceased, and in their stead came a sound like the clacking of castanets. At length, as I moved into a more open space in the forest, I beheld, clear and sharp against the morning sky, the head and neck of a Horned Guan, standing on a topmost branch of a tall, gnarled, moss-shrouded tree.

I have seen among birds few appearances so bizarre as that of the slender neck and black head with its small yellow bill, which opened and closed with a loud clacking, as though the strange fowl tried to intimidate me. With dilated pupils in bright yellow eyes, he stared fixedly at me. His bare throat was scarlet. From the crown of his head rose a tall, slender, truncate spike of bare flesh, the color of ripe strawberries. At intervals the bird, as large as a hen turkey, bent forward, stretching out and lowering his neck, and emitted more of the weird, loud grunts that had drawn my attention to him. Soon he obligingly turned around, and showed me that all the upper parts of his heavy body were black, while his foreneck and lower parts were white. A broad, white bar crossed the middle of his long, black tail. I had difficulty naming the color of his legs, which were between salmon and pink.

I settled down on my heels to record, while still fresh in mind, the impression made upon me by my first encounter with this strange, rare fowl. While I wrote in my notebook, with fingers too benumbed by the early morning chill to form the letters clearly, he vanished without a sound. Doubting that so large a bird could have flown without an audible rustle of wings, I moved all around his tree, craning my neck to glimpse him once more, but in vain.

Not long after this, while I carefully pulled the thread-like rhizomes of a filmy fern from a rock shaded by heavy forest, my Indian guide crept around the corner of the outcrop and beckoned me to follow. I laid down the notebook in which I was preserving the delicate little fronds and advanced, as silently as crackling dead leaves permitted, to a small clear space a few paces away. There, perching on a horizontal limb almost directly overhead, not more than forty feet up, was a splendid Horned Guan, quietly enjoying his noontide repose. Every detail of his odd appearance was quite plain; through my binoculars I could even distinguish the fine, black shaft-streak along the center of each white feather of his foreneck and breast. The bird looked down and seemed to reciprocate the close scrutiny to which he was being subjected. I gazed at him until my neck ached, waiting to see him in action, but he seemed disinclined to demonstrate his mode of flight, so I returned to my ferns.

When I had put away as many of the filigree fronds as I desired and was ready to go, I raised my eyes for a farewell look at the bird, but now at last he had become active. Despite his size, he walked with ease along the slender terminal branches of the tree, plucking the small, green berries at their tips. His hunger satisfied, he stood majestically upon an exposed part of the bough; then he spread his dark wings and glided silently down over the treetops on the slope below.

The warm glow left within me by these intimate meetings with a rare inhabitant of the wilderness was followed by a feeling of chill and loss; for I knew that these much-persecuted guans, peculiar to Guatemala and adjacent Chiapas, were vanishing along with the heavy forests at high altitudes where alone they find a congenial abode—vanishing before any naturalist had described their nest. And if the guan had been so confident in my presence, he would doubtless be equally unconcerned about the approach of a man who intended to harm him; for it is unlikely that this denizen of the remote mountains had learned, like the crow of farmlands, to recognize a gun. Instead of studying the guan's actions with silent intentness, I felt that I ought rather to wave my arms about and shout: "Don't you know me? I am man—man the destroyer. Be wary! Never suffer yourself to be seen by one of my kind! Be alert! Flee! or you and all your race will soon join the ghostly ranks of the Dodo and the Great Auk and the Passenger Pigeon, and these mountains will know you no more."

I passed almost the whole of January at Santa Elena. Whenever the night was cloudless, dawn revealed a heavy frost. Sometimes, when I looked from my window on arising, the boards piled beside the sawmill, the roofs of the Indians' shacks, the bare ground, and the low herbage in the close-cropped pastures, were so white that it was easy to believe that snow had fallen in the night, and that I was in north latitude forty instead of fourteen. I have never seen snow in Central America, but I was told that it sometimes falls on the tops of the highest volcanoes, where it soon melts. One morning at Santa Elena, enough of the hoarfrost to freeze ice cream was collected. At times, sheets of ice an eighth of an inch thick formed along the edges of the rivulet that flowed through the sawmill. In exposed situations, the moist soil along the roads was frequently raised by bundles of long, needle-like crystals.

A frosty night was almost invariably followed by a bright, sunny morning with cold, bracing air. In spots shaded from the morning sunbeams, the frost crystals would frequently remain until noon, but they melted very quickly when the sun reached the zenith and poured down its heat upon them. Since frost never formed in the woodland, but only in spots exposed to radiation into the open sky, it did not anywhere persist much past midday. At intervals, clouds would roll in and veil the mountaintop for days, with occasional light rains, even in the midst of the dry season. Although this cloud-mist chilled one to the marrow, frost was absent as long as it obscured the sky.

As I lay abed those cold January nights, not quite successful in keeping warm even under several heavy Indian blankets, I thought with wonder of the hummingbirds incubating their minute eggs and brooding their naked nestlings out on the frosty mountainsides. One would hardly expect these tiny, graceful sprites to live, and far less to nest, among frosts and chilling winds. But their vitality is amazing, and where food is abundant they manage, by reducing their body temperature and metabolic processes in the manner of hibernating animals while they sleep, to keep alive through the longest and bitterest of the nights on high tropical mountains.

To hummingbirds as a family, altitude is no deterrent. As in other groups of birds, each species has a more or less restricted altitudinal range, but the hummingbird family is one of the few that are almost equally well represented from sea level up to, and even far above, tree-line, as beside bleak Andean snowfields. The bright flowers that provide a substantial part of their nourishment are, at certain seasons, far more conspicuously abundant among the light woods and open meadows of high mountains than amid the dense verdure of humid tropical lowlands; and consequently these little birds are much more in evidence, if not actually more numerous, in the cool uplands than in the thickets and heavy forests of the warm lowlands.

Now, in the early half of the dry season, many kinds of bright-flowered herbs and shrubs blossomed profusely, in defiance of nocturnal frosts. Especially abundant were the shrubby and herbaceous Salvias with scarlet or crimson blossoms rich in nectar. And because their food was more abundant than at any other season, the hummingbirds were breeding, also in contempt of the frosts. Although they had started earlier, in October or November, I still found a few of their nests as late as January.

The Green Violet-ears built their broad, shallow cups of green moss on horizontal branches of cypress saplings that had sprung up thickly, along with a multitude of flowering shrubs and herbs, on slopes from which the forest had been shorn, leaving a few seed trees. The larger Amethyst-throated Hummingbirds preferred to build in the woods, where the profusion of green moss that covered every branch and twig made their small, mossy cups very difficult to distinguish. While the Violet-ears attached dry, shrivelled, little leaves to the outside of their nests, the Amethyst-throat ornamented hers with scattered lichens. Both species used cobweb to bind together their structures and attach them to the supporting twigs, and both regularly laid two elongate, pure white eggs.

I gave much attention to a certain Violet-ear's nest favorably situated, and convinced myself that the mother alone attended the two nestlings, as is true of all hummingbirds that have been carefully studied, with the possible exception of the Sparkling Violet-ear of the Andes. The males were far too busy singing to care for their offspring. Each had his favorite post, usually a dead twig well above the ground on one of the isolated large Cypress trees that remained on the bushy mountainside. Here, from morn to eve, he tirelessly repeated his metallic, rather tuneless *k'chink chink k'chinky chink* that advertised his presence to the opposite sex. Indeed, during the breeding season, to proclaim himself seemed to be the single purpose of the male Violet-ear's existence; by day, he interrupted his "singing" only long enough to chase away an intruding rival, or to descend to the flower beds and sip the sweet nectar of the scarlet Salvia blossoms, and he seemed to begrudge even the time hunger forced him to devote to this.

It made no difference whether clouds hung low and bathed the mountain in their chill mist, or whether the condensed moisture descended in a steady, cold drizzle; after the approach of the dry season set off the Violet-ear's vocal performance, no temporary reversion to wet-season days could make him suspend it. His brave little spirit undaunted by cold and dampness, he continued to sing amid dense cloud-mist that made him invisible at no great distance; and from all around me came the reiterated notes of unseen hummingbirds.

In March, when continued frost and drought had greatly reduced the number of blossoms on their favorite slope, the Violet-ears began to disappear. Then for five months I saw not a single one of their kind. With the return of October and the prospect of sunny days soon to be, the red Salvias seemed to pluck up courage and put forth a few tentative blossoms, which increased slowly in number as the month grew older. By November, the Violet-ears had returned and were singing on their mountainside. One male, whose song was punctuated by a single, clear, ringing note, which served to distinguish him from all his neighbors, performed on the same perches where I had watched him ten months earlier; his peculiar voice left little doubt of his identity. Where he had been in the interval, I can only surmise. These fascinating problems of bird migration, vertical and horizontal, are difficult to solve without resident observers scattered widely over a region. In Guatemala in the early 1930s, I failed to meet anyone who shared my deep interest in the habits of free birds.

All over the Guatemalan highlands, from five to at least eleven thousand feet above sea level, the monotonous little chant of the Green Violet-ear is, to one with senses well attuned to nature, one of the most characteristic of bird notes. Later, in the highlands of Costa Rica and Ecuador, I heard similar utterances from closely related kinds of violet-ears—indeed, the first hummingbird that I found in the Andes was a violet-ear, who drew my attention by notes not too different from those of his Guatemalan cousins to be recognized immediately. This unmelodious "song," no louder nor more likely to alert the ear then the notes of many insects, is one of the sounds most typical of the higher and cooler parts of the American continents, from southern Mexico to Bolivia.

The male Amethyst-throated Hummingbirds sang on low perches in the mossy woodland. Their chant was even more monotonous than that of the violet-ears, consisting of a single, plaintive, squeaky note repeated from seventy-five to eighty-five times a minute. One indefatigable songster uttered six hundred and twenty-five of his melancholy squeaks without a pause.

In a flowery thicket beside the highway, four tiny male Wine-throated Hummingbirds had established a singing assembly. They had chosen stations within hearing of each other's small voices, but no others of their kind could be found within a long distance. Although I have applied the word "song" to the vocal efforts of other hummingbirds for want of a more accurate term, and because their role in the life of the species is the same as that of the more tuneful utterances of other birds, when I refer to the song of the Wine-throated Hummingbird, I do so without reservation. The lay of this diminutive, magenta-throated gem is so impassioned, so intense, so sweetly varied with rising and falling inflections, that if it were a little more forceful without change of tone or phrasing, the Wine-throated Hummingbird would become renowned as a songster. I listened enchanted to his long-continued, ecstatic outbursts of melody.

At the season when the hummingbirds were nesting on the mountaintop, only one other kind of bird, as far as I could learn, laid its eggs. This was the Cinnamon-bellied Flower-piercer, a modestly attired highland relative of the brilliant honeycreepers of lower altitudes. Its short bill is most peculiar, with the upper mandible strongly hooked at the end, the lower uptilted and sharply pointed. While the upper mandible hooks over the base of a corolla-tube and holds it firm, the lower mandible pierces the tissue. Through the perforation so made, the bird draws nectar, doubtless with its tongue, some-times from small florets, sometimes from great trumpet blossoms almost as big as itself. This method of procuring nectar is, of course, unmitigated theft; the flower-piercer does not, like most hummingbirds and many insects that take their sweet drink in the "legitimate" manner, earn its nourishment by pollinating the flowers. Were it not for the activities of the pollinators, the flower-piercer would, before long, find few flowering plants to provide nec-tar for it. Not only does the flower-piercer resemble the hummingbirds in its dependence upon nectar, like them it supplements this sugary diet with proteins derived from tiny insects caught in the air. It was instructive to find these two nectar-drinking birds nesting simultaneously, at a period sepa-rated by months from the breeding season of all the insect-eaters, the fruit-eaters, and the seed-eaters. In the high Andes, where flower-piercers are represented by numerous species instead of the single species found in Guatemala, they also breed at the same time as hummingbirds.

From all over the world, we have a growing volume of evidence that the breeding seasons of birds of all kinds are adjusted to hatch the young at the time when food for them is most abundant. But to accomplish this, the birds must often begin preparations long before this optimum period, often while food is still relatively scarce. In most instances, we do not know just what stimulates the birds to start the reproductive process just far enough in advance of the season of maximum food to ensure that their progeny will have the benefit of it. Might they not sometimes breed in response to an internal rhythm?

Sometimes, in late afternoon, we climbed to the topmost point of the mountain, only a few hundred feet above the house, to watch the sun set over the Pacific. The way led through a cloud forest of low, broad-leafed trees with tortuous branches, all so overgrown and draped with mosses that we could hardly find a spot where the bark was visible. Slender boughs appeared several times their true thickness; trunks lost their firm outlines and seemed soft, unsubstantial columns of green. The ground, too, was overlaid by a rich plush carpet of green moss, which silenced our steps, so that we climbed as though through an unreal forest in fairyland. Amid the moss grew ferns of many kinds: epiphytic polypodies and hart's-tongues draped trunks and limbs; delicate filmy ferns covered the rock faces in the most shady and protected nooks; many other species of ferns rooted among the moss on the ground; but tree ferns were absent at this altitude. Orchids and bromeliads added to the bulk and confusion of the vegetation supported on trunks and branches.

On the nearly flat summit of the Sierra de Tecpán, ten thousand feet above sea level, the character of the vegetation changed. Among the low shrubs that grew densely here were a myrtle (*Myrtus friedrichsthalii*) with small, crowded, dark leaves and little, white, rose-tinged flowers; a shrubby wintergreen (*Gaultheria odorata*) with white flowers borne on a red stem, amid bright red bracts; a white-flowered, bushy arbutus (*Arbutus donnell-smithii*); a blueberry (*Vaccinium confertum*); and other plants of northern affinities.

The rocky edge of this miniature plateau, which fell off abruptly toward Lake Atitlán, was known as "Buena Vista," as it afforded a wonderful panorama of over half of western Guatemala. I like best to remember the prospect from this promontory as I saw it on a blustery evening in late November. As we emerged from the trees onto the stony, wind-swept summit, the mists that blew over it filled the space below, and the sun, hanging low above the lake, was veiled by a pink haze. A moment later, another gust of wind dispersed the cloud, revealing the blue water of the lake far below, with the sun's orb just sinking in an orange glow behind the mountains on the farther shore.

The panorama spread before us was on the grandest scale. Toward the declining sun lay Lake Atitlán, its deep ultramarine water stretching for miles between high, precipitous shores, every irregularity of which, every salient angle on their rugged, seamed expanse, was sharply outlined in the clear evening light. To its left towered the giant cone of Volcán Atitlán, its shapely mass cut darkly into the pink and orange sky. Beside this great volcano stood Volcán San Lucas, a lesser figure with a blunted apex. On the farther shore of the lake, nearer the sun, loomed Volcán San Pedro, second only to Atitlán in majesty. Beyond and to the north, stretching toward Mexico, the high wall of the Sierra Madre met the glow of sunset in a jagged line, piercing it here and there with sharper points. Away to the southeast, in sublime isolation, stood the shapely cone of the Volcán de Agua, and the towering mass of Acatenango blending into that of Fuego. A long, level bank of white cloud, stretching afar from out of the dim east, severed the crest of each great volcano from its foot. Between the two groups of volcanoes, to the west and to the south, the coastal plain spread low and dark, with a faint line of shore beyond, and the long, level horizon of the Pacific, appearing lofty and immensely distant, curving around from Atitlán to Fuego.

Between us and the volcanoes lay a broad and comparatively level plateau, spread out at our feet in a checkerboard where dark patches represented woodland, lighter areas cultivated fields and pastures, all precisely outlined in the still evening air. Scattered here and there lay a number of small towns, whose most conspicuous features were the white mass of the church dominating the cluster of squat dwellings, and the glistening white walls and monuments of the *campo santo* or cemetery. A long, narrow line of highway stretched out toward the base of Agua. Winding through the checkered plain was a deep *barranco*, an immense gash in the earth whose scarped walls here showed green with the vegetation that clung to them, there brown in vertical streaks made by recent landslides.

Thus, for some minutes, the whole panorama would spread in clearest outline before us; then the wind, driving with mournful sighs through the moss-draped boughs of the few gaunt pines and cypresses that stood on the summit, would bear a burden of white mist that spread below us, softening all the outlines, then erasing them wholly, until we stood above a world of cloud, tinged on its western limit with the rose of sunset. The mist that filled, or at least obscured, the valley heightened our sense of its profundity and of our own loftiness—only the moss- and lichen-bedecked dead boughs of a battered cypress tree stood above us. Then the cloud, descending into the warmer air of lower regions, would dissolve again, and gradually the whole vast panorama would take form out of the mist.

Each time it was revealed, the water of Lake Atitlán was a deeper ultramarine, set among mountains of more sombre purple. The warm glow of sunset intensified and spread out until it suffused the whole visible horizon of the Pacific, and its crimson and gold were interrupted only by black masses, so sharply outlined that every tree that clove the skyline was plainly visible, although many leagues away. Above us, the blue vault of the cloudless sky darkened and deepened, until the white crescent of the declining moon grew luminous, and here and there a star shone forth, while far below in Patzum lights began to twinkle. Then the chilling wind, as it drove us from the exposed summit, spread a soft blanket of cloud over all the scene, extinguishing all the earthly lights below us and leaving only the remote luminaries above, by the light of which we led our horses back to the highway, and rode homeward through forests fragrant with the scent of pine.

Four-rayed stars of *Houstonia serpyllacea*, a creeping relative of the Bluets of northern meadows, form a galaxy on close-cropped mountain pastures in the rainy season.

The deep magenta flowers of *Monchaetum tenellum*, a sprawling perennial herb or half-shrub of the melastome family, adorn shady mountain slopes in the Guatemalan highlands in November and December.

17

A Vernal Year

On the Sierra de Tecpán below Santa Elena, the Piras owned another estate, known as "Chichavac," where the vegetation was quite different and the climate more favorable for agriculture. Here, amid pines and oaks and other broad-leafed trees, lived several kinds of birds, rare or absent among the Cypresses of the mountaintop, that I was eager to study. Accordingly, after I had been a month at Santa Elena, I requested permission to reside for a while in the house at Chichavac, about thirteen hundred feet lower. As it turned out, I became so absorbed in following the activities of my bird neighbors at Chichavac that I remained there for nearly eleven months, during which I made occasional longer or shorter visits to Santa Elena, about seven miles distant by the road that wound steeply up the mountain.

After a day of delays, with the young Swede in charge of Chichavac I set off after dark for the horseback ride down the mountain, well muffled against the cold night air. As we trotted down the highway, we caught frequent

Headpiece: Slopes wooded with oaks, alders and pines, pastures, and cultivated fields at "Chichavac," in 1933, before the Inter-American Highway cut through the farm.

glimpses through the trees of the great volcanoes Agua and Acatenango, looming vaguely in the moonlight, far across the wide, misty plateau. The electric street lights of Tecpán shone out far below in a pattern of squares, making the town appear more attractive by night than I found it by day. It was the beginning of February, and the Whip-poor-wills were loudly calling their names, sounding much like their cousins in the North. The farther down the mountain we descended, the more numerous the voices became, until, when we entered the oak woods of Chichavac, they seemed to be quite as abundant as in a Maryland woodland in June. As we rode through an open grove, we heard something scramble up a small pine tree on the roadside bank. My flashlight picked out a small opossum, whose eyes reflected the beam with concentrated intensity.

At Chichavac I occupied an attractive bedroom, panelled shoulder-high with unstained Cypress and ceiled with the same wood, in a pretty cottage of four rooms, with a long, glass-enclosed porch across the front. Ivy clung to the walls of pale Venetian-red stucco, and the roof of native tiles was a lighter shade of the same color. Until Axel came up from the coast, where he worked on a banana plantation, I was the sole occupant of the bungalow; but the overseer, who slept in a room attached to the office, had his meals with me in the dining room, where an Indian boy of ten or eleven years waited on us. An old Indian woman cooked and patted out tortillas in a building apart.

In front of the house was a flower garden, fragrant with roses and English violets, and bright with daisies, periwinkles, and the great, white blossoms of the Floripondio, ten inches long, that hung like so many bells among the dark green foliage of two shapely shrubs, which the Costa Ricans call reina de la noche—queen of the night—because it diffuses its heavy fragrance after nightfall. Beyond the flower beds was the kitchen garden, which in the proper season yielded most of the familiar temperate-zone vegetables. I remember most vividly the delicious peas and the large, succulent artichokes.

In the early morning, while frost whitened the lawn and the rich fragrance of the Floripondio still hung in the crisp air, the garden was the favorite resort of many birds. Then the Spotted Towhees left the concealment of hedgerows to mount high in the trees and sing *cheer tseee tseee tseee*, and the Rufous-collared Sparrows announced, in a voice that was sweetly sad, *We're here too too too*, or *O we're here too too*. Their clear, whistled refrain bore just enough resemblance to the song of the White-throated Sparrow to remind me that its author, who had a white throat and broad black stripes on his gray head, was closely related to the familiar northern bird. Then, too, the mellow whistles of the Yellow-backed Orioles issued from the spreading hibiscus tree, where the graceful gold and black birds sought breakfast among the foliage and in the large, rose-colored flowers. As the house became active, they passed across the valley, in which the white mists of the night were dissolving under the rays of the rising sun, flying heavily with labored wingbeats and pumping tails. Among the oak trees on the opposite slope they continued their singing through the early morning hours, at intervals snatched from the business of hunting caterpillars, but it never sounded more clear and sweet than in the chill dawn.

Although situated in the middle third of the Sierra de Tecpán, roughly between eight and nine thousand feet, Chichavac's two thousand acres were not on a uniform, monotonous slope. On the contrary, the terrain was pleasantly diversified, with gentle, flat-topped ridges and open valleys, but also long, precipitous slopes and deep, narrow ravines, into which cows, reaching too near the edge for some tempting tuft of herbage, more than once fell to their death with the small landslide that they caused. Near the house was a broad, almost level field of many acres, on which maize and wheat were grown, except in the center, which was a low and sedgy meadow, where buttercups, blue-eyed grass, and the small orchids called lady's tresses bloomed in July and August. The rivulet that arose in this marshy ground flowed quietly enough beneath the bridge on the road from Tecpán that skirted the field, then plunged suddenly into a deep ravine that it had cut into the soft earth. Of such varied and, at times, abruptly contrasting elements was the farm composed.

The pleasant alternation of wooded ridge and grassy vale reminded me more of the rolling piedmont region of Maryland, where I had passed my boyhood, than any other part of the tropics, highland or lowland, that I know. Almost always, in other districts, clearings and plantations extend on and on or up and up, with never a patch of woodland worthy of the name, then merge into the tangled growth of abandoned provision plots, through which you must hack and fight your way to arrive at long last in the primeval forest. But the owner of Chichavac had been raised in Sweden, where good forestry is a tradition. Moreover, he needed a reliable supply of logs to burn in the big lime kiln that stood close beside the house. Accordingly, he used his woods without destroying them, and they continued to provide a habitat for the forest birds, at the same time that they yielded a steady supply of firewood. Conditions at Chichavac were, therefore, exceptionally favorable for studying simultaneously the birds of the forest and those of the open fields and bushy clearings. To counteract these advantages, the Indians who worked the farm and the quarry were mercilessly destructive of birds and their nests. Only in the more remote and isolated parts of the estate could I watch birds' nests without fear of betraying them to their enemies.

It was not only the alternation of woodland and meadow that recalled to mind the haunts of my boyhood. The very trees that composed the woods, the herbs that bloomed in the meadows, the birds of forest and field, often reminded me strongly of a more northern land—far more strongly than I have ever been reminded in the highlands of Costa Rica or in the Andes. Among the Cypresses of the mountaintop, especially in stands of sturdy young trees sixty or seventy feet high, I could imagine that I was in a spruce forest of northern New England. But among the oak woods on the lower slopes of the Sierra, the vegetation was more like that of Maryland or Virginia. Possibly, with the Arbutus Trees and many species of birds found also on the West Coast, I should have been even more strongly reminded of parts of California, had I been familiar with that side of the United States. The descent of the mountain brought the same changes in vegetation as a journey southward in the Northern Hemisphere. But despite a general resemblance of the vegetation to that of North Temperate regions, I was constantly meeting plants that recalled to me forcefully that I was between the tropics; and when my walks took me to some point of vantage whence I could gaze across the plateau to the ever present volcanoes rising in the distance, I knew that I could be nowhere but in Guatemala.

Of all the heritages dear from long familiarity that the northerner abandons when he migrates to the tropics, not the least precious is the procession of the changing seasons: winter's gleaming mantle of snow; spring's delicate tints and songs and blossoms; summer's full verdure; autumn's ripening fruits and leaves of many hues. Of the four, spring is to many of us the most eagerly anticipated and the most deeply enjoyed. Yet fully to appreciate the vernal awakening, one must have lived through winter's barren months. One year, when I went north from the tropics just as spring was breaking, instead of the ineffably joyous and wondrous season that I remembered, I found it a disappointing anticlimax. Even after the vernal season had completed its labor of reconstruction, the trees were not so noble and umbrageous, nor the herbs so lush and vigorous, nor the birds and butterflies so flashing and colorful, as those that I had recently left.

The lowland tropics know only two seasons, the wet and the dry. In most of Central America, the woodlands remain green throughout the twelve months, however brown the open fields may become. Even in the most arid districts that I know, the end of the dry season never finds the landscape so barren of verdure and flowers as midwinter in Maryland or Pennsylvania; hence nature's reawakening, when the rains return, lacks the miraculous intensity of a northern spring. Everywhere in Central America, the birds sing and nest most freely after the vernal equinox, and many of the most colorful trees bloom about this time. Of all living things, the birds as a whole are most obviously responsive to the northward-swinging sun.

Here in the highlands of western Guatemala, seven to ten thousand feet above the sea, I found the seasons much more strongly marked than at lower altitudes. Residents recognized only two: the *verano* (summer) or dry season, which usually lasts from mid-October to mid-May; and the *invierno* (winter) or wet season, which occupies the remainder of the year. When I arose on any clear morning from November to April and looked over fields white with frost, I was inclined to ridicule the notion that this was "summer." But the clear, mild days which often followed a frosty morning thawed my reluctance to concede that the weather did in many respects resemble summer. The wet season coincided roughly with summer in the Northern Hemisphere and frost was then absent; but in this part of the year I passed so much time shivering amid the cold, gray cloud-mists that enveloped these mountains for days together, that I was ready to admit the propriety of calling the season "winter." Yet when the sun at last shone forth upon a landscape so fresh and green, with flowers in the meadows and tassels on the maize, I wished to recant my admission—even an English winter is never so green.

Finally, I concluded that the Sierra de Tecpán had, rightly speaking, neither a summer nor a winter. A land without summer, you will exclaim, must indeed be dreary! But nature often gives compensation for what she denies to us; paradoxical as the statement may seem, this high land without a summer was a land of several springs. Why, then, not call it a "land of eternal spring," as the Tourist Bureau in the Capital advertised that Guatemala is? My reason for not taking this course is that we recognize spring, aside from the position of the sun among the signs of the zodiac, by certain terrestrial signs, such as the burgeoning of trees, the renaissance of song among the birds, the greening of the meadows, the opening of blossoms. In the Guatemalan highlands these phenomena do not coincide, as in the North, to make one universal, all-embracing spring, but each has its own particular spring. Accordingly, I believe it most proper to recognize various partial springs: a spring of the trees; a spring of the birds; a spring of the meadow flowers; then, at the year's end, another vernal period shared by the woodland flowers and the hummingbirds. Together, these successive springs occupy practically the whole year.

It must be admitted that these several springs in the course of a single year had very different sorts of weather. But so, in many parts of the temperate zones, does the annually recurrent spring vary considerably from year to year. April may be cold or warm, wet or dry. In any event, on most days during the year that I passed on the Sierra, I found it easier to believe that the season was spring than that it was summer or winter or even autumn, as I knew these seasons in the North.

When I arrived at Chichavac at the beginning of February, the trees were in their vernal phase. The dry season was already somewhat more than three months old; but its early half was not so severely arid as its later half, and occasional showers had fallen. More important, I believe, was the fact that the deep-rooted trees and shrubs drew their moisture from lower layers of the soil, which had been thoroughly saturated by long months of rain, and they were not so dependent on present showers as the more shallow-rooted herbs. Heavy frosts whitened the fields after every clear night, just as they had done at Santa Elena, but they affected only the lower vegetation in the open; at no time of year did they seem to have much influence on the activities of the trees. Despite drought and frost, the oaks (*Quercus donnell-smithii*) with narrow, glossy leaves were renewing their foliage.

I walked through forests of oak, pine, and Alder with mingled impressions of spring and autumn; but this was not surprising in the tropics, where the seasons overlap, and many trees shed their old foliage at the same time as, or only a little while before, they unfold their new leaves. Although some of the glossy-leafed oaks belonging to the black or red oak group had not yet dropped all their old foliage, the majority were covered with fresh young leaves, "as large as a squirrel's ear." On some of these burgeoning oaks, the new foliage was a bright, tender green; on others, yellowish green; on others, tinged with orange; on yet others, bronze. This kind of oak was exceedingly abundant, forming nearly half of the forest on the hillside in front of the house, where it grew mixed with pines, Alders in fresh new foliage, and a broad-leafed species of the white oak group (*Quercus pilicaulis*) still clad in old, hard, dark green leaves. The resulting patchwork of colors was very suggestive of autumn in the North, notwithstanding that the autumnal tints belonged to burgeoning and not defoliating trees. The impression of autumn was strengthened by falling leaves—large, thick leaves of the white oaks that struck noisily against branches in their descent, belated Alder leaves that drifted more softly down. The ground was covered with fallen foliage, dry, crisp, and brown, that crunched noisily underfoot, making it difficult to approach the birds without frightening them away.

Where alders predominated in the woods, the feeling of spring was stronger, less mixed with suggestions of the season of death and decay. Some of these Alders were tall, slender trees a hundred feet high; others, standing in the open, were lower, with full, spreading crowns. Now they were flowering freely, with long, slender catkins dangling from nearly naked boughs—an unmistakable sign of the vernal awakening. Many others of these trees were dressed in the fresh, tender garb of spring; when brightly clad wood warblers—the Black-capped, the Crescent-chested, and Townsend's—foraged among their opening buds, the impression that I was in a northern woodland, and the season late April or May, was difficult to brush aside.

Between the tropics, however, seasonal phenomena are rarely quite so clean-cut and unequivocal as in the temperate zones. Next to that Alder tree which was the very breath of spring stood another that was the symbol of autumn, thinly covered with dark green, dying leaves. As in many tropical trees, the falling leaves of neither Alder nor oaks were brightly colored; they merely darkened from green to brown as they died. Yet catkins dangled amid the moribund leaves of these defoliating trees! Nearby might be another Alder whose lower limbs were leafless, while the upper branches were still well clad in last year's foliage; still others were practically leafless everywhere, their catkins most conspicuous.

In spite of the paucity of flowering herbs beneath the woods, color was not lacking there, for the more deeply rooted shrubs and woody vines continued to bloom through February and much of March. Chief among these were the composites, plants that in the North flower chiefly in late summer and autumn. Thus, in the prevailing conflict among the signs of the seasons, they threw their weight on the side of fall. In this dry period, at least seven species of *Senecio*, coarsely branched shrubs with large and often downy leaves, displayed heavy corymbs and panicles beneath the trees, to represent, at least in color, the goldenrods of the northern autumn. One of the most attractive was *Senecio barba-johannis*, a medium-sized shrub, with broadly oval leaves covered on the lower side with woolly white hairs, and broad, flat clusters of bright yellow flower heads.

The daisy family was also represented at this season by several shrubby or arborescent species of *Eupatorium. E. ligustrinum* was often a small tree twenty feet high, with compact corymbs of white flower heads. A smaller shrub of the same genus, which proved to be a new species and was named for me, displayed white flowers that were wonderfully fragrant. But the most surprising of the dozen shrubby members of the genus that I collected on the Sierra de Tecpán was *Eupatorium araliaefolium*, an epiphyte that perched well above the ground on a tree, which it clasped with a great number of roots that fused where they came in contact with each other, like those of a strangling fig, thus forming a complex network around the supporting trunk. A single root crept along the trunk to the ground and served to draw up the water and salts that the shrub needed, for it demanded of its host only support. The principal branches of this unusual air plant arose directly from the complex of clasping roots; it had no central stem. Some of these stout branches were twenty-five feet long and six inches thick at the base. They bore large, glossy, slightly fleshy, ovate leaves, and, in early March, clusters of white flower heads which told me that I was wrong in taking this plant to be a kind of strangling fig, as at the first glance it appeared to be. Although in the temperate zones the huge composite family is represented chiefly by herbs and low shrubs, in the tropics it produces a greater variety of vegetative types, including tall trees and lianas that spread their foliage over the top of the forest.

Those were glorious days, with the balminess of spring in the air, the freshness of burgeoning leaves on the trees. Early in February, I went by horseback along the road which led down the valley of the brook that flowed in front of the house. As it hurried along between bushy pastures and open woods, the rivulet met others of its kind, until the swelling current rushed down its rocky channel with a loud babble that was almost constantly in the ears of the traveller along this beautiful, winding road.

I wondered if aught of the charm of this scenery filtered into the minds of the poor Indians who passed me, bent under huge loads of pottery or fabrics or the produce of their fields, which they carried on their backs, held there by a rope whose ends were attached to a broad leather band that fitted across their foreheads, which sustained the major part of the burden. They and other pedestrian and equestrian travellers were the only traffic on this road, which led to Paquip, for it was too narrow and rough for wheeled vehicles— except an occasional oxcart, drawn by yokes strapped to the horns of the slow-footed animals. It was disquieting to see heads put to no higher use than the bearing of burdens.

I rode beneath burgeoning oaks and tasselled Alders, with that sound to Northern ears most eloquent of springtime, the loud, mellow *wic wic wic wic* of the Red-shafted Flicker, ringing in the air. Although its associations were with the lowland tropics where spring is a rather indefinite event, the mellow *cow cow cow* of the Mountain Trogon was somehow symbolic of the same glad season. The low, soft *chee chee chee cheet* of the little Tufted Flycatchers, who perched in pairs in the trees along the roadside, would seem to melt the ice of winter in a climate far more rigorous than this. On the banks of the stream grew Capulín trees that closely resembled the sweet wild cherry of the United States, some already in nearly full-grown foliage of the freshest of greens, others with infant bronze-tinged leaves.

At length the road, which had been climbing obliquely up the side of the valley and leaving the growing stream farther and farther below, emerged from light woods at the brow of a long hill. From this point, it inclined steeply downward to join the stream again, passing with many windings between high, clayey banks. To my right was a newly cleared field, a scarcely penetrable tangle, of prostrate pine trees waiting to be burned. The wide valley that opened before me was typical Indian country, bereft of all its forest, save a few sad remnants, to the very summits of the long enclosing ridges. Some of the largest of the remaining pine trees had huge gaps in the base of their trunks, where men had hacked out splinters of *ocote*, the resinous pinewood widely used to light benighted travelers on their way and illuminate the smoky huts.

As though to make amends for its barrenness, the valley into which I rode was filled with the soft warbling of many Common Bluebirds. The sooty thatched roof of a hovel in the foreground was actually attractive amid the delicate pink of the blossoming peach trees that embowered it. Even the remnants of the stalks of last year's maize on the steep, sere hillsides helped to strengthen the aspect of early spring in the North, although it was strange to see the short, blasted stalks rising, in compact clusters of three or four, from the center of a mound at least a foot high. The earth had been heaped up around the still-growing maize plants to prevent the sweeping mountain winds from blowing them over while heavily laden with ripening ears.

At the foot of the descent, the road crossed a tributary stream by a narrow wooden bridge without a parapet, then struck obliquely across the bare ridge on the eastern side of the valley. I followed it only until I came to a tiny rill, choked with grass and watercress, which made a broad swath of emerald at the bottom of a narrow vale, between slopes of naked, dusty soil and brown cornstalks. Here I tied my horse by the roadside and started afoot up the green avenue, treading the soggy turf in which the runnel lost itself, taking care to avoid the swampier spots. My passage disturbed the afternoon repose of a pair of White-winged Doves, who fled with loudly flapping wings from the low tree where they had been resting. Later, they sent their peculiar, sonorous cooing echoing through the valley, bringing memories of their kind among cacti and thorny scrub in other valleys many thousands of feet lower. I tried to paraphrase their distinctive, long-drawn cooing, the nearest to a song that I have heard from any pigeon, but this was the best that I could do: *cuu-cu-c'-c'-cu-cuu-c'-cu-cuuu*.

Beyond the doves, massed in the top of a nearly leafless Alder, rested a flock of about forty Cedar Waxwings, wanderers from the North. A Rough-winged Swallow and several white-bellied Black-capped Swallows wheeled above the barren field. Suddenly, a large part of the flock of waxwings took flight with a loud rush of wings, circled around, and settled in a treetop farther down the vale. Here they were joined by three Gray Silky-flycatchers, who, as is their almost invariable custom, chose the topmost twigs for their perches, and drowned the drowsy lispings of the larger host with their spirited admonitions to *wake-up, wake-up*. It seemed that these slender gray birds themselves approved the classification that places them next to the waxwings, or sometimes in the same family, for they frequently joined these wanderers while they perched.

As the sun swung back to our side of the equator and the drought increased, I found the weather trying. It was impossible to dress appropriately for a day during which the temperature might rise more than forty degrees Fahrenheit. When I went out at dawn, while fields were white with frost and the air held a wintry chill, I needed a woollen coat. A coat and sweater together were not too much if I planned to sit still for hours watching a nest. But by the middle of the morning, I could not with comfort wear a coat in the sun, and I began to peel off my heavier garments. I always carried a knapsack to stow them when not in use. Yet even at midday, the protection of a coat was welcome if I sat in the shade. On April afternoons, the sun beat down through the thin, dry air with such stinging force that lips dried and cracked. The short, brown grass in the pastures became so slippery, after the frost or dew had dried from it, that it would have been as easy to climb steep slopes on roller skates as on leather soles. Rubber soles gave firm footing.

I noticed the season's last frost at Chichavac on the morning of April 2. Then followed six weeks with neither frost nor rain—our brief summer, if we had one. Now that the sun stood nearly overhead at noon, each of the cardinal signs pointed to a different season of the year. The birds, singing in fullest chorus and everywhere starting to nest, proclaimed that this was spring. The maize, pushing its first green blades above the dusty field where it had been sown in March, seemed to have performed a miracle in order to confirm them. I marveled that it had found enough moisture to germinate in that powdery soil; but it had been planted deep, and the inner layers of a mountain that is soaked by almost daily rains for nearly half a year dry out slowly. The forests, clad in uniform full green, indicated that it was now high summer; although a few belated broad-leafed oaks, still nearly naked, betokened spring. The shrubby composites—Senecios and Eupatoriums in the woods, the small-leafed Raijón in the bushy pastures—shedding their plumed seeds, which floated in vast numbers through the dry air, proclaimed autumn with considerable authority. They were supported by the dead leaves, still so dry that they crackled loudly underfoot. The herbaceous vegetation, nearly dormant in the almost flowerless woods and sere, brown fields, pointed just as convincingly to winter. Flowers of all kinds were at their lowest ebb; at no time during the year did I notice fewer. But the birds, singing and building their nests, were right, for the sun had just crossed the equator in its northward swing toward the Tropic of Cancer, bringing spring to the Northern Hemisphere.

The nesting season was at its height during the six weeks of favorable weather between the last frost, at the beginning of April, and the onset of heavy rains, in mid-May. Few species at Chichavac laid eggs before the frosts stopped. Thus the great majority of the birds, with the notable exceptions of the nectar-drinkers that nested early in the dry season and the ground-foragers that waited for the rains, bred while the drought was at its height. But trees and shrubs that flowered early in the year were now ripening their fruits, and the fresh foliage of many trees supported abundant insect life. Accordingly, despite the severe drought, both insectivorous and frugivorous birds found enough food to raise their broods while the weather was dry and days were warm.

I was not sorry that few plants flowered during the final two months of the dry season, because I wished to make my botanical collection as complete as possible, yet I also desired to learn all that I could about the birds. At this period they claimed my attention so fully that I hardly had time for anything else. Before dawn, I was on my way to the woods, with my breakfast of tortillas, boiled eggs, and oranges, and a canteen full of water in my knapsack, and frequently my lunch as well, as I often passed the whole day afield. In the evening, I usually watched some bird retire for the night, as I was eager to learn how the feathered inhabitants of these heights protected themselves from the nocturnal chill. After supper, I worked at my notes until I could write no more, then went to bed with the alarm clock set for half past four or five o'clock in the morning. One advantage of studying birds in the tropics is that one need not arise so inordinately early to watch them start their day's activities, or stay out so late to see them retire, as at higher latitudes in summer. On the Sierra de Tecpán, even in June, the earliest birds did not become active until after five o'clock, Central Standard Time. In this healthful mountain climate, I could go very hard without paying for it later, as in the hot and humid lowlands.

Among the most interesting birds that I studied at this time were tiny gray relatives of the titmice and chickadees, the Black-eared Bush-tits that roam through the more open woods and bushy clearings of the mountains in flocks of from a dozen to a score. Their nests are among the most charming examples of avian architecture: swinging pouches about six inches long, tastefully covered with finely branched gray lichens and thickly lined with soft down, above which is a bed of downy feathers on which the eggs repose. The exquisite structure is entered through a round, sideward-facing orifice just below the hooded top. The black-faced male and the gray-cheeked female cooperate in the exacting task of completing one of these nests, and both take turns incubating the four white, minute eggs for fifteen or sixteen days. After these hatch, other black-faced birds come to help feed the nestlings; the nests that I watched had from one to three of these voluntary assistants. As a reward for their services, the helpers are permitted to sleep in the downy pouches with the parents and nestlings, a privilege that they doubtless appreciate on cold mountain nights.

While I studied these bush-tits, I quite naturally assumed that these black-faced assistants were males who, because of the great numerical preponderance of their sex, could not find mates. Now it appears that some may have been yearling females, because in some races of bush-tits young birds of both sexes resemble the adult male rather than the adult female—a situation found in exceedingly few of the world's many species of birds in which the sexes differ in the adult plumage. All twelve of the fledglings raised in the three nests that I studied had black faces. Contrary to my expectation, neither adults nor young continued to sleep in the nest after the latter emerged. A cozy lodging in dry weather, the downy pouches absorbed so much water that they would have made unhealthful dormitories in the rainy season, which began about the time the fledglings took wing. Thenceforth, all the bush-tits roosted amid the foliage.

Another bird with unusual nesting habits was the Banded-backed Wren, a big, strikingly marked relative of the Cactus Wren that appears to be equally at home in Central America in the wettest lowland forests and on cold mountaintops ten thousand feet high. Such unusual climatic tolerance may be due in part to the protection they receive from their bulky covered nests, each with a side entrance. A family of these wrens that I followed through the year always slept in a dormitory nest, moving at intervals from one to another, probably in the interest of sanitation. As many as eleven grown wrens might occupy the same dormitory. It was amusing, to watch them emerge on cold, wet mornings, when their love of dry warmth conflicted with hunger for breakfast and caused vacillating behavior. The nestlings, who may be raised in an old dormitory nest, are fed not only by their parents but also by unmated helpers—certainly one or two, and probably as many as five at some nests. After the young emerge, the adults patiently help them accomplish the sometimes difficult feat of reaching the doorway of the high nest in which young and old sleep together.

To tell of the Blue-throated Green Motmots that throughout the year slept in pairs in long burrows in which they also raised their young, of the Mountain Trogons that carved nests in rotting stubs, of the Rose-throated Becards that attached great bulky nests to high, slender twigs, and all the other fascinating mountain birds that I studied at this season, would make this chapter far too long. Their story has been told in other books and papers.

In the second week of May, clouds began to gather and to obscure the sky for a larger part of each succeeding day. By the middle of the month, the rainy season was upon us in full force, catching many birds in the midst of their nesting. They continued bravely to attend their eggs and young under adverse conditions, but after raising their brood, scarcely any undertook a second. A few birds, strangely enough, preferred to hatch their eggs and raise their young during the early part of the wet season. They were chiefly ground-foragers that rummaged among fallen leaves and litter for insects and other small invertebrates that seemed to become more abundant after the soil was soaked. Among them were the Ruddy-capped Nightingale-Thrushes, who built their mossy nests in deep ravines reeking with moisture, and hatched their two eggs amid continual rains. Their lovely songs, sweetly sad, helped to lighten the melancholy of the dreary days, fog-dimmed and rain-sodden, that now followed each other in endless procession.

Now spring—a cold, wet, frequently dismal spring—had come to the meadow grasses, and the host of pretty flowering herbs whose roots had for long months rested dormant in the dry soil. Some responded swiftly, others more slowly, to the life-bringing showers: the prevailing low temperatures made their reawakening more tardy than it would have been in arid lowlands to which the rain had come. In two weeks, the plains around Tecpán changed from a nearly uniform grayish brown to a patchwork of bright green pastures alternating with the bright brown of wet soil newly plowed. Pink wood sorrel and magenta verbena bloomed along the roadsides and in the pastures. A little later, violets, both purple and white, pushed up their heads in the open fields and along the banks. The higher pastures were stippled with a multitude of blossoms of humble, ground-hugging herbs, especially the slender, creeping *Houstonia serpyllacea,* one of the chief ground covers, which burst into a galaxy of four-rayed, white stars, so closely set that one could not take two steps without trampling them. Swampy mountain meadows were gay with buttercups, lady's-tresses, and blue-eyed grass. Throughout the rainy season, there was a constant succession of flowers.

During the first few weeks of the wet season, the forests disburdened themselves of their dead wood. Rotting branches and dead, tottering trees, which had barely managed to remain aloft through the dry weather, could no longer bear up their own weight and that of encumbering epiphytes, after their load was increased by the water they soaked up, and they were further weakened by the fungi whose growth was accelerated by abundant moisture. At intervals through the day, I heard the crash of falling branches, and at times the louder rumbling of some large tree as it settled into its final resting place, snapping branches and crushing saplings in its irresistible descent through the forest. I came upon many dead trees, some of great size, which fell to earth soon after the rains began, often bringing down rare epiphytes that would otherwise have been difficult to collect. At this period, it paid to be careful where I set my blind to watch a bird's nest. Falling timber makes the outset of the wet season the most dangerous time in tropical forest at whatever altitude, but the probability of being struck is doubtless smaller than amid the traffic of busy streets and highways.

The rainy season, like a naughty child, seemed to have exhausted itself by its early display of bad temper. In the first fortnight of July, it was not so spiteful as it had been in late May and June. Mornings were usually pleasant and sunny; sometimes a whole day would pass without rain. Those bright days of the *invierno* were, in a way, the most agreeable of the year, for they were without the great fluctuations of temperature that prevailed in the dry season, when a frosty morning was often followed by an afternoon with an excessively ardent sun. The humidity of the atmosphere now moderated the temperature and prevented disagreeable extremes. Moreover, everything was greener in the wet season, and more flowers brightened the fields. Another circumstance, not the least important, which made these sunny days in the rainy months so delightful was their contrast with the dark, chilly days that had preceded them. Yet for me they were tinged with sadness, for nearly all the birds had fallen silent, and I knew that when they resumed their songs I should be far away. Fine days continued to recur at intervals through July and August; but when no rain fell in the daytime, it was almost certain to do so after dark.

It was late September before the maize was in the milk and we began to enjoy corn on the cob, more than six months after it was planted. From March to May, the maize had grown slowly because it had too little water; after mid-May, it had too much water with too little warmth and sunshine. As the grain ripened, many diverse creatures beside ourselves found it tasty. Squirrels, Steller's Jays, and Red-shafted Flickers all claimed their share. But the Indians' half-starved dogs did greater damage, by pulling over the whole cornstalk to reach the ears. To check their depredations, farmers scattered meat poisoned with strychnine through the cornfields, and a number of canines paid the supreme penalty for being hungry. In other districts, various sorts of traps and deadfalls were set in the milpas to catch marauding dogs. Many of those that fell victims to their appetite for corn were such pitiful, famished creatures that death seemed a merciful release for them.

Like the maize, the acorns took much longer to mature in this cool upland region than in the ardent summers of the North Temperate Zone, those of the white oak group requiring about eight months to ripen instead of four or five, as in the Middle Atlantic states. But as the rainy season approached its end they began to mature, and birds that were fond of them flocked to the feast. The three chief acorn-eaters had quite different modes of procedure. The Band-tailed Pigeons, perching in flocks in the treetops, plucked the fruits from their cups and swallowed them whole, making me marvel at the power of their digestion. The Steller's Jay carried an acorn to a convenient perch, where, holding it against the bark with a foot, it hammered with its bill until the shell broke and it could pick out pieces of the white meat. These jays, happy with abundant food, were now more than ordinarily noisy, uttering a variety of harsh and whining calls all day long. I heard their pleasant *sotto voce* medleys only earlier in the year.

The Acorn Woodpeckers, jolly feathered harlequins of the woodland, had still a third method of eating the acorns. They wedged them in crevices on the upper side of horizontal branches, where they could split them open with powerful bills, leaving both feet free. Their hunger satisfied, these woodpeckers often stored fragments of acorns, more rarely whole ones, in convenient crannies in the bark or among epiphytes on high limbs, for future consumption, as told in chapter 2. The jays less frequently hid an acorn in a crotch of a tree, picking up pine needles or plucking lichens from bark to cover and conceal it.

When, after June had passed, I was told that September would be still wetter and more dreary, I could hardly believe it. But it turned out as predicted. What a constant succession of damp, dismal, mist-shrouded days! One storm lasted four days, during which rain fell almost continuously, sometimes fine and driving, sometimes harder and more direct. Clouds hung low, veiling the trees; wind blew in boisterous gusts; the nearly saturated atmosphere was so penetratingly chilly that the thermometer which refused to descend to the freezing point seemed to lie. A bald statement of temperature fails to suggest how cold this thin, damp air can make one feel, especially when rain prohibits outdoor activity. This sort of weather prevailed until about mid-October, when the wet season gave indications that it had exhausted itself. During the five months that it had lasted, there were hardly a dozen days when it rained neither in the daytime nor in the night.

In Guatemala, the *invierno* of 1933 was said to have been the most severe in ten years. When it was over, the *Ferrocarril de los Altos*, which climbed too steeply from the Pacific lowlands to Quezaltenango on the high plateau, was so badly wrecked that it was abandoned; and Panajachel, which had had two highways leading from its deep valley to the outside world, was left with none. A long stretch of one road had slipped bodily down the steep mountainside, while the other was completely blocked by huge rocks and massive landslides. Many weeks passed before a wheeled vehicle could enter or leave this village beside Lake Atitlán. These are only a few examples of the havoc that the hard rains had wrought throughout Central America, and which, in greater or lesser degree, they do almost every year.

The spring that follows a hard winter is no more eagerly welcomed by inhabitants of far northern lands than was the return of bright, sunny days by us who had lived through the long, dreary wet season on the Sierra de Tecpán. The advent of the *verano* was in certain ways reminiscent of springtime, but it was spring at the wrong end of the year, when days were growing shorter instead of longer. No birds, except the little flower-piercers and the mostly tuneless hummingbirds, welcomed with renewed song the arrival of this anomalous spring; scarcely any trees dressed themselves in tender green in its honor. On the contrary, it was the time when leaves began to fall and fruits to ripen. Instead of being the time for sowing, it was the time for reaping; instead of being the season when nights became warmer, it was the time when they became increasingly chill and frosty.

Nevertheless, beneath the trees and in open woodland glades and on bushy slopes the herbs slowly responded to the increased sunshine, until they were bedight with myriad bright blossoms; and many trees had hardly shed their old leaves before they began to burgeon forth again, while others arrayed themselves with flowers. Hummingbirds and flower-piercers built their nests and hatched their eggs in bushes that for months had held no nests that were not old and weathered. It was, indeed, as much spring as autumn and as much autumn as spring; it combined in unexpected fashion certain of the attributes of the two seasons that many people in the North consider the most pleasant of the four.

But just as in the North a few fine days in late February will encourage resting buds to loosen their scales, and start the frogs peeping in woodland pools and puddles, and inspire the Cardinals to sing, making people rejoice that spring has come, when in fact many bleak and snowy days lie ahead; so a few bright days in mid-October cheered us with a premature hope that the dry season had at last returned to these heights. October contained many more days that were dull and cloud-drenched, although less rain fell than in the first half of the month.

November came with clear skies, but it also turned cold. Early on the morning of November 2, I went out under a brilliant moon to visit some sleeping birds, and noticed on the close-cropped pasture grass white patches that proved to be frost. Exactly seven months had passed since I had seen the last frost of the lengthening days. The frosty nights followed hard on the heels of the wet season; probably only cloudy nocturnal skies prevented the formation of frost at an earlier date. The premature cessation of the rains was feared in the highlands, for it might result in heavy frosts in October, destroying so much ripening maize that the people who depended on it faced famine. At altitudes between eight and nine thousand feet in Guatemala, maize requires eight or nine months to ripen—about twice as long as in tropical lowlands. At an altitude of nine thousand feet on Volcán Irazú in Costa Rica, I saw fields in which maize was sown in January and harvested in the following January. This great variation in the rate of development of the same crop is eloquent testimony to the climatic diversity of Central America.

The frost became heavier each night, until, when November was a week old, dawn revealed fields white with hoarfrost, and the vegetation began to feel its destroying hand. The month was very windy, with strong, cold gales blowing from the west or northwest, sometimes for days without intermission, soughing through the boughs of the pine trees and whipping the slender branches of the Raijón bushes in the pastures. The sky was usually clear; but sometimes the wind bore along rain clouds that it drove mercilessly through the treetops, wetting and chilling all the mountain, blotting out the sun, and enveloping everything in a gray mantle, as in the wet season. On such days, a fine, penetrating drizzle might descend; but hard rains, such as had been so frequent in September, were rare. Sometimes we enjoyed a day both bright and calm, a halcyon day that only the Guatemalan highlands and November could produce.

Throughout central and western Guatemala, except on the open summits of the volcanoes and the highest treeless mountaintops, the beginning of the *verano* is the flowery season of the year, and November is preeminently the month of flowers. Now heaven and earth cooperate to cover the landscape with bright bloom; at no other season does the sky contain so much sunshine while the soil holds so much moisture. During the rainy season there had been multitudes of blossoms, especially in the open fields, but they had been mostly, with some notable exceptions, small and retiring florets, like violets and spiderworts, to which one was obliged to stoop in order properly to appreciate them. But now was the season of the tall, bold herbs: Salvias, bur-marigolds, marigolds, Stevias, and many other vigorous composites that displayed conspicuous masses of color and tinctured the landscape.

This was also the time when the most abundant woodland herbs bloomed; in the months just passed, there had been too much humidity and too little sunshine beneath the trees to favor bright blossoms like theirs; soon the soil would be too dry for them. Now, too, many kinds of shrubs came into bloom, adding their brilliance to that of the herbs, and not a few trees extended the floral display still higher into the air. It was chiefly the woods and bushy places that were colorful at this season. Open pastures and meadows, which had their day of glory while the rains continued, were now turned brown and dry by frost and wind. The cornfields, after a brief splendor of brilliant weeds at the very outset of the dry season, soon retained only shreds and tatters of beauty.

The contrast between frosty open field and flowery woodland glade was unexpected; but it was even more surprising to learn that most of the plants which bloomed so freely at this season would be killed by freezing. As November advanced, I found examples of most of the herbs then in flower that had succumbed to frost. One of the first to feel the effects of the cold was the yellow bur-marigold *Bidens refracta*, which of all the plants that then blossomed was the most abundant and conspicuous. In open spots, whole patches of this herb were soon killed down to the ground, all its foliage shrivelled and black. The pretty, yellow-flowered *Bidens pilosa* suffered even more, as it grew almost wholly in open fields. The Salvias, which everywhere brightened the woods and bushy slopes, were no more immune to frosting than the composites. The Bracken Ferns, which had spread their fronds over the open pastures during the rainy season, were killed by the earliest frosts; soon they were as brown and dry as those that cast their thin shadows over a snowy field in the North.

I found herb after herb that could continue to bloom only where protected from frosting, although apparently unharmed by temperature only a fraction of a degree above the freezing point. And the same applies to the majority of the shrubs, whose lower leaves I would find drooping and discolored after a cold night. Later, they shrivelled and blackened.

If the temperature of the lower atmosphere over the mountain had fallen a degree or two below the freezing point and remained there long enough for ice to form beneath the trees, all the forests might have resembled a greenhouse filled with tender plants which had remained unheated throughout a cold January day. But the temperature of the air over the Sierra de Tecpán usually kept somewhat above the freezing point through the nights, and freezing temperatures were neither general nor of long duration. The freezing of the foliage and other aerial parts of plants near the ground on clear nights was almost wholly the consequence of radiation through the thin atmosphere, the effects of which were greatly modified by heat conduction and the downward flow of air chilled by contact with exposed solid bodies. Frost never formed beneath the trees; herbs and shrubs growing in woodland were not harmed by it. Herbs growing high on a steep, exposed slope escaped frosting because the air chilled by contact with their strongly radiating surfaces flowed downward and was replaced by warmer air from higher layers; those growing in a pocket at the foot of the same slope, where the descending chilled air accumulated, suffered severely.

I was constantly meeting with interesting examples of the operation of these several factors and of the exceedingly fine line that separated life from death for many plants, or even individual leaves and flowers. A plant growing beneath the shelter of a small bush might escape frosting and remain green and flourishing, while another of the same kind, situated only a foot away, had been frozen, and stood dead and shrivelled beneath the open sky. The lower leaves of a shrub or sapling tree, or even of an herb only two feet high, might be completely killed by frost, or at least blackened at the edges, while the top of the same plant remained green and uninjured and flowered profusely. Simply by growing in the midst of a compact stand, an herb might win enough protection from radiation to prevent the formation of ice within its tissues, while its neighbors on the edge of the patch were injured or killed. This was especially well illustrated in the case of the abundant, red-flowered *Salvia cinnabarina*, which often grew in close stands in open places.

Just as two weeks of the wet season had sufficed to turn the plains at the foot of the Sierra from brown to green, so two weeks of frost and drying winds were enough to turn them from green to brown again. Nothing remained of the wheat and oats but the stubble; the blades of the maize plants were dry and dead; the weeds in the open fields had fared as badly; and from our height we looked over a vast, nearly level expanse whose prevailing color was golden-brown. Here and there, a rectangle of low, moist pasture was still bright green; and the whole landscape was variegated by the dark green of the oak woods on the ridges.

And now across these open, windswept plains I was about to journey once more; I had fulfilled my dream of following the cycle of the year on a lofty tropical mountain, and my sojourn there was drawing to an end. For twelve months, I had not once descended below an altitude of five thousand feet. Through each of these months some plant flowered, some fruit ripened, some bird sang. Nature, although never as intensely active as in humid tropical lowlands, never slumbered as she does in lands nearer the poles.

Alder Trees (*Alnus arguta*), a hundred feet tall, in mixed forest on the Sierra de Tecpán, Guatemala.

18

Excursions

During my year on the Sierra de Tecpán, I was more stationary than I had been in many past years, or was to be in many future years. I had made the southern slope of the Sierra my Parish of Selborne, and I found enough of interest on the upper three thousand feet of a tropical mountain range to keep me busy for a twelvemonth—or for twelve years, had I been able to remain so long. In the whole year, I did not travel farther from the house where I lodged than I could go afoot in a day. But in the latter part of the rainy season, when I had no daily round of birds' nests to confine my activities within a circuit of a few miles, I made excursions longer than I had taken earlier in the year. I paid more frequent visits to the plateau at the foot of the Sierra, and roamed about more widely there, to become acquainted with the country, collect the plants, and search for kinds of birds that did not occur on the higher slopes.

Although today the Inter-American Highway passes in front of the house

Headpiece: Sierra de Tecpán from the plateau near the town. Thatched Indian huts stand amid cornfields in the foreground.

at Chichavac, at the time of my residence there the highway across the range
was several miles distant from it. To go by car from Chichavac to Tecpán, one
first wound far up the mountain to meet the public road, then descended
about two thousand feet. The pedestrian traveller, however, had the choice
of two routes that were much shorter and made no needless ascent. One was
the road, hardly passable by automobile, that led steeply down the face of
the mountain to Santa Apolonia. In the wet season, this road was churned up
by the sandals and bare feet of so many people, by the iron-shod hoofs of so
many horses and pack mules, that the stretches which were not hard, slippery clay had become deep, oozy mud. My chief objection to this route was
that sometimes one descended with unexpected rapidity and utter loss of
dignity, and reached the foot of the slope with a heavier load than he started
with. The Indians negotiated such slippery declivities with a *caite* or sandal
on one foot and the other foot bare, which gave them a choice of surfaces to
apply to the various kinds of roadway, but this mode of travel hardly appealed to one who wore shoes.

My favorite way to reach the plateau was a little-used footpath through the
woods. I cut across the cornfield, ducking under barbed-wire fences, followed around the contour of a deeply concave slope, a huge natural amphitheatre covered with Raijón bushes, and met the path, which led through
a stand of tall, slender second-growth oaks, mixed with pines and alders and
a few trees of other kinds. This light woods covered most of the long,
thousand-foot slope. It was not so interesting as some of the more mature
forests on the mountains, for the trees bore few epiphytes and the undergrowth was sparse; but even the most monotonous woodland provides much
to detain the observing eye and prolong a journey of a few miles to as many
hours.

After several hundred feet of steep descent beneath the trees, the trail
turned to skirt a small, open, rather marshy area, where a rivulet was born.
Near this spot I enjoyed one of my most intimate encounters with that shy
and elusive creature, the Scaled Antpitta. It was a bird of unique and unmistakable appearance; the body short and plump, nearly tailless; the head
large, with big, dark eyes adapted to life amid dim undergrowth, set off by a
white crescent behind each; the black bill short and strong; the grayish legs
long and heavy, looking as though intended for a bigger bird. The color of
the antpitta's back and wings was brownish-olive, that of its underparts
reddish-brown; its whole plumage blended well with the dead leaves among
which it foraged.

Instead of flying completely out of sight, as these birds usually do when
they find themselves observed by man, this antpitta delayed among some
bushes near the path, giving me a long-desired opportunity to watch it in
action. It did not walk over the ground in the manner of the ant thrushes but

progressed by a series of short, quick hops, made with the feet together. With rapid movements of its heavy bill, it flicked the fallen leaves aside and picked up small creatures that it found beneath them. It was silent, as I almost always found its kind. Hopping and brushing aside dead leaves, it moved off beneath the shrubbery until it was hidden from view. When I tried to follow, it became alarmed, flew up to a low perch, as these birds almost invariably do when surprised, took a rapid survey of the situation with large eyes that made it appear very startled, then plunged into a dense thicket where it was hopeless to try to follow it.

A month later, in the same area, I again met an antpitta. It became aware of me as soon as I saw it, if not before, and retreated with rapid hops, moving by spurts interrupted by pauses, as an American Robin glides over the lawn. While my eyes were fastened on this bird, a low, rather rasping croak drew my attention to another, who stood on a fallen log ahead of me. This second antpitta repeated its queer call, then hurried away behind the first. These two brief utterances were the only notes I have heard from a Scaled Antpitta.

The huge family of antbirds, to which the antpitta belongs, is at home in the warmer parts of the American continents, and few of its members range even a little way beyond the tropics. The antpitta is the only representative of the family that in Guatemala inhabits the altitudinal Temperate Zone. On a bright morning in mid-September, when the season's earliest frost sparkled on the grass of the high, open glades of the Sierra Cuchumatanes beneath the first cold rays of the rising sun, I was surprised to meet an antpitta on the edge of a low copse of junipers on this high tableland, nearly eleven thousand feet above sea level. The black breast, conspicuously streaked with buff, proclaimed it to be an immature bird; since young antpittas seem to wander widely, it may not have hatched at this great altitude. From July to September, I found many of these black-breasted juveniles on the Sierra de Tecpán, always leading solitary lives, as their elders appear to do through much of the year. But the adults, too, were not uncommon among the cypress forests near the summit of the Sierra, and, at the other extreme, they have been found in lowland forest only a thousand feet above sea level.

From the head of the rivulet beside which I saw the antpittas, the path continued downward along the back of a narrow ridge, which on both sides fell off sharply into deep, contracted valleys. Near the foot of the descent stood some huts of the Indians, whose dogs vociferously protested my passage. A field of maize covered the more gentle slope at the base of the mountain. At the lower edge of this field, the highway from Santa Apolonia to Tecpán ran between flowery hedgerows. On the other side of this narrow, earthen road, nearly level fields of wheat and maize and extensive open pastures inclined very gently toward the brook that was my destination.

As I walked across the pastures, I heard the familiar call or alarm note of the Eastern or Common Meadowlark—no, not quite the familiar note, for the call of the local bird was softer than the rasping *zzzrt* of the northern race, but a recognizable version of it. After much searching with my eyes, I picked out three of the birds, foraging on marshy ground where the herbage was low. One rose into the top of a low tree and gave me an excellent view of himself. He was in particularly fine plumage, with underparts of the brightest yellow, the black crescent on his breast bold and heavy. He turned his streaked, brownish back toward me and in this position repeatedly spread and folded his tail, sending out momentary flashes of white from the outer feathers, while he repeated many times the familiar dry trill. In August, the meadowlarks sometimes sang from the top of a fence post or a low branch of a tree. Their clear whistles closely resembled those with which the meadowlarks I used to know in the North called for the "spring o' the year," but they did not seem quite so melodious, possibly because this was not the proper season to be calling so. The Common Meadowlark has a wide range in Central America, where I have found it in habitats so diverse as the alpine meadows of the Sierra Cuchumatanes and near sea level in northwestern Costa Rica.

The brook that was my destination wound through pleasant open pastures, which sloped gently upward to low ridges wooded with oak and pine, or between fields where the maize grew tall and heavy. Although fairly steep, the slopes immediately above the banks were in many places kept wet, even in the dry season, by the groundwater that welled up among the grasses and sedges that covered them. In July and August, these stream-side marshes were ablaze with the tall, red-flowered spikes of *Lobelia splendens*, which much resembled the Cardinal Flower of the North. Slender horse-tails, a rare pipewort (*Eriocaulon benthami*) with heads of minute white flowers, and an orchid (*Habenaria limosa*) with long, heavy spikes of green flowers also flourished here. In September, the one aster that I found in the region (*Aster exilis*) displayed small flower-heads, with white rays and yellow disks, on tall, spreading stems with narrow leaves.

Where the stream swung over against one of the enclosing slopes, it cut cliffs that were often high and steep. When freshly exposed, these escarpments were bare and white, but after the stream had ceased to undermine them, they lost some of their steepness and acquired a covering of shrubs and trees. On these shaded cliffs grew a butterwort (*Pinguicula moranensis*) with splendid, large, rose-magenta flowers, which nodded at the top of the slender, naked stem that sprung from the center of each rosette of light green leaves. The thick, rounded leaves were closely appressed to the bank, and their upper surfaces were covered with sticky glands to which small insects stuck as to flypaper. The bodies of these victims were apparently digested by enzymes secreted by the glands and served to supply nitrates to the insectivorous plant. The thick leaves did not, however, curl over their prey, as do the smaller leaves of the European butterwort.

Some of the cornfields along the stream stretched far. The Broad Bean, frequently planted between the rows of maize, had most attractive flowers, each with a large, spreading, white standard and white lateral petals with a conspicuous black spot. Black and white flowers are so handsome that I regret that they are so rare. When ripe, Broad Beans are boiled until soft and eaten instead of the more common black beans, or else they are roasted and sold in little shops scattered along the roads and in the towns, as well as in the marketplaces. For a half-centavo, one bought a small clay measure of *habas*, as they are called. They have a pleasant flavor and may be eaten like peanuts, only not so rapidly, for they are very hard and can be broken only by those with strong teeth. For a centavo, I bought substance to ruminate upon, walking along the trails, for a whole morning.

Two or three miles below the point where I began to follow it, the stream flowed by a small hamlet of about a dozen adobe houses, then joined a slightly narrower watercourse to form the Río Chayá. The augmented stream coursed through broad, nearly level pastures, emerald green in the rainy season, which stretched to the bases of high, steep, wooded ridges—or ridges they appeared to be, until I climbed to their summits and found that I had merely risen to the level of the plateau in which the deep, troughlike valley had been carved. Riding upon the swift current were many fragments of white rock, of all sizes. In the backwaters and eddies they floated in large numbers, in apparent defiance of gravity. Catching one of the stones and lifting it into the air, I learned the cause of its buoyancy; it was only volcanic tufa, honeycombed with minute air spaces, which made it very light. Yet this is the rock that underlies much of the country and, fragmenting readily, floats down the major rivers to the sea.

Along the banks of the Río Chayá, in August, I first met that engaging little bird, the Black Phoebe. In size, form, and mannerisms, it resembled the gray Eastern Phoebe, but, except for its white abdomen, it was everywhere dull black. It stood upon a log stranded in a large, semicircular embayment in the high bank and darted down to pick tiny creatures from the bare ground. After each descent, it returned to rest and watch on the log or a low hummock of earth. Later, it perched on the fence of a cattle pen farther down the valley and foraged in much the same manner. Sometimes, instead of flying downward, it varied its procedure by darting into the air to catch an insect in more typical flycatcher style. This cow pen was a favorite resort of the Black Phoebe, and on subsequent excursions I frequently saw it here, sometimes in company with another of its kind. At times it deserted the fence to alight upon a boulder that projected above the river, or upon a low branch overhanging the water.

Once, along the headwaters of the Río Chayá, I met a Green Kingfisher, which of all the American members of its family extends highest into the mountains, as it ranges farthest to the north. Along these streams, I looked in vain for the American Dipper, a bird that I glimpsed only once or twice at Chichavac. It was more abundant along the rushing watercourses north of the Sierra Cuchumatanes, which drain into the Río Usumacinta and the Gulf of Mexico.

Not far below the cow pen where I found the phoebes, the river was dammed to furnish water to a millrace, which led for a long distance along the base of a steep slope. An iron grating at the intake of the millrace held back the floating pieces of tufa and other flotsam. The channel supplied water power to turn the wheels of the Molino Venecia, the first of three mills for grinding wheat flour along the upper reaches of this river. The second was a large white building that belonged to the Municipality of Tecpán; no longer operating, its huge waterwheel was rusting away. A few miles farther downstream was the Molino Helvetia, or "Molino Grande," as it was known, for being the largest of all. Flour mills, some of which were operated by electricity, were scattered all over the high plateau of Guatemala, where much wheat was grown.

The Molino Helvetia marked the limit of my explorations downstream. Sometimes I returned to Chichavac by retracing my course; at other times, I branched off from the valley by way of the picturesque Comalapa road and followed devious paths across the plains. This road led upward through the narrow valley of a small tributary stream, its floor open pastureland, its enclosing slopes steep and wooded. Soon this valley contracted into a deep, narrow ravine, through which the little rivulet, screened by the foliage of trees whose summits were on a level with the traveller's eyes, flowed unseen or rarely glimpsed, far below the road, which occupied a shelf cut into the precipitous slope. At the summit of a ridge, I left the Comalapa road for another that led toward Santa Apolonia.

The narrow earthen lane wound between fields of maize and wheat, with here and there an open woodlot, and thatched Indian huts nearly hidden amid the tall maize plants. Blue-crested Steller's Jays, which were legion on the plains, fled before me through the crowns of the trees in the hedgerows; Great-tailed Grackles flocked over the open fields. These are probably the two noisiest birds in Guatemala, parrots excepted, and here where their ranges overlapped they made an incessant din. Little Rufous-collared Sparrows peeped out at me from the bushes with bright black eyes, their crests raised in an inquisitive attitude. Here on the plains they hatched their two or three greenish-blue, brown-mottled eggs in open cups of grass and weed stems, lined with horsehair, well into August, when all the other birds on the Sierra de Tecpán had ceased to breed.

The road that I followed descended a steep slope into another sunken valley of a small stream. On my left I discovered a *barranco* or gorge of astounding proportions. At the mouth, where a tiny rivulet flowed out, I entered the deep trench; it would have been impossible to descend into it, alive, any other way. The towering walls of white tufa rose vertically seventy-five or eighty feet above the narrow floor, but the space between them was in places not more than a third of their height. In this deep gash in the earth, I felt oppressed and insecure, for trees that still bore green leaves lay in the bottom, where they had fallen from above during the recent rains, and several of their erstwhile neighbors seemed to be on the point of following them. It is a peculiarity of the volcanic substratum of much of western Guatemala that, instead of eroding slowly to form open valleys with more or less gentle slopes, as in regions underlain by harder rocks, it erodes rapidly into these *barrancos* whose depth is wholly disproportionate to their width, while their walls are always precipitous. The ease with which detached pieces of tufa are carried away on the surface of the water makes it possible for the streams to perform their work of carving out the valleys with great rapidity, geologically speaking.

Many of the fields between which I passed on that afternoon in early October were yellow with ripening grain, and in some, Indian men and women were harvesting the wheat, a variety with very long awns. The harvesters' methods were most primitive. Each gathered a few straws at a time in his hand and cut them off with a sickle, until he had accumulated all that he could hold; then he tied the handful with a straw into a sheaf and laid it on the ground. This mode of harvesting is as slow as it is ancient, and in a country where labor is not so cheap, it would consume all the farmer's profits.

Later, I watched the threshing done by a method equally primitive. The straw with the heads of grain attached was spread thinly on a circle of hard-packed earth, surrounded by a fence. Three or four horses or mules were led into the enclosure and driven around and around at a trot, to the resounding cracks of a long whip, until I marvelled that they did not drop with vertigo. At intervals, the straw was shifted and the animals given a breathing spell. On a windy day, the grain was winnowed on the same floor where it had been threshed. The mixture of straw and loose grain was tossed into the air with three-pronged sticks, so that the wind could blow away the chaff and shake the grain from the straw. I hold myself fortunate to have witnessed these age-old, worldwide practices, which in more highly industrialized countries one may read about but never see. How different from the mechanical harvesting and threshing that I watched on the farm where I passed my early boyhood!

As I returned from this long excursion in the late afternoon, I met a small party of Baltimore or Northern Orioles, consisting of a male resplendent in orange and black and two or three females or young birds more soberly clad in pale yellow and grayish brown. It was October 9, and these orioles were the first of their kind that I had seen since January. I had the good luck to arrive at the house that evening, in the dark, just as the daily rain started to fall. My visits to the plains more often than not ended in a drenching.

My longest excursion was made at the end of October, when I spent a week at Panajachel, beside Lake Atitlán. At five o'clock in the morning, I set out afoot from Chichavac in the mist and drizzle. When I reached the highway, I was fortunate to meet an acquaintance, who took me in his motor truck up the long incline to the summit of the range. Although this highway was the shortest route between the two principal cities of the country, one could travel along it for hours without encountering any vehicle more rapid than an occasional oxcart. I had arranged to meet Axel at the summit, and since my unexpected ride brought me there early, I had a long wait in the cold cloud-mist that drove fiercely over the crest of the mountain. When finally he arrived, we started off together toward the lake. We had more than twenty miles to cover, but with a drop of almost one vertical mile to give us impetus, the journey did not promise to be arduous.

We followed the highway, shortening its numerous turns by means of footpaths, down to Chichoy, where we turned off the road to the left and headed down the mountainside through thickets of Raijón and other shrubs. When finally we dropped below the clouds, we saw the lake gleaming far below us, beneath the level gray ceiling, while the summits from which we had just descended remained enveloped in the sombre mist. Our narrow footpath led over a low spur, then steeply down along a ridge above a field of yellowing wheat. After reaching the plateau, we walked for miles between fields of wheat and maize, along grassy trails bordered by a bright profusion of blossoms, especially bur-marigolds with yellow flowers, but with a liberal admixture of a species with white rays. The sun had burned through the high ceiling of clouds and its beams brought out the full brilliance of the blossoms. One could walk all day without tiring along such level paths illuminated by such an intensity of pure yellow color.

Our even way came to an abrupt end when, at about noon, we reached the edge of a tremendous *barranco*, which a small river had in the course of ages cut into the yielding volcanic rock that underlay the plateau. This stupendous ravine appeared to be nearly a thousand feet deep; tall pines were dwarfed against its steep walls. The descent, past a small farmhouse that gave its name, "Chuti-estancia," to the whole *barranco*, was relatively easy. Halfway down, we stopped beneath a cypress tree to eat the lunch from our knapsacks. Reaching the bottom, we crossed the stream by jumping from stone to stone, then toiled up the steep trail that zigzagged painfully up the precipitous wall of the gorge.

Once across the canyon, the major obstacle of our journey lay behind us. After a few hundred yards over what remained of the plateau, we began to descend rapidly toward the lake. We passed along a narrow ridge, almost a knife-edge, which fell away in breathtaking slopes on either side. On our right, at points where there had been recent landslides, was a sheer drop of about a thousand feet into the valley below. The cabins of the Indians, almost beneath our feet as we sat on the edge of the precipice to enjoy the prospect, were dwarfed to insignificance; the people themselves, attending their sheep, appeared no larger than ants caring for aphids. A landslide during the last rainy season had barely failed to bury a hut near the foot of the cliff. Swifts and swallows darted and circled over the vast gap in the earth, well below us.

Reaching the end of the knife-edge, we once more beheld the magnificent lake spread out below us, now quite near, glittering in the rays of the sinking sun, which here and there found an opening in the lofty canopy of clouds. On the farther shore, the shapely cone of Volcán San Pedro stood framed between the steep walls of the narrow valley through which the Río Panajachel flows into the lake, and into which we were about to descend. To our left, the summits of the volcanoes San Lucas and Atitlán rose above the shoulder of the enclosing ridges. As we continued the sharp descent through low, light woods, we found two familiar trees that I had not seen as high as the Sierra de Tecpán. One, the Hornbeam, was only a variety of the small tree common in the forests of eastern United States; the other, the Guatemalan Hop Hornbeam, was specifically distinct from the northern tree but closely resembled it. Both were shedding their fruits, as the majority of the northern trees that extend to Guatemala do at this season.

We climbed rapidly down to the wide, boulder-strewn floodplain of the Río Panajachel, which now flowed placidly enough through the broad domain of its destruction, a stream that occupied a small fraction of its flood-time width. Crossing the current on a fallen trunk, we reached the highway, which had been ruined by the recent torrents. A few minutes' walk brought us to the little town of Panajachel, where we arrived at about four o'clock in the afternoon, after an easy journey of a little over eight hours from the summit of the Sierra de Tecpán.

The scenery about Atitlán is on the most magnificent scale. Lying nearly five thousand feet above sea level, the lake is about twelve miles long and almost as broad. It occupies a tremendous depression in the edge of an incredibly eroded plateau about two thousand feet higher. Standing on the lake's shore, one receives the impression that it is enclosed, on the north, east and west, by lofty mountains, which rise steeply from the water's edge in precipitous slopes and rocky cliffs, here and there jutting out toward the water like bold headlands sweeping up from the sea. Only after climbing these apparent mountains does one discover that they are not mountains at all, but the jagged edges of the plateau that almost surrounds the lake.

The southern shore is dominated by the nearly perfect cones of three great volcanoes. San Pedro's imposing figure stands alone to the west with forest covering its truncate apex, evidence that it has long been extinct. Farther to the east rises the equally green cone of San Lucas, with cultivated fields occupying its lower slopes and trees its summit. Directly behind San Lucas, Atitlán, loftiest of the three, towers into the sky. Bare cinder slopes extending far below its crest reveal that in the not very distant past it had poured ashes and lava from its crater, but for the preceding century it had been quiescent. Indians still procured sulphur from its throat. Between San Lucas and San Pedro, a long arm of the lake stretches toward the south. Around the shores and on the slopes above them nestle a dozen Indian villages, renamed by the *padres* for as many saints.

Lake Atitlán has no known outlet, but its water undoubtedly finds its way to the rivers of the Pacific slope through subterranean channels, for it remains fresh. During the wet season the rainfall in this region is very heavy, yet the lake's surface does not rise in proportion to the precipitation. The less sagacious of those who dwell along its shores claim that it has no bottom, which of course means only that they have never tried to sound it with an adequate length of line. In any case, it is extraordinarily deep in proportion to its expanse.

Except on the southern shore of the lake, the only level ground adjoining it lies about the mouths of the short, swift streams that flow into it. Panajachel is situated on one of the largest of these level tracts, formed by the river of the same name. The plain, about a mile in width, is walled by the precipitous slopes and cliffs of the plateau's edge, over which at times pour torrents of water, mud, and rocks, destroying houses and coffee groves at their feet.

Within the valley's sheltering walls a fairly luxuriant vegetation flourished, despite the long, severe dry season. Here I found thriving coffee plantations, shaded by lacy-leaved Grevilleas, or sometimes by Alder trees, an unexpected juxtaposition of plants that made one pause and think. Here grew bananas, sugarcane, Wild Cane, and cannas. Most of the land not occupied by coffee groves was broken up, in an amazingly irregular pattern, into milpas and small garden patches, kept green throughout the year by irrigation. Spanish Plum trees grew everywhere, and excellent oranges. At the beginning of the dry season, weedy pastures were yellow with the same False Marigold that is so carefully cultivated in northern flower gardens. The superb Maxon's Dahlia flourished in moist spots on the hills and about the shores. A profusion of composites, mints, and flowers of many other kinds delighted the hummingbirds.

The vegetation on the slopes above the valley contrasted sharply with that of the level plain. Isolated patches of maize clung to surfaces so steeply inclined that I wondered how men hung on while they planted them; but while the plain was intensively cultivated, the escarpments were mostly neglected. The steepest slopes were open and grassy, or here and there bright yellow with the blossoms of clustering composites. Less rocky and precipitous hillsides (but still amazingly steep) bore a low, rather scrubby woods of oaks, Alder and Arbutus. Birds were not easy to find on these steep hillsides, but they swarmed among the coffee groves, gardens, and weedy pastures of the plain.

Along the northern shore of Lake Atitlán, the transition from the Subtropical Zone to the altitudinal Temperate Zone is exceptionally abrupt and definite, making Panajachel extraordinarily interesting to the student of the altitudinal distribution of plants and animals. It is one of the few places where one can point his finger and say: Here the Tropics ends and the Temperate Zone begins. The unusually sharp delimitation of altitudinal life zones is, of course, caused by the peculiarly abrupt topography. The sheltered plain of Panajachel is Subtropical; the birds confirm the impression that the bananas, sugarcane, and many other plants create. The escarpments that enclose it rise abruptly into the Temperate Zone. Many of the birds of the latter are tempted by the more luxuriant vegetation and abundant fruits to descend into the orchards and groves of the fertile plain, but fewer of the heat-loving birds care to ascend the less bountiful slopes that surround their sheltered retreat.

During my week at Panajachel, where I lived in a little hotel on the lakeshore, I was constantly reminded that here two faunas met. Here Gray's Thrush, ranging up from the lowlands, mingled with the Rufous-collared Thrush and the Common Bluebird, straggling down from the plateau. Here the Spotted Towhee met the White-collared Seedeater, a widespread and common species of the lowlands; here the Red-shafted Flicker of the cool highlands encountered the Golden-fronted Woodpecker so widely distributed over all the warmer parts of the country; here the Tufted Flycatcher saw and heard its heat-loving relations, the Vermilion-crowned Flycatcher and the Tropical Kingbird. Here, too, Steller's Jay of the high country came into contact with the Bushy-crested Jay of middle altitudes. But the two kinds of jays showed clearly what type of habitat each preferred. On the lowest and most fertile part of the plain, between the town and the lake, Bushy-crested Jays were very numerous among the shade trees of the coffee plantations, but here I found only a single Steller's Jay. The Bushy-crested Jays became rarer in the increasingly sterile valley above the town; I saw none at all on the escarpments or on the plateau above, where Steller's Jays abounded.

Migrant birds of many kinds swarmed over the fertile plain. Indigo Buntings, with only faint traces of blue on their brownish winter dress, foraged in large flocks on the weedy fields and were by far the most abundant finches. Rose-breasted Grosbeaks were not rare; and I noticed a few lovely Painted Buntings, the males still wearing their Joseph's coats of blue and golden green and red and black, despite the long journey they had made. Both Baltimore and Orchard Orioles were present, but the latter, who prefer the coastal districts, were much less abundant. Blue-gray Gnatcatchers flitted among the trees, uttering their fine, nasal mew, and Solitary Vireos hunted among the same foliage in a more deliberate and thorough manner.

Ruby-throated Hummingbirds darted over the flowery fields and thickets, more abundant than any single species of resident hummingbird. Individuals with pale throats predominated, but many wore more or less red feathers on their throats, and males with the full, glittering ruby gorget of the adult plumage were by no means lacking. Very truculent, or perhaps only inconsiderately playful, they annoyed other birds from wood warblers to wintering Sparrow Hawks by darting at them again and again, and of course they lost no opportunity to pursue each other. The purple salvias, now blooming profusely, were their delight. Along the shore of the lake, a visiting male Belted Kingfisher dived into water upon which Least Grebes swam peacefully. I did not notice the flightless Atitlán Grebe, which is restricted to this single lake; it seems to occur chiefly on the southern side.

But the most numerous family of migrants was the wood warblers. Tennessee Warblers were present in vast numbers and seemed to be the most abundant birds of any kind. Townsend's Warblers, although numerous, appeared rare by comparison. I saw two Yellow-breasted Chats, which, considering their secretive habits, suggested that they were present in force. MacGillivray's Warblers were still more retiring but advertised their presence amid low, dense bushes by their repeated explosive *tuc*. Black-capped or Wilson's Warblers were abundant, Magnolia Warblers not uncommon. Among the species not so well represented were the Yellow, Nashville, Black-throated Green, and Black-and-White Warblers, the American Redstart, and the Common Yellowthroat. But none of these visiting warblers was prettier than the resident Rufous-capped Warbler, whose bright chestnut head was marked by a white stripe above each eye; its back and wings were plain olive-green; its throat and breast bright yellow. These small, long-tailed birds were abundant in thickets and low woods, both on the plain and the surrounding slopes.

The gem of all the birds at Panajachel was the Slender Sheartail, a wonderful little hummingbird that I have seen only here. A male had his headquarters in a flowery thicket not far from the shore. His narrow, black tail feathers were much longer than his minute green body. A broad gorget of the most brilliant violet covered his tiny throat. His nearly straight, black bill was very slender. I was told that he was the Quetzalito, the Little Quetzal, a name bestowed in allusion to his long and narrow tail. Certainly, if with no change of attire he were as large and conspicuous as the Quetzal, he would be as famous.

19

A Night on a Volcano

Tied to one locality by my desire to make continuous observations on a variety of birds, plants, and natural phenomena, during my year on the Sierra de Tecpán I did not once travel farther than a day's walk from my fixed abode. But when I left the Guatemalan highlands early in January, 1934, I unwittingly embarked upon a long period of almost constant wandering. I proceeded at once to Costa Rica, attracted by its marvellously rich fauna and flora, by the cheapness of living there, and by the accounts that I had heard from other naturalists of the freedom, the democratic simplicity, and the friendliness of a people whose society was not so strongly and harshly stratified as that of Guatemala. It seemed that there, if anywhere, I might fulfill my growing desire to establish a homestead amid unspoiled tropical nature, where I could continue to study and write about the birds and plants. But the small capital on which I had begun my independent studies was already greatly shrunken; I lacked both the money and the experience to

Headpiece: On the road from Quezaltenango to Volcán Santa María. The tall inflorescence of a maguey tree rises on the left.

realize the dream which, seven years later, I did in a measure bring to fruition. After spending two months in unsuccessful attempts to settle in a satisfactory location, I reluctantly decided to return to the United States.

Dejected and uncertain how I should carry on the work on which I had set my heart, I reached Puerto Limón from the interior of Costa Rica with barely enough cash to pay for the lowest-priced cabin on the next ship to New York, only to find that all the accommodations within my means had already been booked. My situation was distressing; if I waited for the boat that was scheduled to sail a week later, my living expenses would so cut into my remaining funds that I would then lack money for even the cheapest accommodation—unless an overdue bank draft finally arrived. It came just in time; I took passage on the next ship, and afterward had reason to be thankful for the enforced delay.

Fellow passengers on this voyage were Professor and Mrs. Oakes Ames. He was America's leading authority on orchids, a generous patron of botanical science, and at that time Supervisor of the Arnold Arboretum of Harvard University. Although I had sent a collection of woody plants and a few orchids from the Sierra de Tecpán to the Arboretum, I had never met Ames. Before we reached New York, he suggested that I return to Guatemala to collect plants for the Arnold Arboretum. Finally, I received a commission to undertake a six-months' collecting excursion for the Arboretum in the Guatemalan departments of Quezaltenango, Huehuetenango, and El Quiché. A ship missed: a way found to carry on—by such rare and improbable chances are our fortunes wrought! Agonizing failures such as I had experienced are perhaps a necessary part of the education of anyone who tries by an independent course to accomplish something out of the ordinary, but they leave psychic wounds that may take long to heal.

In July of the same year, I returned to Guatemala by way of Veracruz, Mexico, and the long rail journey southward to the Isthmus of Tehuantepec, across to the Pacific side, thence along the coastal lowlands of the state of Chiapas to the Guatemalan border. Through-trains, travelling day and night, were scheduled to perform this journey of nearly six hundred miles in thirty-four hours. But I wished to see all the country along the route and planned to travel only by day, stopping off at the little towns at night, and if the vicinity looked promising, tarrying a few days to explore the neighboring country. I enjoyed some memorable days with the birds at Matías Romero and San Gerónimo Ixtepec on the Isthmus and Tonalá in Chiapas. But the trains on this long run through a sparsely populated region seemed to be late more often than on time; journeys scheduled to end before nightfall were sometimes completed after several hours of darkness. Accordingly, I failed to see some interesting stretches of the route.

We arrived hours late at Suchiate, the terminus of the Pan-American

Railroad, where in the dark I was led to a little hotel so filthy that I could swallow no breakfast next morning. Here it was necessary to cross the Río Suchiate, broad, swift, and muddy, to Guatemalan territory. My passport was examined and my baggage inspected on the Mexican side, before I was permitted to embark in a skiff that was poled across the shallow stream to the neighboring republic. Here, beneath some beautiful willow trees at the edge of the broad, sandy flood plain, stood a shelter of palm leaves, in and about which a customs inspector, an officer of the Guatemalan army, several private soldiers, and two Indian women were waiting to welcome me back to the Land of the Quetzal. I presented my passport; then, while the inspector, who seemed too proud to use his hands, looked on, the women began to search my baggage. Never elsewhere have I been subjected to such a minute inspection. Even my letters were opened, one by one; the batteries were removed from my flashlight; every little container in my toilet kit was uncovered; the insides of my shoes were probed—nothing escaped those four sharp eyes. Since I carried a fair amount of apparatus for botanical collecting, I foresaw that this inspection would last through the remainder of the day, while to repack would occupy me all the next day. Presently, I had an inspiration.

"If I have dutiable articles, must I pay here or in the customhouse?" I queried.

"They must be declared in the *aduana*," was the prompt reply.

"Well, my baggage contains dutiable articles, so why not take it directly to the *aduana*?"

So it was released from the clutches of those too-meticulous inspectors, placed in a wheelbarrow, and pushed up the slope into the small frontier town of Ayutla, where the buildings were attractively painted blue and white, the national colors. Here in the customhouse an elderly man without female assistants assessed a small duty on new cardboards to be used for drying plants, but did not even look into my "personal effects."

By rail early next morning from Ayutla to San Felipe, thence by automobile up the steep Pacific escarpment of the highlands, through luxuriant subtropical vegetation, that same evening I reached Quezaltenango. This metropolis of western Guatemala was a quaint town with crooked, hilly streets, and a central plaza where a red lion stood guard before the façade of a cathedral that had been ruined by earthquake. Here, eight thousand feet above the sea, nights are cold and frosty during the winter months. The first objective of my collecting trip was the ascent of Volcán Santa María, whose shapely cone was visible above the low roofs of the town.

On that long train ride through southern Mexico, I had read in a Mexican periodical of adulterated breadstuffs in the capital city. I did not then suspect that I would so soon make their acquaintance in Guatemala, and that they would influence the course of my first excursion. Sunday happened to be my first free day, after I had become settled in Quezaltenango. Wishing to set out early to explore the neighboring hills, I took a few rolls from the hotel before breakfast was served there, bought oranges at a wayside shop, and ate a frugal meal on a rocky hillside above the road. Then I turned my steps toward El Cerro Quemado, a craggy volcanic summit south of the town. But after an hour, I began to be queasy; by midday, I took to my bed, feeling very miserable.

A recurrence of the same gastric disorder, two days later, convinced me that unwholesome bread was in fact its cause. While I was recovering from this second attack, I heard a knock at the door of my hotel room. It was opened by a little *ladino* tailor, sent by an acquaintance whom I had requested to recommend a guide and porter for the ascent of Santa María. He did not impress me as the right kind of guide and porter; he was anxious about payment, and still more apprehensive about the volcano.

Seventeen years earlier, three white youths and their two Indian porters had been murdered while sleeping on the summit, by Indians enraged by the violation of one of the shrines devoted to secret rites which they hide away in remote, inaccessible spots. Everybody in western Guatemala seemed eager to tell me about their most famous murder; the accounts that I heard varied in details, but it appeared that the victims were innocent of the misconduct for which they were killed. In any case, the too-hasty murderers later paid the supreme penalty for their mistake. After all these years, Volcán Santa María was locally considered a dangerous place. Reflection on the circumstances convinced me that it was tolerably safe. Seventeen years is a long time to remember a murder, even a quintuple murder; I am sure that no big city in the North talks about a crime so long, for it always has one newer and more gruesome to occupy the public mind. The fact that this community retained such vivid recollections of its seventeen-year-old murder was, after all, reassuring.

But this reasoning did not occur to me while the little tailor waited in my presence, and I doubt that he would have been able to follow it. He desired me to obtain permission from the civil authorities to climb the volcano—a wholly unnecessary proceeding. Although, at the moment, I felt too sick to go out and seek a less timid porter, I was not too weak to see the humor of requesting official sanction for my intended ascent. So I sent the tailor to buy a sheet of the stamped paper on which all petitions to Government had to be presented, wrote out the request, and directed the man to take it to the *Jefatura*.

Soon my tailor was back again; I must go and sign something—after all my scheming to remain in bed, I must abandon it! At the end of much talk and of twisting much red tape, I was handed a document of most impressive appearance—I have it before me as I write, with both signatures and all four seals—granting me permission to climb Volcán Santa María, on July 26, 1934, accompanied by Santiago_____[the tailor], for the purpose of collecting the rare plants that grow in said locality, to form a herbarium for Howerd [sic] University. But still it was not all finished; I had to *TOMAR RAZÓN POR LA POLICÍA NACIONAL*. This last, I promptly discovered, consisted in signing my name in a huge ledger and paying a quetzal for the privilege. I was not averse to giving my signature for what it might be worth to them, but I balked at paying the equivalent of one dollar, in addition to the more than liberal reward that I had promised the tailor for his services on the morrow. He must, I decided as I began to feel better, come with me without official sanction or not at all. What suprised me most in the whole proceeding was that nobody had seen its ludicrous side. I am sure that no momentous international agreement has ever been signed with greater solemnity.

As I had expected, the timid tailor did not arrive next morning at the appointed hour of six. I waited for him until eight o'clock, then went out to the plaza of the red lion and announced my need to the porters usually to be found there. Soon I had three volunteers for the ascent and chose the most likely, a stalwart, pleasant-faced Indian named Margarito Hornero. Having finally discovered the cause of my ailment, I was quite well this morning, and we were off with a rush. No talk this time about official permits! To the hotel for my blanket and collecting equipment; to a shop for provisions; to Margarito's house, on the outskirts of town, for his *matate* for carrying loads on his back, his blanket, and his brother Pedro, who did most of the heavy carrying—and we were ready to go.

Soon after nine o'clock, we were crossing a broad, level plain, with the whole of Santa María's slender, shapely cone rising up ahead. Its great height made it appear quite near, but we walked mile after mile along the level, dusty highway, before reaching its sloping base. We passed many loads of firewood coming into the city from the surrounding hills, some on the backs of Indians, bending under neatly stacked piles that rose above their heads, others on pack mules groaning under their unwieldy, swaying burdens, still others in creaking oxcarts, which carried the greatest weight with the least effort.

Along the hedgerows were great patches of bright orange, conspicuous from afar, which at first sight I took for brilliant blankets that the Indian women had woven, washed, and spread out to dry. But on closer approach these gaudy patches proved to be small trees overgrown with dense tangles of the common parasitic dodder of the region, called *pelo de león* (lion's mane).

Finally, we started to climb a rocky, sunken road, shaded by Elderberry trees and low oaks. Cleared fields extended upward to an altitude of ten thousand feet, and even above this were abandoned clearings overgrown with shrubs. The way became steeper as we wound upward beneath pine and alder trees, along a narrow trail that was in places hardly more than a tunnel through the dense undergrowth. Above us hung the splendid, bell-like, red flowers of *Fuchsia fulgens*, a straggling shrub that clambered among the branches. Higher still, a small tree with narrow, silvery leaves became prominent in the forest of tall alder trees. It resembled *Buddleia megalocephala*, but it turned out to be a new species that was named *B. hypsophila* by Dr. Ivan M. Johnston.

At twelve thousand feet, we emerged from the dense forest into a region where the only trees were low pines in open stand. Between the scattered pines grew tussocks of coarse grass and lupines with glorious purple and white blossoms. The whole steep slope, stretching upward almost to the summit, was tinted by the lupines: Santa María wore a regal purple diadem. I threw myself happily down among these splendid flowers and reveled in their beauty, while I rested from the long skyward pull. Then I continued upward, slowly, for the way was steep and the atmosphere thin, passing over exposed ledges of rock bright yellow with the blossoms of a prostrate stone-crop.

The windswept summit, where my aneroid barometer registered twelve thousand four hundred feet, was covered with huge, irregular rocks with low herbs growing between them, and here and there a stunted pine tree. There was no trace of a crater; but from time to time pungent volcanic gases, blown up by the wind, greeted my nostrils. In 1902, after a long interval of quiescence, the whole southern quarter of the lofty mountain was blown bodily out by an explosive eruption of titanic force. In the vast gap so formed in the flank of the extinct mother cone, a daughter volcano—or perhaps, more correctly, a son, since it is called Santiago—had slowly been growing up; it was the source of the gases that I smelled. Later, while I collected plants on the coffee plantations on the Pacific slope, I often gazed upon the grim, gray, castellated mass of this young volcano, with a great column of smoke perpetually rising above it. By night, I would sometimes watch glowing streaks of freshly extruded lava coursing down its naked sides. But now, while I stood on the summit of Santa María, this crater was hidden from me by the smoke and clouds that hung above it.

Far off in the west, the great volcanic cones of Tajumulco and Tacaná rose above the clouds that filled the lower intervening spaces. To the east, beyond a profound and narrow valley, I could see extinct Zunil; but all the loftier cones beyond were so wreathed in clouds that I could not recognize them. The sun vanished gradually among towering cloud banks, with no farewell glow. The full, bright moon floated up above Zunil, and threw the great, black, tapering shadow of Santa María over the lower mountains to the west. Through rifts in the clouds gleamed the clustered lights of San Felipe, Retalhuleu, and other towns far down toward the Pacific coast; on the opposite side, the brightly twinkling electric lights of Quezaltenango seemed almost at my feet. Some of the lines that marked the streets were so sinuous that I smiled. Jagged streaks of lightning played among the clouds banked below us, momentarily illuminating the great, opaque masses. The rumble of distant thunder reached our ears. It was a novel and stirring experience to look down, rather than up, on the lightning, as though I were Jove hurling my thunderbolts from high Olympus; it made me feel that I participated intimately in the play of the elements.

While my porters composed themselves to sleep beside the campfire, I lay down on a ledge with head and shoulders beneath a huge overhanging rock that offered a modicum of protection from wind and cold. Just as I was dozing off, the mountain beneath me quivered as though composed of jelly instead of solid rock. Awaking with a vivid sense of the great mass above me, I was on my feet in a trice. My shelter stood firmly enough, but from far below arose the long-continued roar and rumble of a rock slide that the earthquake had started from the precipitous cliffs surrounding the gap in the side of the cone. When all was quiet again, I resumed my hard couch and slept as well as the cold would permit.

In the morning, I enjoyed a superb view over half of western Guatemala. I counted nine volcanoes beside the one beneath my feet. Part of Lake Atitlán gleamed in the distance. The wind was so cold and biting that even wearing a sweater, a jacket, and a woollen undershirt, with my blanket wrapped around me, I could not face it for long without becoming thoroughly chilled. Flocks of hardy Yellow-eyed Juncoes flitted among those glorious lupine bushes. A lone Red-tailed Hawk soared above the mountainside. All day I collected that wonderful flora of the mountaintop, continuing until the waning light made us hurry downward, to walk in darkness across the plateau to Quezaltenango.

After drying the plants from the volcano and writing my notes, I went down to Santa María de Jesús. All the trucks and public cars going that way left Quezaltenango at four o'clock in the morning, to meet the train from Ayutla to Guatemala City. At half past three, the truck in which I had engaged passage shattered the silence of the night as it rumbled over the cobbled street to my hotel. Then, trying to collect another passenger of whose residence he was uncertain, the driver drew a mild rebuke from a policeman by blowing his horn too loudly and too often. By four o'clock, he had gathered all his passengers and started on his way. Daybreak found us at Santa María, thirteen miles down the Pacific slope from Quezaltenango.

The village, consisting of a few dozen houses strung along the highway from Quezaltenango to San Felipe, was beautifully situated, at an altitude of fifty-six hundred feet, in the profound gap between the volcanoes Santa María and Zunil. The slope on which it rested swept grandly upward, with hardly a break in its smooth curve, to the summit of Volcán Santa María, seven thousand feet above. A short distance on the opposite side of the village, the Río Samalá flowed through a deep ravine, where it was impounded by a high dam. The overflow from the dam rushed through a deep and narrow canyon, then plunged over a cliff in a thundering cascade. Below the waterfall, in a deep abyss, stood the hydroelectric plant that generated current for Quezaltenango and, until it was severely damaged by the floods and landslides of the preceding year, for the audacious and costly Ferrocarril de los Altos, which climbed up from San Felipe to Quezaltenango, six thousand vertical feet, in twenty-seven miles.

I stayed in a small inn kept by a German. While arranging my plants in the driers in front of the window of my little room, I could watch the traffic pass along the main highway leading up from the Pacific lowlands. The few automobiles that went by were mostly trucks, busses, and public cars. In the morning, beginning soon after five o'clock, the movement was chiefly downward. In the afternoon, the heavily laden trucks came laboring upward in second or even lowest gear, steam spouting from their overheated radiators, bearing products of the lowlands or imported merchandise. Many stopped in front of the inn to pour cold water into their boiling radiators, replenish their supply of gasoline or oil, or give refreshment to their occupants. But a much greater weight of merchandise, I estimated, passed on the backs of mules and Indian porters, and the latter bore the bulk of it.

It was instructive, and slightly disquieting, to watch these Indians toil upward, bent under heavy loads of the produce of the hot country, among which bananas and plantains were conspicuous, or trot more rapidly downward under great packs of new pottery, cheap furniture, apples and peaches, or other products of the highlands. In the rain that fell almost every afternoon at this season, pack and porter were covered with a rain cape made of long strips from still unexpanded palm fronds, sewn together with overlapping edges, like shingles of a roof. Sometimes a whole family, father, mother, and children, down to little tots who had not been long on their feet, marched by, each, including the youngest, bearing a burden adjusted to its strength. Children of such tender age that, in prosperous societies, no work would be expected of them, toddled along under surprisingly heavy loads. As evening fell, the laden travellers passed by with small bundles of sticks tied somewhere to the outside of their packs, the meager firewood for their bivouac beneath an overhanging precipice, a niche dug into the side of a highway cut, or a crude roadside shelter. Whenever I awoke in the night, I heard the footfalls of burden-bearers passing over the rough cobblestones that paved the road through the village, but the stream was thinner now.

A hundred yards beyond this stream of carriers, across a narrow pasture, I glimpsed a stretch of the trolley wire and tracks of El Ferrocarril de los Altos, tilted upward at an astounding angle for a railroad. A year had slipped by since a train had passed, as rusting rails spiked to rotting ties attested, and nothing had been done to repair the line. Yet a fraction of the labor that went into carrying head loads over this route might have sufficed to restore the railroad, and the water that thundered over the falls, past a hydroelectric plant operating at a fraction of its capacity, could have lifted up all the cargo from San Felipe to Quezaltenango.

By crossing over the top of the dam, I easily reached the flank of Volcán Zunil, covered with wet forest that was broken only here and there, near the base, by small clearings. Its rich subtropical vegetation provided some of the best collecting that I found in Guatemala. Needing a guide to explore the unbroken woods on the higher slopes, I requested the local Comandante de Armas to find one for me. The Comandante, Don Ricardo Flores, replied that it might take a day or two to arrange this, but on the morrow I might go with two of his soldiers to visit some hot springs, called Los Jorgines in honor of General Jorge Ubico, then President of the Republic. A road was being constructed to these springs, which were reputed to possess therapeutic virtue, to make them accessible as a health resort. The soldiers were to blast rocks on the new road, a better use of high explosives than soldiers commonly make. At seven o'clock next morning, I started up the highway with a corporal and a private. They obligingly waited while I collected an herb with splendid purple flowers (*Achimenes longiflora*) that adorned the roadside cliffs, and a shrub with slender, drooping, green stems and a profusion of dull red blossoms (*Rondeletia strigosa*).

Following the highway, then the railroad, then climbing steeply through brush and a cornfield and a pasture, warm work under a bright sun, we finally reached the new road, amid a grove of Alder Trees. It was still a shelf only wide enough to serve as a footpath, cut into the nearly vertical mountainside. We crossed gullies on logs, balancing ourselves precariously high above the treetops, where loss of equilibrium would have been fatal. In spots where the rock had been recently blasted, sulphurous fumes issued from fissures, and the fractured surface was stained yellow by them. Here and there, the rocks were hot, heated by the earth's internal fires. The whole mountainside, a spur of Volcán Zunil, seemed to be underlain by a subterranean furnace. Nevertheless, the narrow shelf along which we passed was to be widened into an automobile road. The hot springs were situated at the head of a beautiful ravine, where ferns, from stately tree ferns to delicate filmies, grew with a profusion that I have rarely seen equalled and never surpassed. Vigorous plants of *Marattia excavata* had finely divided fronds twelve feet long by seven feet broad, springing from stems fourteen inches thick and not much higher.

From a narrow fissure at the base of a cliff sprang a stream of clear sulphurous water, not actually boiling but so hot that, fifty feet from its source, I could not immerse a finger in it. Its taste, although acrid, was not unpleasant; it quickly tarnished a bright silver coin. Into a rock-walled basin, the workmen had conducted part of this hot current and also cool water from a neighboring rivulet. The mixture had just the right temperature for my first bath in water heated by volcanic fires.

After my ablutions, I opened my plant press and collected to the sound of blasting around the flank of the mountain. Among other things, I found a splendid new gesnerid with large, bright orange-red, campanulate blossoms, which was later named *Kohleria skutchii*. I had by no means gathered all the interesting plants in sight when a soldier came to tell me that Comandante Flores was waiting for me along the road; but as it was late in the day and I was tired, I did not hesitate to go. Walking swiftly down the mountainside, we reached the highway, where the Comandante stopped a passing truck that, for a small price, carried us down to Santa María.

Two days later, I started up Zunil with the guide that Don Ricardo had found for me. At first we travelled easily beneath high, closed forest with little undergrowth, but as we climbed higher, the ill-defined trail faded away, the undergrowth of shrubs, vines, and scrambling bamboos became denser, and José, my guide, had to use his machete to open a passageway. While cutting a trail through a thicket of low, slender palms, his dull knife glanced off one of the hard, smooth stems and struck his foot, inflicting a rather long wound that bled a little. As it was then nearly noon, I decided to stop for lunch while José treated his injury with tobacco that he first chewed. We had reached an altitude of eight thousand feet; the summit was still distant; but since we were not prepared to camp, it was time to turn downward and collect along the way. This made our progress so slow that, by half past five in the afternoon, we had descended only five hundred feet below the point where we had lunched. Clouds now veiled the mountain; daylight was fading; it seemed prudent to close the plant press and hurry downward.

Descending rapidly, we soon reached a great tree with plank buttresses, between which José had once spent a night when he lost his way, as he had revealed to me while we climbed upward. As we passed this landmark, I admonished my guide not to let the same thing happen to the two of us. It was soon evident, however, that we had gone astray. I wished to retrace our steps to José's big tree, where we could pick up the trail that he had made in the morning, but he insisted that he could find his way without turning back and wasting the little daylight that remained. After walking a good distance, we found ourselves in a thicket of slender bamboos, on a narrow ridge that fell away sharply in front and on both sides. While I waited with the plant press in the dark, José took my flashlight and explored in three directions, returning each time baffled and breathless.

Convinced at last that we must pass the night in the woods, toward ten o'clock we found a clear, level spot and with some difficulty kindled a fire with wet wood. After heating three tortillas that José had saved from his lunch, we shared them equally—our only supper. I spread my rubber poncho beside the fire; José made himself a bed of palm leaves on the other side, and we lay down to rest. We could hear the clock in Santa María striking the hours, dogs barking, and cars passing along the highway, which seemed quite close but was separated from us by a barrier impassable in darkness. Although we had nothing to cover us against the nocturnal chill, by turning one side and then the other to the campfire, we managed to stay fairly comfortable. By the morning light, we had no difficulty finding our way back to the village, in time for a late breakfast, after which I arranged my specimens in the driers, then slept far into the afternoon. This was the only time in all my travels when I passed a night in the woods because I was lost, although on several other occasions I came perilously close to doing so.

Collecting on Zunil was so rewarding that I might have devoted a month to it, if I had not met Joaquín Glaesemer. A German in charge of a large coffee plantation down on the Pacific slope, he stopped at the inn where I was staying, on his way to the Indian country north of the Sierra Cuchumatanes, to recruit hands to pick the forthcoming coffee crop. Learning that he was familiar with this high, roadless range that I wished to explore, I plied him with questions. From his answers, I gathered that travel there was difficult to arrange, but he kindly suggested that I accompany him. The opportunity was too favorable to miss; I packed up my specimens and returned to Quezaltenango to mail them to the Arnold Arboretum and prepare for the trip across the Cuchumatanes.

My preparation included a visit to Momostenango with Don Joaquín, to buy a blanket. Hiring a car, we travelled by good roads to San Francisco el Alto, where the view over the valley of Quezaltenango was superb, thence for miles through high country covered with coarse bunch grass, and little else, up to ten-thousand-foot crests. The bare, bleak landscape, with grazing sheep, reminded me of the English downs. The wool from these sheep was much in evidence in the plaza of Momostenango, where we found the Sunday market bustlingly active. This was the great emporium for handwoven Indian blankets, great stacks of which stood around the plaza. Brightly colored blankets, still damp from the washing they had received before being offered for sale, were spread on the cobblestones to dry, a treatment that seemed to cancel much of the good effect of the washing. It was difficult to choose from so many attractive designs but, after the customary bargaining, I bought for a quetzal and a half a heavy, intricately patterned blanket that kept me warm for many years. I should have felt guilty paying so little to the patient Indian woman who, squatting on the ground in front of a simple upright loom in a dimly lighted, smoky hut, devoted so many hours, so much skill and taste, to weaving this blanket, but I was sure that the trader who offered it to me had paid less and was not selling it at a loss. Among other things displayed in this crowded marketplace were locally grown English walnuts, very good and surprisingly cheap.

On the following day, Joaquín and I continued northward to Huehuetenango, in a motor truck with an extra seat for passengers behind the driver's seat. The space behind the bench on which we sat was piled high with miscellaneous baggage, including the mail, and above the cargo were piled several Indian men, a woman with her infant, and their belongings. These included several chickens tied up in a net, who squawked loudly when the driver carelessly permitted my suitcase to settle upon them, but it was not removed until I added my protest to theirs. The four of us on the rear bench were hardly more comfortable than the hens in their net, for a large automobile jack occupied most of the narrow space that had been left for passengers' legs. The journey of about fifty miles took five hours and cost a quetzal and a half for each of us, which seemed exorbitant, considering how little space we were permitted to occupy. The ballasted but unpaved highway led over several mountain ranges more than nine thousand feet high and one of ten thousand feet. Except for fringes of trees on the highest crests, the land had all been cleared. The lower fields were planted with maize, while the higher slopes furnished pasturage for many sheep. On the highest summit that we crossed grew tall fir trees, the first that I had seen in Guatemala. When I passed this way again, after an interval of thirty-two years, I looked in vain for them.

Huehuetenango, capital of the department of the same name, was a pleasant, old-fashioned town situated, at an altitude somewhat over six thousand feet, in a broad, rather arid valley with scattered, open woods of oak, Arbutus, and pine. Its climate was among the most agreeable that I have experienced. Here we put up in a clean and comfortable little hotel, kept by a German, and completed the preparations for our strenuous trip over the Cuchumatanes.

20

The Sierra Cuchumatanes and Beyond

At four o'clock in the morning of August 15, an hour after we had intended to leave, the nocturnal stillness of the sedate highland town of Huehuetenango was shattered by the iron-shod hoofs of our horses ringing out on the rough cobblestones that paved the streets. As day broke, we rode into Chiantla, four miles distant by a good road over gently rolling valley lands. Here, in the dim light of dawn, the Sierra Cuchumatanes rose in front of us like a Cyclopean wall. On the farther outskirts of the small mountain town, we found a rough trail that zigzagged up a steep, rocky slope overgrown with flowering shrubs and herbs. As we climbed higher, we found the ascent more gradual, and the slopes, less rocky, were devoted to cultivation. Maize was grown up to ninety-six hundred feet, wheat to ten thousand feet, and potatoes at least five hundred feet higher, where they took seven or eight months to mature.

Little Indian children, seeming very inadequately clad for these cool

Headpiece: Horses and mules roam over the flowery alpine meadows of the high Sierra Cuchumatanes. The ridges are wooded with alder and pine trees.

heights, herded flocks of sheep on the grassy slopes, to the resounding cracks of long whips, which here replaced the shepherd's crook of the East. Green Violet-ear Hummingbirds tirelessly repeated their squeaky songs in the scattered trees, months earlier than I had heard them during the far wetter preceding year, in the more humid mountains farther east. Cinnamon-bellied Flower-piercers repeated their sad little trills as they stole nectar from the bright flowers of shrubs and herbs. The sturdy Indians who carried our baggage surprised me by taking shortcuts up the sheer mountainside, between turns in the winding path, and arriving at the summit as soon as their mounted employers. As we rose higher, more and more of Guatemala spread out grandly before us; the summits of Santa María and other great volcanoes stood revealed above an intervening ten-thousand-foot ridge. Near the top of the long ascent, we passed between low rock walls, on or beside which grew long lines of maguey that had sent up flower stalks of truly gigantic proportions. Behind the walls were the huts and garden plots of the Indians.

I was not prepared for the prospect that greeted me when, after a climb of four thousand feet, we reached the crest of the slope up which we had so long been toiling. I had expected a descent perhaps equally sharp on the farther side, but instead, I looked over a great, nearly level, grassy meadow, stretching off to dim distances both to the right and left. Here and there in the foreground were bare outcrops of weathered gray rock, forming mounds, or even low, irregular hills. At a greater distance were long, steep ridges, supporting woods so open and thin that the whitish rock shone through them, giving these mountains a barren, desolate aspect. Those in the west appeared very lofty, and possibly rivalled Volcán Tajumulco as the highest summits in Central America.

For hours we rode along a nearly level trail, over a plateau whose altitude, as we crossed from south to north, fluctuated between ten thousand six hundred and ten thousand nine hundred feet, according to my pocket aneroid. Upon this elevated base stood ridges that interrupted or enclosed the level areas. We passed over treeless alpine meadows, stretching away, sometimes for miles, in one level sweep to the base of a distant rocky ridge. The ridges alone were wooded, almost exclusively with pine and Alder trees in open stand. The pines were more numerous, and in certain areas the majority had been killed (by a caterpillar, I was told) and stood, leafless and barkless skeletons of trees, awaiting some stronger gust of wind to send them crashing earthward. The dead trees and the croaking of Ravens imparted to these high, empty spaces a melancholy cast, only in part dispelled by the bright blossoms that stippled the green, treeless meadows over which we rode. These alpine flowers became more profuse as we advanced beyond the destructive flocks of sheep belonging to the Indians who dwelt in grass-thatched hovels at the southern edge of the plateau. But even in the uninhabited center of the tableland roamed herds of half-wild horses.

The drainage of this high limestone plateau is largely subterranean. We passed extensive basins lowest in the center. Sometimes they held a shallow pond, but more often the water had drained away by some underground outlet. We noticed many sinkholes, large and small. Beside the trail were deep, narrow pits, where the roof of a subterranean passage had broken through, possibly beneath the weight of a horse's hoof. We had to be careful lest our animals step into them. At midday, we interrupted our journey beside the one stream of running water that we crossed on the whole six-hour ride over the plateau in the rainy season. Here we unsaddled our horses and permitted them to graze while we stretched luxuriously in the flowery herbage. About us grew a remarkable stemless thistle with spiny leaves that lay flat on the ground, forming rosettes two feet or more in diameter, in the midst of which, only an inch or so above the ground, were large heads of white, purple-tinged florets. It turned out to be a species of *Cirsium* new to science, to which Dr. S. F. Blake gave my name. In the cliffs walling in the lush, narrow vale were shallow caves and recesses in which travellers slower than ourselves found shelter for the night. This uninhabited spot was called Tunimá.

After our midday rest, we continued to ride at about the same altitude, but over a part of the plateau where the rocky ridges were more crowded and the intervening meadows less extensive and level. Beneath the pine and Alder trees on the ridges, a magnificent purple dahlia grew profusely in company with a beard tongue (*Pentstemon gentianoides*) with spikes of splendid, large, reddish flowers that in size and shape resembled foxgloves. In the middle of the afternoon, for the first time in hours of riding, we fell below the ten thousand five hundred-foot level and began to descend the northern face of the range, passing through noble forests of Cypress and Fir, now sadly disfigured by fire. To the right of our path, an impressive cascade poured down into a profound valley. Soon we reached the oak woods, then open slopes devoted to cultivation and pasturage.

Riding down into the narrow, picturesque valley of the Río Cocolá, we found lodging in the long, low, red farmhouse of the Hacienda Quesil. Here, amid a swarm of boys, dogs, pigs, and chickens, all of whom wandered freely in and out, we enjoyed a substantial supper without inquiring whether the kitchen conformed to sanitary standards. The meal over, we spread some dry grass on the dirty floor of a nearly empty room, covered it with our blankets, and slept soundly despite the hardness of our beds. This farm of about five thousand acres belonged to an English company that operated some of the largest and finest coffee plantations on the Pacific slope. Its chief value to the company was as a source of hands for coffee picking. The Indians who built their rude huts and planted their milpas on this estate might be evicted if they failed to heed the annual summons to make the long journey afoot to the coffee plantations when the berries ripened. Each year, as picking time approached, whole families migrated on a large scale from this remote hinterland down to the rich plantations.

Resuming our journey early next morning, we crossed two high deforested ridges and before noon reached Soloma, a large Indian village of low, thatched huts almost hidden amid the maize plants that grew tall in its fertile soil. Here we put up in a white-walled house fronting the central plaza. Our host, Celso Díaz, was the *habilitador* or recruiting agent for Don Joaquín's plantation; he received a commission for sending Indians there to pick the crop. On the following morning, Don Joaquín, Don Celso, an Indian named Gregorio, and I went afoot to visit one of the few remnants of unspoiled forest in the district. Passing through milpas, then up long, bushy slopes where sheep were pastured, after a hard climb we reached the crest of a ridge. Here our trail led through thickets with many shrubs and small trees of the Mexican Bayberry, which sometimes grew to be forty feet high and bore small, wax-covered fruits much like those of the Carolina Bayberry. Finally, at an altitude of over nine thousand feet, the shrub gave way to an old forest of tall, moss-draped trees.

At the edge of this forest, both Don Celso and Gregorio halted and refused to accompany us farther. This, they told us, was an enchanted forest, devoted to certain magical rites called *brujería*. Those who dared to penetrate the forbidden domain would lose their wits, as had happened to a poor unfortunate who had returned from an excursion to the mountaintop with a disordered mind. Moreover, the *brujos* or magicians might cut us to pieces with their machetes if they found us there. To test the strength of Gregorio's conviction that the place was enchanted and dangerous, I offered him fifteen centavos—a whole day's wage—if he would accompany us into the forest, but he could not be induced to proceed a step farther.

Don Joaquín wished to examine the scene of the Indians' *costumbre* and I to see the vegetation, so we continued upward without our guides. Like most travellers in the wilder parts of Guatemala, we carried revolvers and felt able to protect ourselves if threatened. Passing along a neglected trail beneath ancient broad-leafed trees, we suddenly emerged into a narrow, grassy opening on the mountaintop. Here stood three, small, rude, thatched shelters, all facing in the same direction. In front of each lay a rough, flat stone, covered with cold, blackened ashes. Beneath the shelters were sockets for holding votary tapers. At the back of each, interlaced pine boughs screened what was evidently a shrine; but deeming it imprudent to touch or remove anything, we did not learn what emblems or images were hidden there. Later, high on Volcán Atitlán, I came unexpectedly upon another small clearing with rough stones, bespattered with tallow from candles burnt in secret rites, resting on a bed of pine needles. The contrast between this pitiful little shrine and the great cinder cone that rose above it, a silent monument to nature's vast, heedless power, brought melancholy reflections on the weakness, credulity, and folly of man.

After satisfying our curiosity, except in regard to the contents of the three thatched shelters, Joaquín and I returned without incident to our companions at the enchanted forest's edge. Now they told us of a long subterranean passage that led to a magic cave, and a small lake or pool on the mountainside, exciting our curiosity anew. We were on the point of turning back to seek these novelties, but our host insisted firmly that it was time to return to his house for lunch, which we did. Later, Don Celso excused himself for not accompanying us by explaining that if the Indian elders of the village learned that he had trespassed upon the forest which belonged to them, he might be fined or imprisoned.

To Christianize the Indians was a principal goal of Spanish policy in the newly conquered western lands. Acceptance of the Cross was considered equivalent to submission to the Crown. But the highland Indians of Guatemala mistrusted promises of heavenly bliss by white strangers who had brought them earthly misery, first reducing them to bondage, then substituting for chattel slavery two centuries of serfdom under the *encomienda*. To placate the white conquerors, those who dwelt near the centers of authority went through the forms of Christian ritual; but everywhere the aborigines persisted in the belief that the magic of their ancestors was more potent than the imported variety. In remote mountain fastnesses throughout the Guatemalan highlands, they continued to practice their ancient rites, doubtless in degenerate form. The massacre on Volcán Santa María, of which I told in the preceding chapter, is an example of what might befall those who molested their secret shrines. Had the New World remained isolated from the Old for another millenium or two, it might have developed, without borrowing from the older cultures of the Eastern Hemisphere, an advanced civilization all its own, which might well have been admirable and would certainly have been interestingly different from any that we know. Now it is doubtful whether the aboriginal inhabitants of the Americas will ever become what they might have become had they been spared the tragedy of nearly five centuries of subjection and degradation.

The saddle on the horse that I obtained for the next day's journey had stirrups much too short for me and they could not be lengthened. After riding a few miles with my knees drawn up, I decided that it would be much less tiring to walk, so I sent the horse back to Soloma with Don Celso, who had accompanied us part of the way. We crossed another high ridge that had been shorn of all forest except a fire-scarred remnant at the top. Near the pass over this ridge was a ten-thousand-foot summit called Los Chivales, where, we were told, Indians came to catch birds. On dark, cloudy nights in August and September, when hosts of small migrants are travelling southward, the aborigines kindled large fires on the exposed peak. Confused by the conspicuous blaze, the winged wanderers flew straight toward it, as to a lighthouse on a foggy coast. Then the waiting men beat them down with clubs and gathered up their small, bright bodies for the pot. Diminutive wood warblers, which migrate by night and at this season, were probably the chief victims of this cruel practice, which I hope is now prohibited.

At dusk, we reached the Indian village of San Miguel Acatán, where we were hospitably received in the home of Don Efraim Molino, a shopkeeper who also recruited labor for the coffee plantation that my companion managed. Appropriately for one so named, Señor Molino operated a flour mill, all the heavy machinery of which had been carried across the Sierra Cuchumatanes from Huehuetenango on the backs of mules and Indians, by trails such as we had followed. Don Efraim had strewn the floor of his dining room with fragrant pine boughs, a pleasant custom of the country that was supposed to drive away the fleas so troublesome in elevated regions of the tropics. A good supper was decently served on a clean tablecloth, and, for a pleasant change, nobody wiped his mouth on its edge or, at the meal's conclusion, swished water through his mouth and spat it out noisily on the floor. Although my bed that night was only a mattress, I had clean white sheets for the only time on the whole trip.

At San Miguel, I left my agreeable companion to continue his journey alone, for he assured me that the country ahead was, if possible, even more thoroughly shorn of its original vegetation than that through which we had just passed, and I was eager to resume my collecting. While he continued down the valley, I retraced my steps, travelling slowly and gathering such plants as I could find along the way. I delayed a few days at Soloma, then, passing by way of the Finca Quesil, started the return crossing of the Sierra Cuchumatanes.

I anticipated with pleasure collecting all those wonderful herbaceous plants on the high plateau, but fate decreed otherwise. After a night in camp among the pine and Fir trees at the northern edge of the tableland, symptoms of influenza began to appear. Doubtless I had been imprudent to bathe at dawn of the preceding day in the icy water of the Río Cocolá. That southward passage over the meadows remains in memory as the longest journey of my life. My horse was now laden with specimens and equipment, as were both of my porters; there was no choice but to creep along on foot, at intervals resting on a rock, or even throwing myself down in the herbage, to recover strength.

When, at long last, we reached the squalid hovels on the southern side of the range, darkness was falling and my men urged me to halt for the night. But I could not face that prospect; I had become obsessed with a single idea, a soft, clean bed in the Hotel Gálvez in Huehuetenango, and I crept stubbornly on, reaching Chiantla that night and Huehuetenango the next morning. For a week I lay on such a bed, attended by the departmental doctor, and acutely aware of certain organs of whose presence within me I had until then no direct evidence. The proprietor of the hotel, Don Rodolpho Apel, was most kind during my illness.

While convalescing from this attack of influenza, I explored the valley of Huehuetenango with my plant press. A few miles from the town, the ancient ruins of Zaculeu stood on a tongue or peninsula of flat land bordered on three sides by ravines or barrancos. The site reminded me of that of the ruins of Iximche, near Tecpán, although the latter was on a more extensive table between deeper canyons. Evidently such a situation was chosen for a citadel because it was not too difficult to fortify and defend.

The central feature of Zaculeu, visible from afar across the valley, was a pyramid that rose in eight superposed stages or platforms, each smaller than the one below, leaving a broad shelf or terrace, along which one could walk freely, around the next higher stage. The summit, occupied by a rectangular court surrounded by a parapet, was reached by two very steep, parallel stairways on the side of the pyramid toward the center of the ruins. I estimated that the nearly square structure was about a hundred feet on a side by forty feet high. The stone walls that supported the earthen core of the structure had been covered, a few years earlier, with mortar to strengthen them—and make the pyramid more conspicuous by its gleaming whiteness. Scattered over the peninsula were smaller pyramids and mounds that had not been repaired; their crumbling masonry was overgrown with bushes and weeds. Cattle wandered among the ruins, cropping the scant herbage. When I revisited Huehuetenango with my wife three decades later, the largest pyramid had been more extensively restored, but with the loss of its aspect of venerable antiquity it had lost much of its charm.

The brooks that flowed through the ravines on opposite sides of Zaculeu united to form a shallow creek, ten or fifteen feet wide, bordered by willows and alders. As I followed the stream downward through an open valley between gentle hills, I found many Sabinos. These trees, closely related to the Bald Cypress of the southern United States, stood not only along the banks but sometimes in shallow water, with the current flowing all around them. They also grow in the rocky beds of swift mountain torrents, such as the Río Salegua, where one wonders how they ever got started and how they can withstand the pounding of the floodwaters. The thicker trunks of these Sabinos, sometimes as much as ten feet in diameter, were most irregular, with ridges and deep embayments, and not far above the ground they separated abruptly into a number of upright divisions or trunks. Moreover, the trees with very thick trunks were no higher, about ninety feet, than others with boles only two or three feet in diameter at breast height and the simple, gradually tapering trunk that one expects in coniferous trees.

These facts, and the examination of transitional stages, convinced me that the very thick trunks were formed by the fusion, or natural grafting, of several trees that grew up close together. Possibly the same is true of the famous big tree of Tule in Mexico, a member of the same species, whose trunk, some fifty feet in diameter, is reputed to be the thickest in the world. In this same valley below Zaculeu, I was surprised, but by no means delighted, to meet an old acquaintance that I never regarded as a friend—Poison Ivy.

By mid-September, I was strong enough to undertake a second expedition to the Sierra Cuchumatanes. I was slightly ashamed as I rode up the steep trail above Chiantla upon Col, the white horse I had hired, while Silverio Agustín, my porter, trudged behind, bent beneath a heavy pack. But I tried to allay my scruples by reflecting that such was *el costumbre del país*—which Silverio himself would doubtless have accepted as sufficient justification for any practice—and that, after our arrival in camp, I must be fit for concentrated hard work, while Silverio would be required only to make the camp fire, boil rice and sweet mountain potatoes, and heat the tortillas. Nevertheless, after we reached the plateau, I made a pack animal of my horse and walked with Silverio, an arrangement that expedited our progress. Col earned forty centavos a day for his owner while Silverio received only fifteen, the usual wage, both being fed at their employer's expense.

It was pleasant to walk over those high, flower-stippled alpine meadows on this sunny afternoon. When we reached the stream at Tunimá, I climbed a notched log to examine the ledges and niches on the gray cliffs where travellers passed the night. Finding them already occupied by porters and muleteers, with the prospect that more would crowd into them before nightfall, I decided to seek a cleaner and more secluded campsite. Higher in the valley, I found a tributary rivulet that issued in full volume from the rocks at the mouth of a deep and narrow ravine. The entrance to this small canyon was guarded by an ancient, gnarled Juniper Tree, fifty feet high, with a twisted trunk two feet thick, that was rooted in a cleft at the base of the cliff. Pushing through the thicket of much smaller Junipers that covered the bottom of the ravine, I proceeded inward until I found a ledge, halfway up the rocky wall, that was roofed by the projecting cliff above. This sheltered platform was wide enough for both of us to spread our bedding, for the woodpile and camp fire, and for storing our equipment. After the fire dried the dripping mosses on our ceiling, our camp site remained dry even in the rain. A pool of clear water among the rocks on the bottom of the ravine supplied all that we needed for drinking and cooking; the Junipers provided plenty of excellent firewood. By buying potatoes from the Indians who were carrying them to market over the trail a quarter of a mile away, we made our provisions stretch for four days, at the end of which we returned to Huehuetenango with as many specimens as I could properly care for.

On the morning after our arrival at this pleasant camp, September 13, I noticed frost on the green and living herbage of a nearby open glade. Here was a flourishing stand of thistles, the tallest of which were shoulder high. Their large heads of purple florets attracted many hummingbirds of four species. Between visits to the thistles and other flowers, Green Violet-ears sang persistently in the surrounding trees. After the rising sun had warmed the air, another visitor to the thistles, the Broad-tailed Hummingbird, gave spectacular flight displays; with loudly buzzing wingbeats he swung back and forth through a gigantic U with vertical arms forty or fifty feet high. Sometimes I could detect a demure female Broad-tail perching inconspicuously at the very nadir of his cometary orbit. The big Magnificent Hummingbirds and the little White-ears were less abundant here. Two diminutive Wine-throated Hummingbirds repeated their sweetly varied songs on a bushy slope, at an altitude of eleven thousand feet.

Red-shafted Flickers, Hairy Woodpeckers, blue-crested Steller's Jays, Golden-crowned Kinglets, croaking Common Ravens, Common Meadow-larks, Eastern or Common Bluebirds, Brown Creepers, and Spotted Towhees gave a distinctly northern aspect to the bird life on this frosty morning. Savannah Sparrows were numerous along the stream that flowed through the wet meadow, where I surprised one carrying a caterpillar, evidently for a fledgling that I failed to find amid the dense herbage. On my first crossing of the Cuchumatanes, in mid-August, I had heard the buzzing song of this resident race of a northern species but could not pause to look for the songsters.

Even now in mid-September, the widespread Rufous-collared Sparrows still sang freely. At this late date, one pair fed two newly hatched nestlings, in an open cup of grasses and rootlets, covered on the outside with green moss, lined with horsehair, and well hidden amid low herbage. But the most delightful songsters on this cool, bright morning were the Audubon's Warblers that flitted through the scattered pine trees, the males in full, resplendent nuptial plumage. Doubtless they, too, nested on these heights. Among migrants, I saw a Solitary Sandpiper along the stream, a Northern Water-thrush in the ravine where we camped, and a single Townsend's Warbler. Amid all these birds of northern types, the young Scaled Antpitta that lurked amid the Junipers seemed distinctly out of place.

The herbs in the meadows were already drying up and setting seed; flowers were less abundant than I had found them a month earlier. Nevertheless, I found enough to fill my papers. Most abundant were the composites, including dahlias, cosmos, everlastings, fleabanes, stevias, and a new species of hawkweed. Mints were also well represented, especially by the sages or salvias, beloved of hummingbirds, of which I collected five species; the parsley family, by sweet cicely and two kinds of eryngo. Geraniums bloomed on the hillsides, and buttercups brightened a shallow pool in the alpine meadow. I left this fascinating plateau with a great desire to return for a month or so, earlier in the year when more birds were nesting, but, like so many other cherished projects, this was never fulfilled.

The country around Huehuetenango was delightful for horseback riding, with roads and trails usually mudless even in the rainy season, few cars to bear down disconcertingly upon the rider along a narrow highway with a precipice on one side, grand mountain panoramas, and a climate usually spring-like. After collecting for six weeks on the great coffee plantations on the Pacific slope (of which I shall presently tell), I returned early in November to Huehuetenango to arrange my longest equestrian journey, which was to take me into the wild country of northern El Quiché. Again with Silverio and Col, we passed eastward into the arid valley of the Río Blanco. Here, at Aguacatán (the place of avocadoes), where the Indian women dressed attractively in white bodices and dark blue skirts, I learned what happens to some of the water that sinks out of sight on the lofty plateau of the Sierra Cuchumatanes. It issues in crystal springs from the foot of long, rocky, sterile slopes that sweep up and up to high, tree-fringed summits.

The Río San Juan, a stream twenty to thirty feet wide, is the surface-flowing continuation of a strong subterranean current, which emerges through a fissure in the rocks with a turbulent upwelling that reminded me of ocean breakers thrusting themselves irresistibly through narrow coves and recesses on a rockbound coast. A few yards from the point where it leaps forth into the daylight, this river, born full-grown, flows full and deep through a fertile, intensively cultivated plain whose verdure is its gift. Many Black Phoebes and a few Amazon Kingfishers frequented its course. In this and similar areas, the Indians, using primitive methods of irrigation, raised onions and other vegetables in flourishing gardens that contrasted sharply with the surrounding barren slopes. And thus the rains that sink through the porous rock on the high plateau at last serve the purposes of man.

The newly opened highway that we had followed from Huehuetenango ended at Aguacatán. From this point onward, we travelled over a rough trail, passable only by horses, mules, and men afoot. This led us down the valley, past some ancient Indian ruins, across the Río San Juan and a smaller stream, then up a long, rocky slope with scant, scrubby vegetation. Here Rock Wrens were more abundant than at Huehuetenango and Aguacatán, where I often watched them hopping over the roofs of unglazed tiles. My attention was drawn to them by a pleasant chirping that made me look upward, for it sounded like the twittering of swallows. Then I would look around and espy a pair of these brownish, spotted birds standing on a pile of loose rocks, twitching their bodies up and down like dippers on a streamside boulder. But instead of diving into the water, the wrens dived into crevices among the rocks, doubtless to hunt insects and spiders. They sang with loud, bell-like notes and seemed to remain paired at all seasons.

On this same stony mountainside, a month later, I saw more Common Ravens than I have ever found elsewhere. Six that hopped around together on the ground seemed to be playing. One circled in the strong breeze with a Marsh Hawk, apparently just to enjoy the exercise. Later I had nine in view at once. Bolder or tamer than I have ever known the Common Crow to be, one raven stayed on its trailside perch while I rode by within twenty-five feet of it.

On the higher slopes the soil, although shallow, became less dry and rocky. Here the trail led past Indian hovels, strongly tilted fields of ripening maize, and pastures where thinly clad children herded sheep with noisy whips. Here, toward its eastern end, the Sierra Cuchumatanes was lower, with a pass at only eighty-five hundred feet. As so often happens in Guatemala, in crossing this divide we passed abruptly from an arid to a humid region, and, after ascending by a dry and stony trail, we descended the northern side of the range over slippery mud.

Continuing downward to the big Indian village of Nebaj, we entered a region with so many trees of familiar northern kinds that it was easy to imagine myself twenty or twenty-five degrees of latitude farther north. I collected specimens of most of these trees along a delightful two-mile stretch of the well-named Río de las Cataratas below the village, where with many a leap and plunge it rushed loudly down a rocky channel, then threw itself over a hundred-foot cliff into a wild, verdant gorge of quite different aspect. Between the tree-lined bank and wooded enclosing slopes were long, level glades covered with emerald sward that made walking a pleasure.

Here, in November, at an altitude of about six thousand feet, Ash, Hornbeam, Hop Hornbeam, and Maples of two kinds were shedding their winged or bladdery fruits; four or five species of Oaks dropped their acorns; Holly was laden with scarlet berries and Cornel with blue; fragrant Mexican Bayberry was covered with little waxy fruits. Willow, Sweet Gum, Alder, and Sumac also flourished here. The foliage of one kind of maple and certain trees of the Sweet Gum was bright red, but most of these trees of northern types were now shedding their leaves without brilliant autumn tints, as is, with few exceptions, the way of trees in the tropics.

A pair of long-legged, lead-colored American Dippers foraged on the wet streamside rocks, or plunged fearlessly into the surging, pellucid water where it was most broken, to the ceaseless music of the many cataracts.

When this lovely valley was flooded with the most brilliant sunshine without becoming uncomfortably warm, I was so filled with joy that I wished this one perfect day would never end.

While collecting in this valley, I suffered an eruption on my hands, similar to that caused by the familiar Poison Ivy of the United States, which I attributed to a species of Sumac, a small tree, that I had gathered a day or two before. Later, in Costa Rica, I again had a slight rash on my hands after making specimens of a sumac, a tall but very slender tree of the mountain forests. Of all the thousands of kinds of plants of tropical America whose juices have moistened my skin in the course of my collecting, the sumac, and the Cashew with its seed pods, are the only ones that have caused dermal eruptions. The sumac is a close relative of the common Poison Ivy and the Cashew is a more distant member of the same family. The plants of tropical America are no more noxious to man than those of the temperate regions of the continent. Nettles there are in considerable variety, including some that are formidable, but unlike the insidious juices of certain species of the genus *Rhus*, they give prompt warning of their defensive array of poison-filled syringes, and one rarely suffers more than a few sharp pricks from them.

At Huehuetenango, I had been given a letter of introduction to Don Guillermo Samayoa, one of the wealthiest citizens of Nebaj. When, a few days after my arrival there, I presented the letter to him, he rebuked me mildly for collecting plants on his property along the Río de las Cataratas without permission. Then, thawing, he suggested that I visit the Finca Chailá, a farm that he owned on the Río Copón, in the northern part of the department of El Quiché. His son, Oscar, a boy of seventeen then on vacation from the agricultural college in Guatemala City, was about to join an older brother who took charge of this very isolated farm, and I could go with him. When Silverio heard of this intended trip, he came to tell me that he had just received news that his wife was sick, and he must return promptly to her. Although I doubted this story, I did not insist that my porter accompany me, since I had engaged him to go only as far as Nebaj and Solzil. However, Don Guillermo persuaded Silverio to go with me to Chailá, promising that he could return promptly if he did not like it there. When I offered to send a message to Silverio's wife, advising her that he would be absent about two weeks longer, he assured me that this would be unnecessary. I heard no more about her illness.

After breakfasting with the Samayoa family, I set forth on a wet morning for the wilder country to the north, mounted on Col, and in company with Oscar, who rode a shaggy mule with a long and tireless stride. We travelled through humpy limestone country covered by fields of ripening maize and scrubby pastures, over new roads intended for motorcars, which no car could yet reach because many roadless miles separated this district from other highways. Under the rain, the soft, unconsolidated earth of these roads had become soft mud into which the hoofs of our poor animals sank deeply. About noon, we rode into Cotzal, a most unattractive village of slovenly, neglected houses scattered over a hillside, with muddy or rocky passageways that served as streets between them. Here the chief occupations, aside from growing corn, seemed to be flea-hunting and rope-making. The females of the village squatted by twos in doorways and on narrow porches, diligently searching for vermin in each other's stringy black hair and putting to their mouths whatever they found.

To watch the rope-making was more interesting and pleasant. The cordage was made of maguey fibers by teams of three. While one man stood with a wooden twirler in each hand, an assistant added fibers to each of the strands that he was twisting, taking care to keep them of uniform thickness. As his strands lengthened, the operator of the twirlers walked backward across the yard and out into the roadway, where, fortunately, traffic consisted only of pedestrians and an occasional horseman. The cordage made in this simple manner was strong, durable, and cheap.

Continuing northward from Cotzal, in the afternoon we entered a beautiful valley where, at an altitude of nearly four thousand feet above sea level, enormous Sycamore Trees, a hundred and fifty feet high with trunks seven feet thick, stood above the stream like columns of white Parian marble that some titanic Phidias had carved and adorned with verdure-bearing boughs. Even taller, approaching two hundred feet in height, were the Mexican Elms and the Walnuts, which had already dropped their fruit and were now almost leafless. Huge oaks (*Quercus skinneri*) were strewing the ground with great acorns hardly smaller than the walnuts. At Nebaj, I had found mature acorns (of *Quercus exaristata*) about the size of garden peas. I calculated that the large acorns had two hundred and fifty times the bulk of the small ones. The forest that once covered this valley must have been truly magnificent, but nearly all the original trees had been felled to make way for the coffee plantations of the Finca San Francisco, where I was hospitably entertained. The three or four hundred thousand pounds of coffee that this isolated farm produced annually had to be transported on mule back across the Sierra Cuchumatanes to Huehuetenango or Quiché, a three-day journey, before it could even be loaded on motor trucks for the long haul to a railroad. Cypress trees were being planted on a large scale to replace the vanished hardwood forest.

Passing still outward, over the abrupt Cerro Putul, where first I glimpsed that magnificent bird, the Quetzal, we entered a subtropical region of humid mountain forests burdened with epiphytes, where magnolia trees grew, and Slate-colored Solitaires sang their sweetly pensive verses, which reminded me of a little mechanical music box that I had once owned as a child. Riding along the forest trail, I rummaged in the depths of my mind until I recalled the simple tune that the box played when I wound it up, a melody that I had not heard for nearly a quarter of a century. Then it occurred to me that perhaps nothing once remembered is ever quite forgotten but is merely stored away in a dim corner of the mind, until revived by an appropriate stimulus.

Descending the steep flank of a forested ridge in the "Zona Reyna" of northern El Quiché, we stopped to lunch beside one of the most enchanting rivulets I have ever seen, and while our Indian porters built the inevitable fire for heating their tortillas, I followed it upward through the woods. The steeply descending channel was a series of roughly semicircular basins, each deepest on the upstream side, from which the bottom sloped gently upward to the smoothly rounded, level rim. Each basin was brimful of the most pellucid water, which flowed slowly over the rim to the next basin below. Some of these pools were so broad and deep that I might comfortably have bathed in them, but the journey ahead was so long that we could not linger. The sand-colored rock, evidently travertine, that formed the streambed had been built up by the slow precipitation of the calcium carbonate dissolved in the limpid water. Living roots were embedded in it, yet it was certainly too hard for their growing tips to have penetrated. When we passed this rivulet on the return journey, it was reduced to a mere trickle by a week without rain, yet it was the only stream that we crossed in a whole day's journey. We were still in limestone country with largely subterranean drainage.

In this remote region, only the long, almost level-topped ridges were wooded. The broad valleys, which not long before had evidently been extensively planted or used for pasturage, were now choked with dense second-growth thickets. The trail through these thickets was often a deep, mud-bottomed trench cut through the riotous verdure, or even a narrow tunnel, along which I rode with my head affectionately pressed to Col's white neck, to avoid the fate of Absalom, as in this barberless region my hair was growing long. Perhaps scarcity of surface water, along with lack of roads, was the reason for the large-scale neglect of this "Zona Reyna," which had acquired considerable fame as a land of great agricultural promise and had been parceled among his friends by President Reyna Barrios. At one of the widely spaced farms where we passed a night, I was ashamed to lead my overworked horse to drink at the stagnant pool, covered with green scum, that was the only water available for him.

After four days of strenuous travel over rocky or muddy tracks, where it was often better to lead one's horse than to ride, we came in view of the Río Copón, the bluest stream that I have ever beheld, winding its impetuous way through the last foothills on the border of the vast, hot, forested plain of

El Petén, then a scarcely broken wilderness that covered a third of Guatemala. Viewed from an elevation of three thousand feet on the last ridge that we crossed, this plain was a sea of treetops that stretched away and away to a distant horizon as level as that in a seascape. Great, isolated masses of white cumulus clouds drowsed at midday over this ocean of dark foliage, with their flat undersides all on a level, as though they floated upon an invisible fluid. Their shadows were great, irregular, black blotches upon the velvety carpet of treetops.

Although far inland, at the Finca Chailá I was back again in the Caribbean lowlands, among Montezuma Oropéndolas, Melodious Blackbirds, White-tipped Brown Jays, Royal Flycatchers, Rufous-tailed Jacamars, and many wintering Catbirds, in the world of vegetation and birds with which I had already become familiar at points nearer the sea. But the botanical collecting was poor. The big, substantially built stone house that we had come so far to reach stood in the midst of a vast area of low, impenetrably tangled second-growth thickets, remote from the splendid forests that we had laboriously traversed.

After three days at the Finca Chailá, I bade farewell to my young companion Oscar and started back with Silverio and Col. The horse seemed reluctant to desert the lush lowland grasses for Huehuetenango's lean pastures; although ordinarily docile, he balked at the first little stream that crossed the trail, and it required much persuasion, fore and aft, to make him pass. As we rose into the highlands, where the dry season was beginning, the wayside was adorned with a profusion of flowering salvias, purple and blue, that would have gladdened the heart of the most melancholy traveller. Proceeding slowly and collecting along the way, we reached Huehuetenango after a month's absence. Here Don Rodolfo, the hotelkeeper, had a sad tale to tell. When Silverio failed to return after two weeks, his wife had come every afternoon to the hotel, to weep for her absent husband, who, she firmly believed, had been eaten by the *tigre* or jaguar. In vain Don Rodolfo tried to convince her that Silverio would return with me. I remembered then how Silverio had rejected my offer to send a message to his wife.

The ancient gnarled Juniper (*Juniperus* sp.) at the mouth of the small canyon where we camped on the Sierra Cuchumatanes.

21

Contrasts in a Plant Collector's Life

Dividing my time between the mountains of northern Guatemala and the coffee plantations of the Pacific slope, I led a life of contrasts during my six months of botanical collecting. North of the Sierra Cuchumatanes, where the population was almost wholly Indian and the only roads were rough trails, I lived with few comforts, sleeping on the ground, the floor, a hammock, or a hard wooden platform, eating such fare as I could find, although occasionally, as at Nebaj and the Finca San Francisco in northern El Quiché, comfortable lodgings were available. On the Pacific slope, I received unforgettable hospitality on great coffee plantations, where it was delightful to live among cultured people, amid magnificent scenery, and yet find the collecting even better than in some of those northern valleys where the aborigines had destroyed all the forest, leaving a desolate region in which neither nature nor man had much to offer for the spirit's nourishment.

I twice visited the coffee plantations on the Pacific slope, from late Sep-

Headpiece: The Indian village of Soloma, amid fields of ripening maize, north of the Sierra Cuchumatanes.

tember to early November in the wet season, and from mid-December to late January in the dry season, each time staying five or six weeks and moving from plantation to plantation. At beautiful "San Diego Miramar," near Colomba in the Department of Quezaltenango, a wartime major in the British army showed me paintings and pen-sketches of distinction, which he had taught himself to make during rainy afternoons on his plantation. In a room above the *beneficio* where the coffee beans were processed, I saw for the first time how an etching is printed. In his three children, I found interested disciples in the cruder art of preparing botanical specimens. At huge "Helvetia," I spent rainy October afternoons discussing metaphysical questions with a lady of Norwegian birth who spoke excellent English— without which our discussions would have been impossible. At "Mocá," we ate breakfast on the vine-covered portico of a mansion almost palatial, looking out beneath royal palms upon a long lake whose still water reflected the great shapely cone of Volcán Atitlán, with snow-white egrets resting along the shores, ducks, grebes, and wintering coots swimming in the shallows. To protect these lovely adornments of the pond from the slingshots of the Indian plantation hands required constant vigilance.

At San Diego, I was enchanted by the coffee plantations under natural shade. After clearing away the undergrowth and thinning the canopy, the coffee bushes had been set beneath the original forest trees, instead of trees specially planted to provide the shade they need. These naturally shaded plantations proved a rich collecting ground for trees. Since thinning and absence of tangled undergrowth left a more open woodland, it was much easier to see whether tall trees bore flowers, many of which were green and inconspicuous and could be distinguished only by examining the canopy through field glasses. Moreover, the less crowded trees could bear their flowering branches nearer the ground, where they were more readily reached.

The great variety of trees in these plantations, some flowering, others fruiting, attracted multitudes of birds, under conditions peculiarly favorable for watching them. Here I enjoyed good views of elusive denizens of the treetops, such as the Green Shrike-Vireo, which in unbroken forest I had often heard but seldom seen. I resolved that if ever I owned a coffee plantation, it should, if possible, have natural shade, despite the opinion of experts that planted leguminous trees are better for the coffee bushes. (This wish was never fulfilled; years later, when I planted a little coffee, it was shaded by banana plants, whose fruit I needed.)

Although I had been told that in September and October I should find no trees in flower in this region, this was far from true. Generally trees with showy masses of brilliant blossoms display them during the dry season, but many with small, inconspicuous flowers were in bloom during the rainiest months.

In the district south of Volcán Santa María, the ground was almost everywhere covered with a layer of whitish pumice, in places nearly a foot thick. At San Diego, coffee, shade trees, and fruit trees were planted in the bottom of pits dug through the volcanic ash to reach the rich soil beneath. In these conditions they flourished, and the verdant plantation hardly revealed that, thirty years earlier, it had been devastated by a volcanic eruption. If not too severe, such an eruption may be a blessing in disguise, for a layer of disintegrating volcanic ash fertilizes the soil. Hence volcanic regions often support a more productive agriculture than tropical areas where the red clay soil has been leached by countless centuries of torrential rains, with no additions from plutonic sources. A lava flow, on the contrary, brings enduring desolation to the district it covers.

On my first visit to Mocá, a tall tree with pale red flowers was everywhere in blossom. After searching in vain for inflorescences that could be reached, I reluctantly felled one of the smaller of these trees, still a growth of imposing size. Collecting the five or six twiglets that I needed for specimens, I left the remainder of that fine tree to wither and rot. The thought of what I had done oppressed me. A troublesome question kept revolving in my mind too persistently to be brushed aside: Had I performed a laudable act in the service of science, or had I been guilty of vandalism? After long pondering, I found myself unable to define precisely the boundary between science and vandalism; but I suspected that some of the things done in the name of science—murderous and wasteful collecting; vivisection by tyros unfit to make the beneficial discoveries that alone might justify the suffering they cause; the removal and scattering of ancient monuments—verge dangerously upon this boundary. I resolved to walk more carefully in the future.

After this, I procured the services of a wiry highland Indian trained to prune the shade trees in the coffee plantations at Mocá. He began his work with me by stating roundly that he could climb *any* tree. Disliking braggarts, I was careful that first day to send him up the most difficult trees I could find. But he justified his extravagant claim, and thereafter I was more considerate. He worked almost entirely with lines, throwing a thin cord over a lofty branch, with this drawing up a stout rope that he doubled over the bough and secured at its nether ends, then climbing up the rope with hands and bare feet, sailor fashion. When the lowest limbs of a great tree were beyond reach of his lines, he would ascend into the top of a neighboring tree, bridge the gap between the two with his rope, then pass across to the taller tree. It was a pleasure to watch him at work; he took such delight in his skill and hummed a tune as he climbed his ropes, hand over hand, into the crowns of huge trees over a hundred feet high.

Later, in Costa Rica, I employed a boy who could scramble to the crown of the tallest tree if it happened to be draped with stout woody vines, as so many trees of the tropical forest are. Since that day at Mocá when I could not decide where to draw the line between science and vandalism, I have sacrificed no big trees, whose fall brings destruction to a wide area of forest, and few small ones. I do not believe that my collecting suffered from this forbearance. So often I devoted hours to obtaining the flowers of some lofty tree or vine, only to find afterward more readily accessible flowers of the same kind, that I tended increasingly to pass the more inaccessible specimens by, or to collect them only in special cases, as a last resort. Nothing is more exasperating than to spend much time or money for something that we might have with little effort.

This Pacific slope of Guatemala, in the mountain zone of frequent mist and rain, is the home of the splendid *Sobralia macrantha*, whose delicately pinkish-purple blossoms are among the largest of orchid flowers. The most vigorous plants that I saw grew in a low thicket at the top of a steep roadside bank on the upper part of Helvetia plantation, at an altitude of four thousand feet above sea level. The creeping rhizomes of these orchids formed a mound at least four feet in diameter, from which sprang scores of slender, leafy stems, each about seven feet high. The flexible shoots bent gracefully outward, those on the lower side of the mat leaning over the edge of the bank, and bore the superb blossoms at the tips, usually only a single full-blown flower on each stem. The vase of these orchids that adorned the dining room table in the manager's house was a joy to behold.

I spent Christmas and the New Year with Major Cecil Hazard and his family at San Diego. The dry weather had begun, and nature was in her most festive array to celebrate the holiday season. Many of the most beautiful of the tall trees of various kinds that shaded the coffee plantations were in full bloom, making a glorious display with their lofty crowns covered with pink, white, mauve, or yellow flowers, some delightfully fragrant. Many of these trees shed their foliage when they flowered, thereby making their bright colors still more conspicuous. Among the blossom-laden boughs, murmurous with the humming of myriad insects, flitted great numbers of gaily clad birds, including orioles, honeycreepers, and wintering wood warblers and vireos.

A branch of one of the cypress trees planted in the yard, cut to provide a Christmas tree for the children, was most attractive when adorned with tinsel, colored spheres, and the usual knickknacks. On Christmas Eve, the Indian men and boys who worked on the plantation assembled in front of the big house; the women had also been invited, but with characteristic shyness they failed to appear. Then candles were lighted on the Christmas tree; rockets, invariable accompaniment of any celebration in Latin America, shot up into the air; the children twirled electric sparklers; and a marimba band played in the garden. Next Major Hazard distributed cookies, candies, cigars, and little surprises to all his numerous tenants. After this, the marimba, followed by all the Indians, was carried to the laborers' cabins, where it was played until dawn. The big house again became quiet, and we all retired early.

After the New Year, I went from San Diego to Mocá, hospitable home of the Gordon Smiths. On my earlier visit in October, I had climbed Volcán Atitlán to the base of the cinder cone, and now in January I repeated the ascent, to seek plants that had come newly into flower. Starting before daybreak with an Indian guide, for two hours we rode upward through the great plantation, along a winding road bordered by lacy-leafed Grevillea trees from Australia, which had been planted for windbreaks. In the lower reaches of the plantation the coffee was shaded by the *chalum*, a species of *Inga*, but at the top, around five thousand feet, it grew unshaded. Here the coffee bushes were more robust, with more compact masses of glossy, dark green foliage, than those growing under shade at lower elevations. They were rooted on the precipitous slopes of deep *barrancos*, between which were knife-edge ridges barely wide enough for a footpath. To retard the erosion of the naked soil of these high plantations, a species of yucca called *Isote* was planted in close-set, horizontal rows, forming a living palisade that caught some of the earth washed down by the rains.

Reaching at sunrise the upper edge of the plantation at an altitude of fifty-six hundred feet, we sent back our horses and continued upward on foot through the primeval forest. The obscure path led along a razorback ridge, between tremendous gorges where Brown-backed Solitaires, the *guarda barrancos* or "guardians of the ravines" of the Guatemalans, sang turbulently to greet the new day. Their wild piping, a cascade of chiming notes, differed greatly from the pensive, restrained song of the Slate-colored Solitaire that I heard in the forests of northern El Quiché. Perhaps I can best convey the contrast between the songs of these closely related birds by calling the Slate-colored Solitaire an artist who has studied classical music in a conservatory, his brown-backed cousin a wild, untutored mountaineer, who expresses his abounding vitality by a spontaneous outpouring of ringing notes. His is the true voice of the untameable *barrancos* where he persists in the midst of encroaching cultivation.

For twenty-five hundred vertical feet, we followed upward along the same narrow ridge, through broad-leafed forest that yielded few flowering specimens for my collection. Passing a narrow opening where stones bespattered with tallow betrayed that here Indians practiced secret rites, we finally emerged from the woods onto the base of the cinder cone. Before us, it stretched bleakly up and up into the clouds that hid the volcano's summit; to our left, it continued downward as a long, broad tongue invading the forest. The first few steps over the cinders demonstrated the difficulty of climbing a slope where they rested at the limiting angle of repose. The pieces of volcanic ash were of two colors, dark gray and reddish-brown, and all sizes, but mostly between that of a walnut and that of an orange. When I tried to climb over this loose, rounded material, it rolled downward and I slid down with it, my feet embedded up to the ankles. Only with the aid of a stick could I hold myself erect and make any progress. The thought that I might have to struggle upward for three thousand feet over this treacherous footing extinguished all desire to reach the cloud-veiled summit. In any case, my work ended where the vegetation stopped.

The bleak desolation of the cinders was relieved by long fingers of green vegetation that reached up into it from the forest below. But how could flowering plants get rooted and live in such an unstable, inhospitable substratum? The hardy pioneer that bravely invaded the naked slag where nothing else grew and prepared the way for other vegetation was a member of the buckwheat family, *Muehlenbeckia volcanica*. I was uncertain whether to call it a shrub or a vine. Its slender stems pushed through the loose pieces of cinder and spread over the surface, forming low, compact, green mats, amid which I had to search carefully to find the small green flowers and the little black fruits.

Where the slope had been stabilized by this pioneer, a variety of other plants flourished. In late October, the most splendid was Maxon's Dahlia, which grew eight feet high and generously displayed large flower heads with broad, delicately purple rays and bright yellow centers. A variety of other flowering composites then adorned these ascending tongues of verdure, along with melastomes, clusias, and figworts. Even a terrestrial orchid with brownish flowers, *Epidendrum polyanthum*, was rooted among the cinders. In areas of older vegetation, low pine and Alder trees stood above a luxuriant carpet of mosses and ferns into which we sank deeply at every step. It was much easier to climb upward along the fingers of vegetation, whose roots bound the cinders together, than over the bare, shifting slides, and we ascended several hundred feet up the cone before the gathering clouds released their moisture and sent us downward.

The coffee plantations on Guatemala's Pacific slope, from the Department of Suchitepéquez westward to Mexico, occupy a zone of abundant rainfall that was once covered by a heavy rain forest of dicotyledonous trees, such as I had found on Volcán Zunil. The pine woods that cover so much of the Caribbean slope at all altitudes are rare in this region. Here, below the higher levels of the volcanoes, I found only one stand of pines, which was situated in the Finca Mocá, on a long strip of rocky ground beginning at about three thousand feet and stretching upward for a thousand feet or more. The transition from the pine woods to the mixed forest of broad-leaved trees, or the coffee groves that had replaced them, was abrupt; evidently a narrow tongue of volcanic material had created here a soil more suitable for pines than for other trees, or for coffee. Here the pines (*Pinus pseudostrobus*), noble trees at least a hundred feet high with trunks three feet thick, grew in open stand, with only a slight admixture of other arboreal species. The ground beneath them was thickly carpeted with moss and fallen pine needles.

Beneath the light, open canopy of the pines, herbs bloomed more profusely than in the denser dicotyledonous forests nearby. Some were of species that elsewhere I found only at considerably higher altitudes. A terrestrial orchid with bright orange flowers (*Epidendrum ibaguense*) grew abundantly here, its long, leaning stems propped up on wiry roots. Elsewhere I have found this orchid chiefly on the sides of highway cuttings, on bare slopes, in open marshes, and on arboreal ants' nests. Great clusters of Sobralias likewise flourished on the stony soil beneath the pine trees. Here, too, in January, I found wintering Townsend's Warblers, although I saw none among the surrounding broad-leaved trees. As I watched them flocking through the pines with Black-throated Green Warblers and an occasional Hermit Warbler, I seemed to be in the highlands, where they abound at this season.

On my first visit to Mocá, in October, I glimpsed a small bird so elegant that it reminded me of a bird of paradise. In January, after I had finished collecting and packing all my specimens for shipment and was waiting for my ship to sail, I had more time to seek these splendid little birds. In a thicket of low, tangled, scarcely penetrable second growth, I found a party consisting of about half a dozen of them, of both sexes. The males were attired in glossy black, with a bright red cap and a mantle of long, light blue feathers on the back. Their two, slender, central tail feathers were longer than their body. Their legs and toes were bright orange. The females were plain olive-green, wholly devoid of the males' adornments. By their short bill, compact body, and brisk flight, I recognized these birds as manakins, the most elegant representatives of this ornate family that I had seen. I did not hear the bright, cheery *toledo* of these Long-tailed Manakins, which at a later season rings through the woodlands on the Pacific side of Central America, from Guatemala to northwestern Costa Rica. Not until many years later did I have the good fortune to watch the elaborate courtship dance, in which two males jump up and down in perfectly synchronized rhythm.

A tall silk-cotton tree (*Bombacopsis Fendleri*) at the edge of the laboratory clearing on Barro Colorado Island.

A cayuca newly carved from the great trunk behind it, with its makers, on Barro Colorado Island, Panamá.

22

Barro Colorado Island

At the end of January, 1935, I finished collecting for the Arnold Arboretum and was free to resume my studies of the birds. I wished especially to learn more about those of the lowland forests. Although on my journeys through Guatemala I had traversed great tracts of splendid forest at all elevations, most were far from houses and sources of provisions. I knew only one place where I could find, without delay, living accommodations and facilities for study in the midst of unspoiled forest—Barro Colorado Island in the Canal Zone, to which I had already made two short visits. Accordingly, I took passage on the *Acajutla*, a small coasting steamer, from the Guatemalan port of Champerico to Balboa.

At Champerico, an unprotected port, the steamer anchored far from shore and the cargo, mostly coffee, was carried out in lighters. The only passenger, I was lowered from the high wharf to a small boat in a chair swinging from the end of a cable attached to the steam crane. The leisurely voyage of eleven

Headpiece: Barro Colorado Island from Gatún Lake in 1929.

hundred miles down the Pacific coast of Central America took ten days. We stopped to load coffee at most of the small ports along the way, sometimes tying up at the wharf, but as often dropping anchor well offshore. While we waited at anchor, I was impressed by the regular alternation of the dry-season winds. In the mornings, the breeze blew strongly off the land, and the ship, tugging at her anchor chains, kept her bow toward the shore. Around noon, when the sunbathed land became warmer than the ocean, the wind shifted, now blowing shoreward and swinging the ship around on her moorings until she pointed toward the open sea. As the land cooled after sunset, the heavier air again flowed strongly from the highlands toward the ocean, and the steamer again swung around to bring her bow toward the shore. One morning, when I made the mistake of writing on the cool deck instead of in the stuffy cabin, a sudden stronger gust of wind scattered my papers, blowing into the water a sheet of records that I was sorry to lose.

As I came on deck on the third morning of our voyage, the *Acajutla* was entering the Gulf of Fonseca. The scenery was magnificent. The spacious bay was rimmed around the landward sides by jumbled mountains, and to the seaward guarded by high islands, some of which were extinct volcanic cones. Behind the port of Cutuco rose the double-peaked mass of Volcán Conchagua, over four thousand feet high. Cultivated fields stretched halfway up its steep slopes, above which they were wooded. Off in the northwest rose Volcán San Miguel, with a great cloud of smoke continuously pouring forth from its broad crater mouth. A number of lesser, inactive volcanoes were scattered around the circumference of the Gulf.

While the ship spent two days at Cutuco loading coffee that was slowly delivered by rail, I wandered through the open woods of low, thorny trees that surrounded the little port. Many were leafless after several nearly rainless months, yet a number were in flower despite the severe drought. Among the most conspicuous of the birds were wintering Scissor-tailed Flycatchers, which I had long desired to see. By dozens, they perched on the telegraph wires and the exposed ends of the higher branches, whence they made long, graceful sallies to catch flying insects. Their chief diversion was chasing Crested Caracaras and White-throated Magpie-Jays. Sometimes two or three of the flycatchers joined in the pursuit of their much larger victim, buffeting its back. Years later, on a trip by bus from Chiapas to Costa Rica, I was impressed by the abundance of wintering Scissor-tailed Flycatchers throughout the dry Pacific side of Central America. Other migrants numerous in the light woods beside the Gulf of Fonseca were Baltimore Orioles, Orchard Orioles, Yellow-throated Vireos, Yellow Warblers, and Western Tanagers. Despite the drought, there was evidently no lack of food for the birds.

After sunset, we crossed the Gulf to Amapala, Honduras, where seventy-two heavy bars of silver, from the Rosario Mines far in the mountainous interior, were hoisted aboard in slings and locked in the ship's strongbox down in the hold. Then we continued southward, calling at Corinto, Nicaragua, and Puntarenas, Costa Rica. Up to this point, the land along which we had coasted was, now in the midst of the dry season, brown or grayish-green, in striking contrast to the full verdure of the Caribbean littoral at all seasons. Southward from the Gulf of Nicoya, however, this Pacific coast became greener. The shore along which we passed was covered with by far the heaviest forest that I had seen since our voyage began. Where the mountains receded a little from the beach, leaving some level ground, thatched huts stood among palm trees. Inland, the mountains that rose ever higher to the lofty, cloud-veiled summits of the Cordillera de Talamanca were everywhere, as far as I could see, covered with heavy, unbroken forest. I little suspected that, before the year's end, I should begin to explore these forested mountains of southern Costa Rica, and eventually make my home in a valley among them.

As the *Acajutla* entered the Panamá Canal on the afternoon of the tenth day of our voyage, the few passengers for Balboa were taken ashore in a launch. At the landing, I was met by James Zetek, custodian of Barro Colorado, who passed my baggage unopened through the customs—a pleasant change from the troublesome inspections to which it had been subjected at certain frontiers. After buying a few things at the Canal commissary, I took the afternoon train to Frijoles station, in the middle of the isthmus, where a launch awaited me. As we crossed Gatún Lake, we saw the *Acajutla*, which passed from view behind a wooded promontory of Barro Colorado Island just as we stepped upon it, shortly after sunset.

The impounding of the water of the Río Chagres, as part of the construction of the Panamá Canal, formed Gatún Lake, which flooded one hundred and sixty-five square miles of mostly forested lowland. The higher elevations within this area became islands, of which the largest was Barro Colorado, comprising nearly four thousand acres of hilly terrain, covered partly by primeval forest and partly by old second growth. Although only slightly more than three miles in diameter, Barro Colorado has more than twenty-five miles of deeply indented shoreline. Doubtless the birds, mammals, and other animals surrounded by the rising water were concentrated on the shrinking areas that rose above it; but whether this resulted in a permanently high population density of certain species, or whether territorial behavior and food requirements soon reduced the density to its normal level, is a question that appears not to have been satisfactorily answered.

In any event, Barro Colorado, with a rich representation of the animal life of the region living in unspoiled forest, was in 1923, by decree of the governor of the Canal Zone, made a reserve for the preservation and study of its flora and fauna. The Institute for Research in Tropical America undertook its development as a place where scientists could live and work in favorable conditions. A large two-story frame building, with living quarters and laboratory space, was built on a narrow spur between ravines, a hundred and ten feet above the boat landing at the head of a narrow inlet of the lake, from which it was reached by a long flight of concrete steps. Smaller buildings, for cooking, for lodging, for a library, grew up around the central edifice.

Miles of trails were cut through the forest, named for outstanding naturalists or men who had contributed to the development of the island, and marked at frequent intervals. These clean paths made it easy to observe the forest's varied life without getting lost, as one so readily does in trackless woodland. Unfortunately, as happens in reservations everywhere, these trails, intended only for those who came to watch and to study, served the purposes of poachers who came to destroy, despite the island's isolation by water and its protection by government.

First impressions are often the most vivid and enduring. Perhaps I can best convey the charm of Barro Colorado, its surprising contrasts, and its impact on newcomers to lowland tropical forests, by drawing on notes written just after my first visit, when, on my way from Almirante to Tela in 1929, I spent two days there, between ships. At the experiment station beside the Changuinola Lagoon, where I had passed the preceding six months, I was far from primeval forest, to which my visits were few and brief. Here on Barro Colorado I was in a narrow clearing pressed closely on three sides by heavy forest. As twilight deepened on my first evening on the island, the surrounding woods became vocal with the calls of unfamiliar birds, some melodious, some weird and startling, and frogs began their nocturnal chorus. Out in the lake, a red and a white channel light flashed into view. With a loud clanking from throttled engines announcing its approach long before it slipped into sight from behind a wooded headland, a low-lying freighter passed in front of the laboratory on its way to the Caribbean. After the vessel rounded the second of the forested capes that limited one's view of the canal, the clanking died rapidly away. Stillness reigned in the dark forest until it was interrupted by the low rumble and deep-throated whistle of a freight train on the railroad across the lake.

Early next morning, I set forth along one of the trails that radiate from the laboratory clearing. An easy walk took me to a wooden tower on the highest point of the island, four hundred and fifty-two feet above the lake. Climbing to its top, I looked over the green crowns of the myriad trees that covered the northern half of the island, through which I caught glimpses of the ship canal which passed around three sides of it, coming close to the ends of its longer peninsulas. For a long while I stood there in the sunshine, gazing over the island-studded lake with its intricate shoreline, beyond which low, wooded hills, dappled with the shadows of the great, white cumulus clouds that floated low above them, stretched far away into Darién.

Watching ship after ship glide past the bays in the island's shoreline, I thought of the long centuries during which a substantial part of the commerce of the Americas, and men hungry for wealth and power, had flowed across the Isthmus of Panamá in sight of this hilltop where I stood. Doubtless, even before Columbus, the aborigines had traded across this convenient neck of land. Then came Balboa, the first European to behold the Pacific's eastern shore, and rapacious Pizarro on his way to destroy the Inca empire. For centuries, strings of heavily laden pack mules transported Peruvian gold through those hills, from the old city of Panama to Porto Bello, to freight the galleons that bore it to Spain. Finally, North Americans built the Panama Railroad, which carried from the Atlantic to the Pacific coachload after coachload of adventurers impatient to reach California's goldfields. And now the Canal, with its continuous stream of iron ships bearing the modern world's infinitely varied commerce among the six continents!

Of all the multitudes of people who passed this way through so many centuries, scarcely any paused to give more than a fleeting glance to the marvellous vegetable and animal life around them; of all the incalculable wealth that flowed across the isthmus, scarcely any had been dedicated to the study of its teeming tropical life. Now, at last, a sustained effort was being made to disclose the secrets of tropical nature. Happy the island set apart for this high endeavor! Fortunate the man, gifted with patience and understanding, loving nature more than gold, who could turn aside from the streams of trade endlessly passing and tarry here to study Life!

On the next morning, I followed the Thomas Barbour Trail. I had scarcely left the laboratory clearing when loud wing beats drew my gaze upward into the crown of a great tree, where I saw my first Crested Guans, long-tailed, dark birds the size of turkeys, with red wattles pendent from their throats. Boldly colored toucans flew overhead, their enormous bills giving the impression that they carried some large, vividly tinted fruit. Here and there the thick bracts of a *Heliconia* or *Renealmia* growing beside the trail brought a touch of bright color to the sombre interior of the forest. Interspersed among the mighty broad-leafed trees stood massive Corozo Palms bearing enormous bunches of orange fruits beneath their crowns of erect pinnate fronds, and Chonta Palms whose tall, slender trunks were propped well above the ground on thorny stilt roots. The frog-like croaking of Rainbow-billed Toucans floated down from the treetops, and from away in the distance came the mournful cooing of a Gray-chested Dove. While I stood delighting in the wealth of bird life, a distant, low humming grew into a roar that drowned the notes of all the lesser birds, as an airplane, hidden from me by dense foliage, flew above the treetops. This alien sound was followed by the deep bass roars of a troupe of Howling Monkeys, excited by the aircraft's passage.

After climbing a steep hill, I stopped to examine an epiphytic fig tree whose earthward-growing roots, fusing together where they met, completely enveloped the trunk of a smaller tree, which at the first glimpse appeared to spring from these foreign roots. A short way beyond this, I frightened up some trogons who had been eating something on a trailside stump. Then, as I continued onward, the air became heavy with a strong, musky odor. A rustling of leaves, a few low grunts, and a Collared Peccary, with the bristles on its neck standing erect, stepped from the undergrowth to the open trail, regarded me intently for a moment, then scampered off through the forest. I heard several companions who remained unseen in the undergrowth. The air still held a trace of their scent when growls from close beside the path drew my attention to a Coatimundi. With long muzzle lowered, it advanced a few steps toward me. Suddenly I noticed a second coati beside the first, then both ran off together. These long-snouted, long-tailed plantigrades are more diurnal than their close relations, the raccoons, and I often saw them singly or in small parties, usually in trees.

The trail turned and began to drop downward. At the end of a long, straight stretch, I glimpsed the lake. Hurrying onward, I reached the shore just in time to see the *Japanese Prince* glide slowly past Colorado Point, hardly more than a stone's throw from where I stood. Passing through the stumps of dead trees that projected above the shallow water near the bank, the waves of its wake broke at my feet, as the freighter continued onward to Pedro Miguel Locks and the Pacific Ocean. Across the canal rose the radio towers at Summit. The transition from the ancient world of nature to man's modern world was startlingly sudden.

The backwash from passing vessels was eroding this shore, exposing the slippery red clay that gave the island its name and large rocks that stood above the shallows. Here I noticed several Gray Basilisk Lizards. The larger ones were about two feet long, with prominent crests on their heads and along their backs and tails. When I chased these wary old lizards, they ran back into the woods, but I succeeded in making some of the younger ones demonstrate what I wished to see. With foreparts elevated, they ran over the water to take refuge on the exposed rocks and stumps. Only their long hind toes touched the water, so lightly and rapidly that surface tension, or perhaps the water's inertia, prevented their breaking through and becoming submerged. The long, elevated tail balanced the lizard's raised foreparts. On land these saurians also ran on their hindlegs alone. If for any reason they broke through the water's surface, they could swim beneath it like frogs.

While I played with the basilisks and made the acquaintance of a flycatcher new to me, the *Bethore*, lying low in the water with its heavy mineral cargo, slipped by toward Gatún Locks, reminding me that, after all, I was living in an iron age, governed by clocks. I had promised to return to the laboratory for lunch at noon and had only a half-hour to retrace my steps. As I hurried back along the trail, counting the hundred-meter markers that slipped by, its charm had vanished, for wild creatures do not often reveal themselves to those who rush. Just as I was about to enter the laboratory clearing, however, a rustling in the branches overhead made me halt and look upward, to meet the gaze of a party of Squirrel Monkeys peering down at me.

Although hardly bigger than Gray Squirrels, these marmosets appear to possess a full measure of the common primate trait of curiosity. With agility surpassing that of squirrels, they jumped through the boughs to look inquiringly through gaps in the foliage at the fellow primate who rushed along the forest trail. Unlike most New World monkeys, they did not use their long, furry tails for grasping. They were attractive little creatures, with the black skin of their faces showing through a sparse covering of short gray hairs, and contrasting sharply with the crest of long, white fur on the crown and forehead. The back was gray, the nape bright chestnut, the chest and forelimbs white. Two of these titis, as they are called in Panamá, were confined behind the main building. As I stood beside their cage, reading about them in Goldman's *Mammals of Panama*, one stuck its little hand through the meshes of the wire and continued to pluck at the book as long as I held it there. I was sorry that on the morrow I had to sail from Cristóbal for Puerto Barrios.

In mid-December, 1930, I returned to Barro Colorado for a sojourn of five weeks, and after arriving from Guatemala in early February, 1935, I remained until June. During the whole of my second visit and most of my long third visit, Frank M. Chapman was also present. During his later years, he regularly passed the dry season on the island, where he made some notable studies, especially those of the Wagler's or Chestnut-headed Oropéndola and Gould's or the Golden-collared Manakin, and gathered material for his two delightful books on Barro Colorado, *My Tropical Air Castle* and *Life in an Air Castle*. Curator of Birds in the American Museum of Natural History, Dr. Chapman was then, without much doubt, the most distinguished ornithologist in America, widely known for his popular field guide, *Handbook of Birds of Eastern North America*, his editing of *Bird-Lore* (which later became *Audubon Magazine*), his numerous books and magazine articles intended for the general reader, his work as a conservationist, and such substantial scientific works as *The Distribution of Bird-Life in Colombia* and *The Distribution of Bird-Life in Ecuador*. His position in the prestigious American museum brought him into contact with many prominent men, including President Theodore Roosevelt and John D. Rockefeller, the elder, about whom he had interesting tales to tell. Aside from Herbert Friedmann, the great authority on cowbirds and other parasitic birds, whom I met on several brief, helpful visits to the National Museum in Washington, where he was Curator of Birds, Chapman was the only ornithologist whom I knew somewhat intimately until years after I began to publish in this field. I have never been so closely associated with any other ornithologist as with him, during nearly five months together on Barro Colorado.

Although Chapman was a famous scientist forty years my senior and I was scarcely known, he often invited me to accompany him about the island. On an early walk together, he helped me to distinguish the notes of three of the five kinds of trogons found there. Years later, I studied the nesting of these trogons, chiefly in Costa Rica. A short distance beyond the trogons, we noticed a tiny, short-tailed, brown bird walking toward us over the ground with a careful, mincing gait. It was the first Nightingale Wren recorded on Barro Colorado.

On the following afternoon, Dr. Robert Enders, the mammalogist, and I accompanied Chapman in the launch around the shore of the island to Drayton House, one of the cabins that stood at the end of certain of the trails, equipped with food and bedding for those who wished to pass a night or two in the midst of the forest. We helped him to carry his heavy photographic equipment to a small stream where he had seen Tapirs' tracks on the muddy bank, and to stretch across the rivulet the wire that he hoped would release the shutter for a flashlight photograph of the animal as it passed in the night. Then we returned to the laboratory, leaving Chapman to sleep alone in the wilderness, as he preferred.

Although he had promised to return the next morning, he did not appear until late in the afternoon, when, fearing an accident, we were preparing to go in search of him. He rejoined us in high spirits, after a day of adventures. Paddling along the shore in a cayuco, he had stepped on a rotten pier, which collapsed, spilling him into the water along with his binoculars, gun, and machete. While he recovered himself, his field glasses, and his gun, the dugout canoe floated away. He failed to retrieve his machete. After landing, soaking wet, he became entangled in a dense thicket of bushes and wild plantains, with only a small sheath knife to cut a path—a most inconvenient predicament. Working his way laboriously along the shore, he finally caught the derelict cayuco and returned to the cabin to dry out. While sitting on the porch, drying, he saw his first Sungrebe or American Finfoot swimming in the lake, in full view for about five minutes. He finally rejoined us minus his machete, his field glasses flooded with water, without a picture of the Tapir, yet he assured us that his day had been most enjoyable; he relished this deviation from the sheltered lives that we habitually lead. It was good to find so youthful a spirit of adventure in a man of sixty-six. I took apart his binoculars, dried and cleaned them, and made them as good as ever.

A few days later, Chapman and I went afoot to pass a night in the cabin at the end of the Drayton Trail. As we passed through the heavy forest in the center of the island, he pointed out my first Scaly-throated Leaf-tosser, a small, dark brown member of the ovenbird family, busily throwing dead leaves right and left on the forest floor with which it so well blended. Then a sudden, brief shower made us take shelter beneath a leaning trunk and a clump of palms. Finally reaching the cabin, we embarked in the cayuco moored in front of it and started to explore the cove which, branching and rebranching, wound back into drowned valleys and ravines.

The narrow channel of calm, open water ran between shallows covered with cattails, sedges, and other marsh vegetation, beyond which the high forest rose like a wall of almost solid verdure. Gaunt, almost branchless, dead trees standing in the water reminded us how the lake, with its many arms, had originated. Their naked trunks were riddled with the holes of nesting birds. As we paddled along, careful to avoid the numerous stumps and snags that lurked just beneath the surface and menaced navigation, we surprised a Sungrebe swimming ahead of us—probably the same that Chapman had seen while he dried after his immersion. We came close enough to distinguish its short red bill, the bold white line above each eye, and the black stripe down either side of its neck. It swam rapidly until, finding that we were overtaking it, it rose easily into the air and flew a short way low above the water, into which it dropped at a safer distance from our vessel. This was repeated as we bore down on it a second time. Finally, the bird entered a narrow, palm-bordered inlet and was lost to view. This was the most intimate encounter with a Sungrebe that either of us had enjoyed.

We continued up the narrowing inlet until we found ourselves in a cul-de-sac where many small palms grew at the water's edge, beneath tall trees which arched overhead. Then, in the dusk, we returned to the cabin for supper. As we were finishing our hasty meal, a deep, indescribable sound from the surrounding forest sent cold chills creeping along my spine. The first explosive utterances were followed by low moans, as though of a soul in agony. This, Chapman explained, was "Juan," the Bird of Mystery of which he had written in *My Tropical Air Castle.* Later, while I stooped over the end of the landing float and washed the dishes, the same blood-curdling calls seemed to emanate from the tree above my head. Although, knowing that they came from a bird rather than some man-eating animal of the "jungle," I was without fear, these half-grunted exclamations again brought spinal shivers. Their author remained unseen amid the dusky foliage. I do not know whether Chapman ever succeeded in identifying "Juan," but I am fairly certain that he had no desire to remove the bird from the realm of mystery by attaching a dry scientific name to it.

Shortly after sunset, a nearly full moon rose above the trees on the opposite shore and flooded with soft light the broad expanse of Gigante Bay, studded with the skeletons of many dead trees and scattered rafts of floating vegetation, that opened in front of the cabin. Far away, above the hilly mainland shore, the twin ruby lights atop the radio towers at Darién shone like two red stars. Again we embarked in the cayuco. As we pushed out from the shore, a large bat flew so low above my companion's head that he ducked violently, giving me a momentary vision of spilling into the lake. In my flashlight's beam, the bat was light gray.

As we drifted quietly along the shore, we repeatedly saw this or a similar bat skim the surface of the water with its hindparts, producing a clearly audible rippling sound and stirring up little waves visible in the moonlight. Sometimes, too, we heard the bat strike the water with a louder impact, but we could not see just what happened on these occasions. Chapman recalled that, more than thirty years earlier, in one of the passages between Trinidad and the South American mainland, he had shot a bat of similar aspect above the sea, and found the remains of fish in its stomach. The bats that we watched were probably Hare-lipped or Mastiff Bats, well-known for their habit of fishing, which they sometimes practice even in broad daylight, capturing small fish that have been stunned or wounded by diving pelicans. Apparently they seize the fish in their feet, which are exceptionally large for bats, with strong claws; but it has also been suggested that they scoop up small fry in their interfemoral membrane, which stretches between the hindlegs and includes the tail.

During my last visit to Barro Colorado, Chapman promised that if I spent the night with him at Fuertes House, another cabin beautifully situated at the head of a narrow inlet of the lake, I would hear the Common Potoo, a nocturnal bird that resembles a large goatsucker but belongs to a different family. While we waited, listening, the full moon rose above the trees that grew close around the cabin, but we heard only the chorus of frogs from the marshy cove below us. Finally, despite the irritation caused by the numerous redbugs and ticks that I had picked up on my walk across the island, I fell asleep.

I still slept soundly when a call from Chapman woke me. Sitting up on my cot, I looked out the window upon a night flooded with the yellow light of the sinking moon, reflected in countless spears of light from the glossy foliage of the surrounding trees, wet after a passing shower. The frogs still continued their chorus; and from the distance came, subdued but clear, the most melancholy notes that I had ever heard from any creature. The wildness of the setting, the pale moonlight that accentuated the darkness of the shadows beneath the great trees around us, heightened the mournful quality of an utterance that in any circumstances would have sounded forlorn. The soft, plaintive notes were so beautifully modulated that they at once brought to mind a phrase from Shelley's *Adonais*, "most musical of mourners." It was not difficult to imagine that out there in the moonlit forest wandered a beautiful maiden who had just lost her beloved, or a mother whose dear children had been snatched away, or, as Waterton the Wanderer had fancied a century earlier, Niobe wailing for her blasted children before she herself was turned to stone. *Poo-or me, oh, oh, oh, oh*, the mysterious voice seemed to cry, the notes strongest at the beginning of each phrase and falling away toward the end. As the Poor-me-one (as the bird is called in Trinidad) repeated its plaint, we noticed that two potoos were answering each other in the distance. I did not see a potoo as it called until, many years later, a pair incubated their single egg on an exposed knothole in a tree behind the house in Costa Rica where I write these lines.

At the time of my third visit to Barro Colorado, Chapman, then seventy, had recently undergone a serious operation. Although he could not walk far, he continued to explore the forest of his beloved Barro Colorado. Unable to climb the long flight of stairs in front of the laboratory, he rode up in the little car, pulled along rails beside the steps by means of a winch and cable, in which freight was hauled from the boat landing to the buildings. He did not like to be seen on this baggage truck, and one day, when a visitor snapped a photograph of him taking this ride, he became quite angry.

Dr. Chapman was a pleasant, considerate, and exceptionally stimulating companion, but one aspect of his conversation perplexed and occasionally distressed me. At one time, he would speak about birds as though they were sensitive little people in feathers; at another time, he would tell of the hundreds of thousands of bird skins under his charge at the museum, and of plans to increase this already huge collection. Not only did this seem inconsistent, it even raised a suspicion of insincerity. If birds were indeed the feeling creatures that his conversation sometimes represented them to be, then it was surely wrong to shoot them—especially when one recalled that collectors never retrieve all the birds that they hit, but inevitably leave many to die lingeringly of their wounds. But, as I have grown older, I have become more poignantly aware of how difficult it is to achieve perfect moral consistency in this world where interests clash on every side and good and evil are so perplexingly intermingled.

The great tragedy of biology is the difficulty of acquiring certain kinds of information about living things without harming them. Too often, to advance his studies in such fields as physiology, anatomy, or taxonomy, the zoologist must mutilate, torture, or kill the animals that he professes to love. I have been content to learn what I could about birds without harming them, but others, because of professional commitments or scientific curiosity so strong that it overruled compassion, have taken a different course. To be fair to bird collectors, their effect upon populations of widespread, flourishing species is negligible, but this hardly softens their impact upon the unfortunate individuals chosen to become specimens.

Perhaps, if I had tried, I might have enlisted Chapman's support, and that of the wealthy museum which he represented, of my studies of tropical birds. But I hesitated to become dependent on a museum, lest it require me to collect the bird specimens of which museums seem never to have enough, and this I could never bring myself to do. Stubbornly independent, I knew that I must go my own way and work out my own destiny, difficult and perplexing as that sometimes was. Many years later, after Frank M. Chapman had died and a fund to support ornithological studies had been established in his name at the American Museum of Natural History, I received from it research grants that enabled me to extend my studies into new regions, but these grants were given without any obligation to collect.

Toward the end of February, one of the most beautiful lianas of the forest, the Purplewreath, came into bloom. Its flowers, each consisting of a violet-like purple corolla in a larger lavender calyx, were displayed over the high treetops, where we ground-dwellers could rarely see and enjoy them; but fallen blossoms littered the forest floor in many places. This ornamental vine of the verbena family is widely planted in Central America; in the municipal park of Retalhuleu, Guatemala, I stood long admiring a circular arbor that was almost solidly covered by the pretty lavender flowers. In late April, the Purplewreath bloomed again, even more generously than in February.

At the height of the dry season, the Palo Santo brightened the woodland here and there with masses of red of a lovely pale shade. I wished to examine the flowers of this arboreal member of the buckwheat family, but they were invariably borne near the tops of trees that were often seventy-five or eighty feet high. The slender trunk, covered with light gray bark that peeled off in ragged flakes like that of the guava and the eucalpytus, branchless for many yards above the ground, gave no encouragement to a climber. Moreover, the hollow stem and branches of these trees were usually inhabited by colonies of stinging ants, more formidable than those that infest the Cecropia tree. Since I was reluctant to fell so beautiful a tree, I was unable to examine its inflorescence, until one day I found a large branch that had just fallen. No longer in flower, it bore developing fruits, each enveloped by a bladdery calyx with three spatulate lobes an inch and a quarter long. These calyx lobes retained their light red color, making the tree as colorful in fruit as in flower, or perhaps more beautiful, as is likewise true of the burío. When the ripe fruit is shed, the calyx lobes act as a parachute, twirling it around as it floats slowly downward, often to a distance from the parent tree.

The branch of the Palo Santo that I so opportunely found was about two and a half inches thick and had been gnawed all around by some animal until it broke off. But what animal could have done this? Monkeys, anteaters, and sloths, the most abundant fairly large arboreal animals in the Barro Colorado forest, hardly fell under suspicion, as they are not bark-eaters and their teeth are poorly fitted for such work. I was pondering this problem as I walked along the trail when a blackish hawk, as large as a Turkey Vulture, suddenly flew up from the ground ahead of me. As it quickly vanished over the treetops, I noticed that its broad tail was crossed by several narrow white bars, but I did not see enough to identify the raptor. As I advanced a few steps more, I found a freshly killed porcupine, which had not been there when I passed a few hours earlier. A gaping wound in the animal's scalp had evidently been inflicted by the hawk's beak. From one side of the victim's body the quills and fur had been torn, revealing short gashes in the naked skin.

Without much doubt, the porcupine had obliged me by cutting down the branch and then had been struck dead by the hawk. But can a raptorial bird kill and dismember a porcupine without suffering cruelly from its spines? This is a question to which we have no answer, but in Guyana a porcupine's skull was found at a Harpy Eagle's nest. The porcupines of Central America, belonging to the genus *Coendou*, have prehensile tails and pass much of their lives in trees and vine tangles, where I have sometimes found them resting by day.

Of all the trees in this forest, the Guayacán made the most arresting display. Usually leafless when in full bloom, these great trees, of hardest wood, were densely covered with trumpet flowers of the purest gold, like sunlight condensed and intensified. Those on the hills across the lake, often miles distant, stood out from the deep green forests through which they were scattered like brilliant yellow beacons in a dark night. In 1935, when the early months were wet, the Guayacáns favored us twice with their delightful display, in February and in late April. In some cases, at least, the same trees participated in this double flowering, but I believe not the same branches. A tree of the related Roble de Sabana also covered itself twice with its delicately pink trumpets, in mid-March and again in April. In the center of the Isthmus of Panamá, these trees of drier country were much less abundant than the Guayacáns.

Among the few plants that brought a touch of color into the lower levels of the heavy forest was the Quassia, a shrub or small tree up to twenty feet high, which in December and January displayed large, attractive racemes of long, pink flowers. At about the same time, another small tree, *Clitoria arborescens*, adorned the forest undergrowth with white blossoms delicately tinged and streaked with pink. Like the more familiar flowers of *C. ternatea*, they are resupinate or inverted; the enlarged standard, nearly three inches long, is at the bottom of the corolla rather than at the top, as in the Sweet Pea and most other leguminous flowers. Although one looks in vain for violets at low altitudes in tropical America, the family is represented there by *Hybanthus*, a shrub with pretty white flowers. In the forests of Barro Colorado, *H. anomalus* grew abundantly, its slender branches rarely reaching a height of fifteen feet. Only one petal of each flower, the lowest, was enlarged, but this was an inch and a half long, marked at the base with pale yellow. In February, and again in late April, these white petals shone out in the dim underwood of the high forest.

Among the most stirring sights and sounds of the island was the drumming of the Crested Guans. In January we heard it frequently in the twilight of both morning and evening, but never while the sun was in the sky. Beginning by moonlight, the guans sometimes boomed for an hour before sunrise. It was fascinating to watch these big birds walk with graceful ease straight forward along a slender branch a hundred feet in the air, like a tightrope walker. From such a lofty station they began their drumming flight, which often carried them across the laboratory clearing from the forest on one side to that on the other side, giving us an excellent opportunity to see how it was done. After taking a few measured wing strokes to start it on its way, the big bird beat its spread wings very rapidly, making the loud booming sound. Then, after a short, silent glide on spread wings, it drummed again in the same manner, after which it continued by ordinary flight to the next treetop. Nearly always the guans drummed twice on a single flight. These splendid birds are so persistently hunted for their flesh that they become rare and wary in all settled districts, and, unlike the related chachalacas, they seem unable to survive apart from primary forests. Accordingly, to see six in the top of a tall, nearly leafless tree, eating berries, as I did beside the laboratory one evening in January, was a rare privilege.

Very different from the booming of the guans were the liquid, rippling notes of the Purple-throated Fruitcrows, large, stocky black members of the cotinga family. Although their calls were among the characteristic sounds of the forest, they were not easy to see, for their straggling flocks usually remained high in the treetops. One morning, while I laboriously cut my way through the formidable hooked leaves of a dense stand of Wild Pineapples, I heard their pleasant notes sounding here and there in the forest canopy far above me. Presently a lone male dropped down to alight in full view on a branch of a Cecropia tree. As he lowered his head to preen his breast, the tufted feathers of his throat patch stood out on either side of his neck, forming a two-pointed shield of richest purple, a gorget equal in splendor to that of a hummingbird but of course much larger.

Soon the rest of the flock settled down in the middle layer of the forest, like so many small crows. Some fluttered their tails rapidly from side to side and made an unpleasant noise, like that of a man clearing his throat as he prepares to spit. Other notes sounded like the scolding of a jay, but mixed with these harsh utterances were more of the melodious liquid notes, elemental music that I loved to hear. These birds have a habit of moving sideways along a branch, like the Black-fronted Tody-Flycatcher. They eat many small fruits, which they sometimes gather while perching but more often pluck from an exposed branch while they surge past on the wing.

On Barro Colorado, as elsewhere, I lost no opportunity to learn about that neglected side of the life of birds—how they sleep. As I wandered through the darkening forest on an evening in late February, hoping to follow some bird to its roost, a sharp *penk* drew my attention to a flock of Collared Araçaris in the treetops high above me. Soon one of them tried to enter an opening in the lower side of a thick horizontal limb of an immense tree, a hundred feet above the ground. The orifice was so narrow that these slender, middle-sized toucans could barely squeeze through it. Watching their dark, colorless figures silhouetted against the last glow of daylight in the sky, I saw them flutter below their narrow doorway and frequently turn back, to try again and again before they gained a toehold on the doorway. When at last one succeeded in clinging there, it squirmed laboriously in, its long tail projecting stiffly outward after its body had vanished, then slowly following it inward.

The natural hollow in the branch was evidently fairly spacious, for six of the araçaris slept in it, as I confirmed by watching them emerge next morning. While the ghostly hooting of the Great Rufous Motmots filled the dusky forest, a great beak was thrust out from the narrow aperture and a slender body fought its way after the beak. One by one, five bedmates left the cavity in the same labored fashion. Then all six flew off together.

At the end of March, eggs were laid in this hole. I could not see them, but whenever in the daytime I stood in sight of it and clapped my hands loudly, a big, black and white bill was stuck out. After peering down with bright yellow eyes, the araçari always squeezed through the orifice and flew off through the forest; despite the great height of its inaccessible hole and my distance from it, the bird felt unsafe in my presence. Now a single araçari slept in the cavity with the eggs. I could not learn where the others slept, but it was evidently not far off.

Before the middle of April, five araçaris were sleeping in the high cavity; one of the original six had vanished. They were now taking in food, evidence that the eggs had hatched. What is more, all five were feeding the nestlings; one morning, soon after they became active, they all carried in food in such rapid succession that I was certain that I had not counted the same one twice. At first, they brought chiefly insects, held prominently in the tip of their great bill, and laboriously squeezed into the hole to deliver each meal. As the nestlings grew older, more and more fruits were brought to them, and also seeds of the Wild Nutmeg, each prettily embraced by its bright red, coral-like, branching aril, visible from afar. After digesting off this soft, spicy envelope, the equivalent of the mace of the true Nutmeg, the nestlings would regurgitate the hard brown seed, about three quarters of an inch long by half an inch broad. When the young araçaris were about five weeks old, the attendants delivered their food while clinging below the doorway, thereby avoiding the laborious squeeze in and out. But while hanging back downward, they could not regurgitate all the items they brought in their throats or still deeper in their alimentary tracts. After passing in the more accessible ones, they flew up to a perch to regurgitate the remainder, then returned to the doorway to deliver it.

The attendance of a nest by five grown birds raised some interesting questions. Was this a case of communal nesting, as in anis, or were a single breeding pair helped by non-breeding yearlings, as in White-tipped Brown Jays? There were no obvious age differences among the five attendants, who all looked alike to me. To know the number of eggs or nestlings would have helped to answer my questions, but the hole could not be safely reached even by the most intrepid climber. The best I could hope to do was to count the young as they left their high nursery, or better, as they returned to sleep in it, as I expected they would do.

When the nestling araçaris were about six weeks old, one or another spent much time looking through their high doorway. They called *pitit* like the adults, but their voices were weaker. On the following day, at least one of them left the nest. When I arrived in view of the hole that evening, the attendants were showing it how to return for the night. While one hung upside down below the doorway, the fledgling came and clung momentarily to this bird's back, but it could not enter because the older bird was blocking the way.

The nest tree was rapidly losing its foliage, leaving the araçaris exposed to the open sky, and they all seemed to be extremely nervous. They had reason to be. While I was slipping into my blind to watch the birds without alarming them, a big White Hawk swooped down and seized one in its talons. The victim wailed piteously, like a heartbroken child. Although the other araçaris had dispersed as the raptor struck, they promptly rallied and pursued as it flew across a deep ravine with their comrade, whose cries grew faint in the distance. I was reminded how toucans of another kind had gathered around Walter Bates when he seized a member of their party that he had wounded by gunshot in the Amazonian forest. But the araçaris availed as little against the much bigger hawk as did the Brazilian toucans against the naturalist with a gun.

I was not at first sure whether the raptor had carried off an adult or the fledgling that had been trying to reenter its nest. When it was almost dark, five araçaris cautiously returned to the nest tree and darted into the narrow orifice so quickly and expertly that I was certain that they had done this many times before. A young bird had emerged from its protecting nest only to fall promptly into death's waiting arms, as too frequently happens.

Before this episode, I had sometimes seen a White Hawk fly over the forest with a snake or a lizard dangling from its talons, but I had not known it to attack a bird. This particular individual, however, had its appetite whetted for araçaris' flesh, and during the next two days it hung around their nest, hopeful of securing another. Despite this persecution, the adults continued to bring food to the young still in the nest, but with the utmost wariness. At times, for no reason evident to the watcher in the undergrowth far below them, they were seized by sudden panic and all darted frantically off together, calling *pitit pitit pitit* as fast as they could. Probably they had seen the hawk while it was invisible to me.

There had been at least three eggs and nestlings, the last of whom left the hole forty-six days after I first saw the attendants carry in food. Possibly, however, only the oldest member of the brood had hatched when feeding began, and the nestling period was a day or two shorter than this. After the alarm of the hawk's attack died away, the grown birds resumed their old habit of sleeping in the high hole, but I failed to learn where the surviving young lodged. Without the raptor's interference, I have no doubt that the whole family would have continued to occupy this dormitory, which would then have sheltered eight araçaris each night. I believe that at this nest a mated pair were helped by three non-breeding birds, probably yearlings, but I could not prove this.

The Collared Araçaris were not the only birds with helpers at the nest that I found during this season's work on Barro Colorado. In a tree beside the main building, four Plain-colored Tanagers attended two nestlings. The relationship of the four adults was not clear; possibly a pair bereaved of their young found an outlet for their thwarted parental impulses at the nest of a neighboring pair. And on my very last morning on the island, I watched a young Golden-masked Tanager, only three months old and still in dull greenish juvenal plumage, help its parents to feed its younger siblings in a later nest. In subsequent years, I have repeatedly found immature Golden-masked Tanagers bringing food to later broods of the same season. Sometimes, too, three individuals in full adult plumage attend the same nest, which holds only its normal brood of two nestlings.

A small colony of Yellow-rumped Caciques (*Cacicus cela*). The water of the cove in which the dead, epiphyte-burdened tree stood failed to protect the hanging nests from an invading snake.

Female Crimson-backed Tanager incubating on Barro Colorado Island, Panamá.

23

A Tragedy in Gatún Lake

Although in other localities I had watched a few nests in lowland tropical forests, Barro Colorado was the first place where I studied birds wholly within lowland forest or a clearing so narrow that it was strongly influenced by surrounding woodland. Here I was made vividly aware of how difficult it is for birds of tropical forest to raise their young. It also became evident that a principal enemy—probably the foremost enemy—of nesting birds is the snake. Birds, paleontologists tell us, sprang aeons ago from reptilian stock. Thus they are distantly related to snakes; but if these so different classes of animals are in fact derived from a common ancestor, they are separated from it by a series of generations inconceivably long—longer, in all probability, than that which separates man from the most primitive of his brother mammals.

If we think of the ancestral reptiles as four-legged creatures somewhat

Headpiece: Gatún Lake from Barro Colorado Island. Sixteen years after the forest was drowned, many trees stood, gaunt and bleached, in the backwaters, providing sites for hole-nesting birds.

similar in form to contemporary lizards or crocodiles, then snakes have been derived from them by specialization with degeneration, birds by a creative advance long and complex. While the ancestors of snakes were losing their limbs, their external ears, and their eyelids, the forerunners of the birds were gradually transforming their forelegs into wings, their scaly vestiture into feathers. They were also developing the capacity to regulate their body temperature, and learning to sing, to build nests, to incubate their eggs and care for their young. At the same time, or probably later, many acquired advanced social habits, associating in flocks, choosing mates to which they remained faithful throughout the year, and even evolving complex group systems of reproduction. Meanwhile, snakes were incapable of comparable developments. On the whole, they acquired no accomplishments that their four-limbed, primitive ancestors did not already possess. They learned neither to produce melodious sounds, nor (with the exception of the Indian Rock Python) to incubate their eggs and care for their young, nor to associate in creative partnership with others of their kind.

The natural world probably exhibits no conflict more fierce and unrelenting than that between bird and serpent. For ages these two—the one aerial and vocal, the other creeping and silent—have preyed on each other; their strife continues undiminished to this day. Each, it is true, has many other enemies, and each devours smaller or weaker individuals of its own class. But, at least in the tropics, I believe that snakes are the chief agents limiting the population of birds, while birds are largely responsible for preventing snakes from overrunning the country.

They work, of course, in quite different manners. Birds wage a bold warfare, some of them catching and eating large, active snakes; whereas snakes prey chiefly upon the eggs and immobile nestlings of birds—those hostages they have given to Fortune. How many adult birds are surprised and swallowed by snakes in the night while they sleep on their nests, or amid foliage, we have no means of knowing, but I suspect that this does not often occur. By day, the winged creature is too alert and mobile to be often captured by its creeping enemy. It would be incautious flatly to deny the tales one sometimes hears of snakes "charming" birds—paralyzing them by some mysterious power of fascination and then seizing them—but I am skeptical of such stories. In many years of bird-watching, I have never witnessed such a thing, nor have I known any reputable, level-headed person who has.

In the forested regions of tropical America, adult birds appear, on the whole, to have few enemies, at least where man does not persecute them severely. I have rarely seen a hawk capture a bird. Few other forms of violent death have come to my notice, and in regions where snow never falls and food is available at all seasons, death from cold and starvation cannot be common. All the available evidence points to the conclusion that the birds of tropical America live, on average, much longer and more tranquilly than those of higher latitudes, which each year must either undertake a long migration or live through months of cold and dearth, both of which courses are fraught with great peril. Accordingly, they can maintain their population by laying smaller sets of eggs and raising fewer offspring each year.

But when we turn from the adults to their eggs and nestlings, we find a very different situation. I had not studied Central American birds very long before I began to appreciate how difficult it is for them to reproduce their kind. Again and again, I have returned to a nest, after an interval of a day or less, only to find that the eggs or nestlings had vanished. At best, as on the Sierra de Tecpán, slightly more than half the nests of which I kept records produced at least one living fledgling. At the other extreme, within the forest on Barro Colorado, only five out of thirty-five nests (one in seven) escaped destruction until the young were ready to depart. Because of the small proportion of nests that avoid premature disaster, the student of life histories in the lowland forests must be endowed with a fortitude that survives repeated discouragements. But I have derived a consoling thought from the great number of nests that I have seen destroyed: if it is so exceedingly difficult for these birds to reproduce, yet their population does not dwindle, it follows that the adults must be fairly long-lived.

Sometimes, as with Amazilia's first nesting (see chapter 7), I have found the despoiled nest swarming with ants; in such cases, one knows immediately the agent of destruction. In wet lowlands, ants take a heavy toll of nestlings, but even the ravenous hordes of army ants seem unable to injure eggs with uncracked shells. Rarely I have seen a graceful Swallow-tailed Kite hover on wing beside an exposed nest in a treetop to pluck the eggs or nestlings from it; or a mouse has converted a low nest into a home for itself, disregarding the eggs; or a woodpecker's high hole has been torn open in a manner that could be accomplished only by powerful mammalian claws; or great-billed toucans, ogres to all the smaller birds, have devoured the nestlings. But most frequently, the nest has been emptied of eggs or helpless young without a clue left to the identity of the robber: the structure remains uninjured and its surroundings unaltered; the contents have cleanly vanished. In the majority of such cases, I believe, a serpent has been the destroyer.

For four months on Barro Colorado Island, I followed the fortunes of a pair of Crimson-backed Tanagers, who feasted on bananas provided for them by Dr. Chapman and nested in the hedge behind the main building. In February, they successfully raised their first brood of two, and in March they built a second nest in another part of the hedge. After the nestlings of this second brood were feathered, my attention was arrested, early one afternoon, by the loud, nasal calls of the parents. I hurried to the nest, found the occupants safe and sound, and saw no reason for the parents' alarm, so I withdrew to watch from a distance. Soon the adults renewed their cries of distress, and their grown offspring of the first brood, arriving on the spot, joined their voices in protest. Other kinds of birds, including a pair of Streaked Flycatchers, a pair of Green Honeycreepers, a female Variable Seedeater, and a Southern House Wren, gathered around the excited tanagers. I ran back to the nest, where I was horrified to find a large, gray, yellow-speckled snake, a Mica, with one of the nestlings already in its mouth. Picking up the first stick that I could find, I fell upon the serpent and killed it, too late to save its poor victim, who dropped lifeless to the ground.

After removing the snake, I tied up the tilted nest and replaced the surviving nestling, who had jumped out during the excitement. Before the stars waned next morning, I went out and found it sleeping peacefully, then returned indoors to wash and dress. But these matutinal preparations were interrupted by cries from the parent tanagers. Knowing now what to expect, I snatched up my machete and rushed forth—too late. Another snake was over the nest, and only a fat bulge a few inches behind its head revealed what had become of the nestling. This serpent paid the same penalty as the first.

Soon afterward, the bereaved parents built a third nest in a different part of the clearing, close beside the cabin in which I lodged. One afternoon in May, when the nestlings of this third brood were well grown, the continued cries of the parents, of the two young tanagers of the first brood who still remained with them, and of a variety of other birds drawn by their excitement, apprised me that this nest was also in peril. A careful search through the surrounding herbage failed to reveal the cause of alarm. Finding no lurking enemy, I returned to my writing, but the birds continued ill at ease, and their reiterated cries prevented my concentrating on my notes. Finally, concealing myself near the nest, I saw that the tanagers perched much in a low rosebush and cocked their heads to peer down into the dense herbage beneath and around it. Here I discovered and killed another large Mica that had been lurking nearby for at least three hours. The tanagers had been so upset by this long interval of nervous tension that for the next hour they neglected to feed their hungry nestlings; they returned again and again to the rosebush, to look down at the spot where their enemy had lain and assure themselves that it was no longer present.

Near the end of a long, slender arm of Gatún Lake, reaching far inward between two of the many wooded ridges of Barro Colorado Island, stood a low, decaying trunk, a lone remnant of the forest that had been drowned when this valley was flooded by the impounded water of the Río Chagres. Although its own leaves had long since withered and all its branches had fallen, the stub was verdant with the foliage of a variety of orchids, ferns, bromeliads, and even shrubs that found here a place in the sun; it was draped and embraced all around, from the water up to its broken summit, with the roots of its aerial garden. Bees had built their hives and wasps had attached their nests amid this tangled mass of epiphytic vegetation, and a flock of Yellow-rumped Caciques had started to weave their swinging pouches at the ends of the branches of the shrub that flourished at the top of the dead trunk. This stub stood a hundred feet from the nearest shore and rose about twenty-five feet above the water.

At dawn, when the placid water of the lake was ruffled by no waves larger than the ripples that spread from the prow of my cayuco carved from a solid log, I would paddle up the cove to watch the caciques at their work. They were black birds with patches of bright yellow on their wings and rumps and beneath their tails. Their strong, pale yellow bills tapered to a sharp point. The females closely resembled the males but were smaller and less glossy, with eyes less intensely blue, or even brown on some. These hens went quietly and efficiently about the weaving of their long, pendent nests, with slender vines and narrow strips torn from palm leaves and tough vegetable fibers brought in their bills from the forest. When the thin, strong fabric had been completed and the rounded bottom closed, they lined the pouch with soft vegetable fibers and silky down from the bursting seed pods of the Barrigon Trees on the shore. To weave and line a nest took from six to eight days.

As in other colonial-nesting members of the oriole family, the males, less numerous than the other sex, helped neither to build the nest, to incubate the eggs, nor to feed and brood the young. They were brilliant vocalists who seemed to be fully aware of their talent. Perching on the coveted top of the trunk, or close by a nest, or even clinging to its side, they uttered a variety of beautiful, clear, liquid phrases to which I never tired listening. While delivering their bright songs, they bowed slightly forward, vibrated their relaxed wings, shook their tails, and raised the yellow feathers of their rumps, making themselves more conspicuous. They spent much time chasing each other, but it seemed to be all a game, as they never fought and rarely so much as touched each other. If a male cacique occupying the coveted station at the top of the trunk saw a second male flying toward him, he invariably relinquished his perch to the latest claimant.

By early May, when thirteen female caciques had completed their nests and settled down to incubate, the colony became quieter. Two brown-eyed hens, probably younger than the others, abandoned their nests unfinished. Soon weak cries issuing from some of the swinging pouches, and food in the bills of the returning owners, told me plainly that the eggs had hatched. For a few days, the number of nestlings increased and the colony prospered. But then, watching from the cayuco tied to a submerged stump at the head of the cove, I noticed that some of the nests were no longer visited by the hens. Daily the number of abandoned pouches increased, until half were unoccupied. Since none of the nestlings was yet old enough to fly, it was evident that something was preying on the eggs and young. But it was impossible for me to reach and look into the nests hanging from the ends of slender branches at the top of a rotting trunk inhabited by many bees and wasps.

Since, in all my hours of watching by daylight, I never saw anything molest the caciques, it was a fair conclusion that the nests were plundered at night. Accordingly, seated with a companion in the cayuco tied at the head of the cove, I kept vigil as twilight faded into darkness. The caciques had retired into the forest to roost, leaving, as is their custom, the surviving older nestlings alone in their pensile cradles. In the dusk, the mournful call of a Common Potoo floated out of the woods beside the inlet. Presently the big, dusky, nocturnal bird emerged from the trees and flew over the water, coming toward the caciques' colony, then swerving off to the opposite shore. In form and mode of flight it resembled a large, slow-flying hawk, a similarity that suggested predatory habits.

Could it be that this potoo, this larger relative of the goatsuckers, which I knew only as a stirring voice sounding through the moonlit forest, preyed upon the nestling caciques? I was reluctant to believe that a creature that sang so feelingly had habits so disagreeable; yet this was my first clue to the mystery of the nestlings' disappearance. Pushing the cayuco farther back into the shadow of the trees on the shore, where we would be less conspicuous to the penetrating vision of a night-bird, we waited breathlessly. Again the dusky bird winged toward the colony, but flew above it without pausing, to vanish into the forest. It did not again appear.

We continued to watch in the darkness, hearing only the liquid calls of the frogs along the shore, and once the grunts of some peccaries off in the forest. We waited until the waning moon, rising late, floated up over the crests of the trees on the eastern ridge. But among the caciques' nests all remained quiet, and at last we reluctantly paddled away through the moonlight, the mystery unsolved.

The following evening, I returned alone to guard the caciques' blighted colony. As day waned, the young birds in one of the nests cried out loudly and impatiently shook their swinging nursery, but their mother did not bring food to them. All the other caciques' nests seemed to be deserted now. A pair of yellow-breasted Rusty-margined Flycatchers were bringing insects at short intervals to their young in an oven-shaped nest of straws, situated among the roots of the epiphytes that covered the trunk, a few feet above the water. Still supposing that the attack on the nests would come from the air, I made a berth for the canoe among the tall, dense marsh grasses at the head of the cove, where I could wait partly concealed, viewing the nests in silhouette against the sky. As daylight waned, I loaded my revolver and placed extra ammunition and the flashlight where I could reach them promptly.

On the farther shore of the lake, at the end of the long, forest-bordered vista down the cove, a massive cumulus cloud rose above the darkening hills, with the tints of sunset playing around it. The roseate glow of early sunset deepened to purple, which in turn slowly faded and left the cloud a dull, leaden mass, sharply outlined against the clear amber afterglow in the open spaces of the sky. The birds had all retired to their roosts, and bats began to flutter back and forth above the water; brilliant Pyrophorus fireflies threaded through the surrounding forest; mosquitoes began to buzz and beetles to boom among the lush grasses around me; the frogs along the shores called in liquid unison. Off in the woods, the potoo uttered a few soft notes, but it did not reveal itself above the water, as on the preceding evening.

The afterglow gradually faded from the west; the stars now shone brightly overhead, with the Great Bear standing up above the mouth of the cove. The caciques' nests and the foliage among which they hung coalesced into a single dark mass that formed the effigy of a rearing horse with a bushy tail, black against the starlit sky. Every nest and every leaf visible in silhouette remained perfectly motionless; peace seemed to reign in the devastated colony.

Time slipped by. Since no sign of menace came from the air, I pushed the cayuco out from its bed of grass and paddled toward the colony, into which I threw my flashlight's beam. Almost the first glance revealed a Mica stretched in a sinuous line along a branch that supported several nests, now hanging empty and deserted. Holding the torch in my left hand and the revolver in my right, I fired upon the snake, repeating the shots until the magazine was empty. Neither the sudden blaze of light nor the revolver's loud reports seemed to make the slightest impression on the reptile, which, slowly and deliberately, continued to slither along from nest to nest, each of which it had already plundered and now found empty. Meanwhile, the cayuco drifted slowly about; it was difficult to keep the long craft in position, load and aim, all by the light of the torch that filled one hand. Again and again I fired, until I had shot off all my twenty cartridges, and the revolver's barrel became too hot to touch. But a snake, even at twenty-five or thirty feet, is a slender target to hit with a short-barreled revolver in a shifting light; when I had exhausted my ammunition, the ravager continued its search among the nests it had already desolated.

As a last hope, I threw an extra paddle, then three spare flashlight batteries; but these missiles failed to reach their mark. The neighboring shores provided no stones; I had nothing more to throw. But I shouted, hissed, whistled, splashed water, all without effect. Helpless now, I watched with a sort of horrid fascination as the snake, heedless of all else, continued its search for prey. Finally, it stuck its head into a nest at the end of a branch and withdrew it with a bulge behind its mouth. Then it opened its jaws widely and vomited up two eggs, which fell into the water with a splash and sank. Evidently the ravening snake cared only for nestlings, there in the midst of plenty!

About this time, a steamship, aglow with many lights, slipped steadily through the dark water of the canal channel beyond the cove's mouth. If anyone aboard had looked to the port side, he may have noticed a beam of light directed into a low tree by the distant shore, but he could never have guessed the tragedy that the beam revealed to me.

When it had finished its search among the pouches that hung closest together, the snake slid across to the opposite side of the colony, toward the only nest that I knew to be occupied, by nestlings about ten days old. I had fired my last shot and hurled the last detachable object that I could spare, and still death advanced relentlessly toward its goal. The wasps and bees whose nests were plastered or suspended all over the tree did not sally forth to punish the invader, which slipped too smoothly and silently over the boughs to arouse them from their night's repose; it did not occur to me until later that I might have stirred them to fury by hammering against the trunk. What I had already seen was sufficiently distressing; I had no desire to watch this nocturnal drama to its tragic end. Turning the bow of my small vessel toward the open lake, with heavy heart I paddled away into the darkness of the night, leaving it to cover an act which it would have been painful to me to behold.

Next morning, I returned to make another attempt to dislodge the Mica from the blighted colony. Even if the last of the caciques' nestlings had been destroyed, I might at least save the young Rusty-margined Flycatchers in the domed nest among the roots near the water level. As I paddled along the wooded shore, I reflected that the snake, easily swimming across the water on which the caciques had foolishly relied to protect them from the attacks of terrestrial enemies—evidently the chief peril of their colonies—had climbed to the top of the stub by way of the roots draping it all around. Then, after pillaging some of the nests, it hid in a deserted pouch by day, to sally forth and plunder others while darkness shielded it from the attacks of hawks and prevented any defense by the purely diurnal caciques. So it would continue until the last egg and nestling in the tree had been devoured.

But as my cayuco approached the head of the inlet, I beheld, to my amazement, the Mica hanging, head downward, beside one of the caciques' abandoned pouches. A nearby nest, contrary to my expectation, still cradled living nestlings, and their mother, devoted bird, was carrying food to them, less than a yard from the motionless body of the destroyer. On the neighboring shore, I cut a long pole and pulled down the snake, which had already begun to putrefy. In its posterior half, I counted three bullet holes, through one of which the viscera extruded! In its death writhings, the serpent had tied a tight knot in the hinder part of its slender length. With a lack of sensibility that one would hardly expect to find in any living thing, the ravening snake, which might easily have dropped into the water and swum unharmed away, not only continued its search for prey after having been thrice struck by bullets, but even swallowed two eggs when mortally wounded. Probably it was the effect of these wounds that caused it to disgorge them.

The nestlings that I had supposed to be doomed when, my last shot spent, I left the serpent advancing toward them, also miraculously escaped destruction, and their mother continued faithfully to attend them. Other female caciques, bereft of their young, still came to the tree bearing food, which after an interval they swallowed, or carried off again in their bills. Still others came to peer into empty nests, or cling to their sides. Despite all these poor birds had endured, their attachment to their nests and young remained unshaken. Nor did the idle males utterly desert the stricken colony. Until the last fledgling flew, they came at intervals to deliver their mellifluous songs among the swinging empty pouches.

Years before, in Honduras, I had watched a Mica, six or seven feet long, climb upward through the tangle of vines that draped around the lower part of the trunk of a tall, isolated tree, from the ends of whose spreading boughs hung scores of the long pouches of a colony of Montezuma Oropéndolas, larger relatives of the caciques. But a pair of yellow-breasted Great Kiskadees, one of the biggest and boldest of the flycatchers, feeding nestlings in the crotch of a neighboring tree, would not tolerate such an insatiable nest-robber so near their covered nest. Again and again, they darted close by its head, uttering angry, threatening notes as they swept past. They persisted in these feints until the big snake turned and slid downward among the vines. Meanwhile the oropéndolas, feeling secure in their high treetop, merely looked on from its lower boughs. What a tremendous destruction of eggs and nestlings would have ensued, had the Mica reached its objective! Now I understood why an oropéndolas' colony is nearly always established in the top of a tall tree with a smooth trunk that cannot be scaled by a snake and boughs that stand clear of all neighboring vegetation.

In a later year, while bathing in a Costa Rican stream, my attention was suddenly arrested by a commotion in some bushes overhanging the bank. A slender green snake had attacked the nest of a Scarlet-rumped Black Tanager, and the orange-breasted female, uttering cries of distress, was attempting to drive it away. Nothing surprised me more than this display of spirit by the tanager, for in the presence of man this is one of the shyest of birds at its nest. Not waiting to witness the outcome of this unequal struggle, I seized the nearest stick and knocked the serpent into the river, too late to save the two pretty blue eggs spotted with black. When mortal bullet wounds will not extinguish the rapacity of a snake, how can the bill of the bravest small bird turn it aside from her nest?

Accordingly, it is difficult to resist the impulse to help the birds against these inveterate enemies, whatever views about the balance of nature one may hold, and however distasteful it may be to destroy even the meanest of living things. Doubtless, to go about killing every snake that we find would be an indefensible practice, if only because the interactions of the several forms of life are so complex and poorly understood that we might, in the long run, injure what we wish to protect. Yet to refuse to come to the aid of a bird whose nest is menaced by a serpent would be to suppress a strong and precious impulse, as though we failed to help a kinsman or friend savagely attacked by a stranger. For birds, not only by their warm blood but also by their capacity for attachment to a mate and devoted service to their progeny, are so much more akin to ourselves than cold-blooded snakes, which never even enter into relationships that, in ourselves at least, evoke love and loyalty, that we spontaneously feel closer to them, and they seem to have a better claim to be defended by us.

24

A Wanderer's Harvest

After leaving Barro Colorado in June, 1935, I made a short visit to Costa Rica, then continued northward to Guatemala, where I passed six weeks studying the birds of the Pacific slope on San Diego coffee plantation, near Colomba. Here I had the only attack of malaria from which I have ever suffered, evidently contracted while passing through the lowlands; but it was promptly and permanently cured by injections of quinine. I had intended to remain in Guatemala; but irritated by what I considered shabby treatment by the Guatemalan department of immigration, I left in mid-August for Costa Rica, where I have lived almost continuously through the subsequent years. I have not regretted this sudden change in my plans, for, with the exception of one brief but bloody civil war, Costa Rica has been peaceful throughout the last four decades, while Guatemala has been torn by fierce political and social conflicts. Moreover, I found in an isolated Costa Rican valley an exceptionally favorable place to continue my studies of tropical

Headpiece: "Whiteface," a male Groove-billed Ani (*Croptophaga sulcirostris*), incubating in a sprayed orange tree. The white paint mark on his face distinguished him from his mate.

nature with slender means. As has happened repeatedly in my life, a distressing reverse had favorable consequences.

As I review the journeys I have made and the adventures I have had in several decades of wandering off the beaten track in tropical America, I find that my journeys were neither very long nor hazardous, nor my adventures much out of the ordinary. It was better so. Although hairbreadth escapes from death and extreme hardship may be expected in the exploration of remote, uncharted regions, it has been truly said that the modern naturalist or scientific traveler rarely has such experiences except as the penalty for careless preparation or taking foolish risks. Although my travels led me into none of those predicaments that are so agonizing to live through yet make such gripping tales, they were certainly not exempt from discomforts, anxieties, and disappointments.

We set forth on our journeys to see more of the earth's beautiful and wonderful things than we could find at home, yet we cannot avoid encountering much of its ugliness, misery, and horror. Beauty and ugliness; pleasure and pain; elation and depression; ardor and exhaustion; life and death—these are the two faces of the same coin; we might even lack names for them if they were not brought into high relief by their opposites. We undertake our travels to gather the glittering coins of beauty and joy, but too often we find them lying with their dark faces up. What could we expect? Unless we were fools, we knew before we left home that every coin has two faces.

Fortunately for our faith in life, the coin's glittering face makes, as a rule, the most lasting impression on the mind; just as the brightest object most strongly affects the photographic film. Thus our journeys frequently appear more enjoyable in retrospect than while we were in the midst of them; and memories of past delights lure us ever and again to embark upon fresh travels, and incidentally to expose ourselves once more to the old annoyances and disillusions. Nevertheless, whether they bring him pleasure or pain, a naturalist's travels are a necessary part of his education. They give him breadth of vision, as his long and studious sojourns, his observation of the gradual growth and development and decay of organisms through the slow procession of the seasons, bring him depth of understanding. To me, these sojourns have yielded more contentment and knowledge than my wanderings, and have rewarded me more richly. I have traveled chiefly to make them possible.

I have, by inclination, been an adventurer in little things. My red-letter days have not been made by planting my banner on the summit of some forbidding peak hitherto unscaled, nor by stumbling upon some ancient city lost in the forest's depths. Among them I include the day when, riding down an abrupt mountainside in northern Guatemala, I first saw a magnificent living Quetzal and heard the sweet strains of the Slate-colored Solitaire; the day years later when, after long waiting, I found my first Quetzals' nest; the day when, riding through the rain along the side of a profound Andean gorge, I saw my first glowing orange male Cock-of-the-Rock resting beside the trail. These days match in memory that on which, with my head stuck through the car window as the train speeded away from Veracruz, I beheld my first tropical snow-peak looming ahead like some whiter and more sub-stantial cloud; and that when, the sky clearing after days of rain and mist, the snow-crowned cone of Sangay, plumed with a long streamer of smoke, stood revealed above the vast forests of the Ecuadorian Oriente.

But a place still more sacred in memory is held by the afternoon when, dragging myself from bed with a raging pain in an infected leg, I hobbled down the hill at Lancetilla to watch a nest of the Groove-billed Anis, and saw a juvenal of the first brood, hatched in July, helping to feed and protect his younger brothers and sisters hatched in September. My memorable discov-eries of this sort have been as numerous as I could reasonably expect; I could not tell about all of them in this book, but most are to be found in other writings.

The most substantial harvest of these years of wanderings and sojourns off the beaten track, the reason and justification for them, is the bird lore I have garnered. I find enduring satisfaction in having built up, by long vigils and innumerable visits to nests, fairly complete pictures of the habits of several hundred kinds of birds, beautiful, shy creatures whose life stories had never been told. Moreover, it is most gratifying to recall that such insight into the lives of birds as I have won has been gained without the intentional sacrifice of a single life, without the willful destruction of a single nest. I have from the first regarded my bird-watching as a game—a serious and important game that can tax one's strength and skill to its limits, but still a game—which has definite rules, and in which the odds are always in favor of the birds, where they rightly belong. And these rules prohibit the destruction of occupied nests, even, for example, cutting down a high, inaccessible nest that cannot otherwise be examined, or digging out and leaving exposed those tucked away in narrow crannies and deep burrows.

As in other spheres, I have found that, in the long run, it pays, in objective results no less than inward satisfaction, to abide conscientiously by the rules, not making exceptions for special advantages. In complying with the rule not to tear open active nests in deep holes or crannies, I have been forced to devise various methods, more or less ingenious, for peeping into them by means of mirrors and electric lights. With these I not only see the shapes and colors of the eggs; but since I do not disturb them, I can make continuous observations, discover how long they take to hatch, and follow the develop-ment of the young, as likewise the behavior of the parents. Similarly with high, inaccessible nests; although the rules prevent the measurement and description of the lifeless eggshells, they ensure (so far as it depends on me) that the nest remains undisturbed, available for the far more interesting study of the habits of the living birds who attend it. With attentive watching of these high nests, one can even learn the approximate length of the incubation period, and how long the young remain within. And if one is patient and searches long enough, in the end he is likely to find a nest of these height-loving species that he can reach without unduly risking his bones.

In chapter 10, I told how at Lancetilla I studied the habits of the Groove-billed Anis, black members of the cuckoo family that often nest communally. It happened, however, that my first anis' nest belonged to a single pair. I wished to know which of the two was the male and which the female. By sticking a small paintbrush above the nest where they would rub against it, I gave these birds white marks that served to distinguish them; but neither by appearance, size, voice, nor habits did they reveal their sexes. The professional ornithologist might have answered this question in five minutes by shooting one or both members of the pair and performing an autopsy. But after watching this affectionate pair of birds throughout their first nesting, I regarded them as friends, and felt about as ready to take their lives as I would be to murder the man whose fine hospitality I had accepted.

Instead of this unfeeling course, I resolved, when these anis prepared for their second brood, to watch and learn which of the two laid the eggs. This endeavor cost me several long, tiresome days, sitting on a hard box in a stuffy little burlap tent beneath a hot sun, with most of the time nothing but an empty nest in front of me. But in the end I was successful and saw one of these anis lay an egg—not the one that I had expected to do so, but the other, whom for three months I had been calling the male! Her mate, named "Whiteface" from the white paint mark on his cheek, had incubated every night and been by far the bolder in defending the nest, sometimes buffeting the back of my head when I looked into it, and for these reasons he had always been designated by the feminine pronoun. Now I was certain that he was the male.

To have made this surprising discovery without any sacrifice of life greatly enhanced the joy it brought me. The destruction of one of these anis or any of their offspring, even to satisfy scientific curiosity, after I had been a silent witness to the whole course of their first nesting, would have been a crude way to terminate a pleasant association and have left an ugly blot upon memory. But if these considerations could not reconcile the most coldly intellectual scientist to three days of ennui and discomfort and in themselves barren, I have another that might have weight with him. Some of the most interesting revelations that this family of anis vouchsafed to me came while they were raising their second brood—and dead birds raise no second brood.

Moreover, by these long vigils I learned that anis lay their eggs around midday, not early in the morning as many other birds do and I had expected them to do, and at irregular intervals. Later, when I studied the Smooth-billed Ani on Barro Colorado Island, this tediously acquired information enabled me to learn the sexes of the paint-marked birds with a minimum expenditure of time and patience, by employing the same method of watching the laying of the eggs from concealment. In this species, too, the male took charge of the nest by night, and was its more zealous defender.

Although I have lost no opportunity to learn facts about birds and other living things, to trace the whole pattern of their lives in all its details, such information, satisfying as it can be, is at best superficial and not what I most desired to win. What we really need for our enlightenment and guidance is insight into the inner or psychic life of the beings of all kinds that surround us. To have this insight might solve some of the mysteries of this so baffling universe and bring us closer to an understanding of its secret springs; for this hidden realm is the abode of all realized value, of everything that, as far as we can tell, gives significance to the existence of the universe as a whole or its least component. But this is just the aspect of things that is most securely hidden from us. Whether we use our unaided senses or our most delicate and sensitive instruments, we see only the outer shell of the objects, living and lifeless, that surround us. We speculate about life on Mars or planets that revolve around distant stars, yet we remain profoundly ignorant of the inner life—the only life that really matters—of the creatures that surround us on this earth. We know phenomena, not the intimate nature of things.

It seemed to me that it should be easier to penetrate to the inner life of birds, which are warm-blooded like ourselves and in their social and familial relations resemble us in various ways, than to that of animals more remote from us in our systems of classification or more different in their manner of life—than to that of reptiles, fishes, or insects, for example. Unfortunately, studies of animal behavior, or so-called animal psychology, shed no light directly upon the quality of the inner life of their subjects; they reveal only sequences of activities from which we draw speculative inferences. Or such studies do lift a corner of the opaque curtain that hides mind from mind only when supplemented by imaginative sympathy, that precious spiritual power to which we owe all that we know of the inner life even of those most like ourselves. But the insights of imaginative sympathy are not capable of scientific confirmation. They are not for that reason false; they merely remain unproved. Sometimes, in the warm glow generated by intimate encounters, I have seemed to share the very feelings of the creatures I watched; but as this glow faded, doubts would arise. Nevertheless, I am convinced that these moments of sympathetic participation possess some residual virtue. At least, they have made me distrustful of all too-simple and too-mechanistic in-

terpretations of the behavior of the higher animals. Although the veil that
separates the mind of a bird or non-human mammal from my own remains
opaque, I have gained profound respect for what lies behind it.

My continuing ignorance of the psychic quality or inner life of the creatures around me, of their true nature, makes me reluctant to harm even the
least of them. When we kill any living thing, we do not know exactly what we
destroy. Birds, for example, seem to live more intensely than we do: their
vital processes are more rapid, their senses keener, their reactions swifter; it
may well be that they also feel more deeply than we do. Although the
life-span of the smaller birds is much shorter than ours, they may crowd
more real living into a day than we commonly do in a week or a month. Their
relatively uncomplicated life amid trees and flowers and sunshine may be
more joyous and satisfying than ours in a complex artificial world beset by
perplexities and misgivings. They may know intuitively much that we try to
discover by the slow, analytic methods of science; as when, without charts or
instruments, they find their way to a definite goal over vast expanses of the
earth's surface. In short, their lives may be, in various aspects, more perfect
than our own; and who would wish to be guilty of destroying something more
perfect than himself? Even to suspect that the earth supports many creatures
who live joyously gives us greater confidence that the world process (of which
organic evolution is a phase) has not gone miserably astray and is an antidote
to pessimism.

Whatever the quality of birds' inner life may be, outwardly their lives are
beautiful. Much as I have loved and sought truth, I have loved and sought
beauty even more—not only sensuous beauty, but likewise moral and
spiritual beauty, the beauty that ancient philosophers equated with the
good. Beauty is always, in a sense, truth, although if incautious we may draw
false inferences from it; but truth, in a world so full of harsh and ugly facts as
this, is often the antithesis of beauty. Although one may contend that to
know the truth, even about ugly and evil things, is in itself good and valuable,
such knowledge can hardly avoid being tainted by its loathsome object. But
to learn the truth about beautiful things, by long and patient effort to disclose
their carefully guarded secrets, is one of the most satisfying of pursuits.

Birds, as I have said, are among the most beautiful of living things, and
their lives, or at least those of the less fierce and ravenous kinds, are nearly
always beautiful, as a whole and in their details. To learn pleasant facts about
beautiful things is altogether delightful. This is what makes the study of their
habits so richly rewarding, at least to one with my particular temperament.
Certainly, for our survival, we need to know about many things that are dry,
unpleasant, or revolting. But so many brilliant minds, supported by wealthy
institutions, are dedicated to these investigations, that it can do no great
harm if a few "world losers and world forsakers" devote themselves to the

pursuit of the beautiful truths that enrich us spiritually even if they contribute nothing to our survival in a competitive world. Yet the study of bird life must be regarded as more than an innocent hobby; by contributing generously to our understanding of basic biological problems, it can help to make our own lives saner and more secure.

Not only did the years of wandering about which I have told in this book fill my mind with treasured memories and my notebooks with valued records, they prepared me for the undertaking on which I had set my heart: to establish a homestead in or beside unspoiled tropical forest, where I would enjoy leisure and independence to study and write about nature for decades. This was not easy to accomplish, especially by one with small resources. Until I had seen much of the tropics, and learned to deal with its people, and outgrown the spells of nostalgia that at first oppressed me, and hardened myself to isolation, and discovered in which of the many varieties of climate that tropical America offers I could live most healthfully and efficiently, and accumulated a little capital, I was not prepared to realize this dream. How it was finally accomplished, and what rewards it yielded, has been told in another book.

The course I have followed has been long and devious; sometimes I have faltered; but always the call of a bird has lured me on, too imperative to be resisted. From the sunless depths of lowland forests, from the shores of pellucid mountain torrents, from high and bleak windswept plateaus, from the heart-straining slopes of equatorial snow peaks, I have heard the calls of the tropical birds. In a thousand different tones they have sounded, harsh and liquid, dull and shrill. Following them, I have known hunger and fatigue and loneliness, illness and the torments of countless insects, and occasionally a chilling sense of abandonment. Yet they have rewarded me with rare beauty and great joy and the incomparable elation of discovery. Whether I followed them eagerly or, as at times, hesitantly, always these calls were irresistible.

> We are what suns and winds and waters make us;
> The mountains are our sponsors, and the rills
> Fashion and win their nursling with their smiles.

>   Walter Savage Landor

Index

Abbey Green Plantation, Jamaica, botanists' headquarters at, 44–46

Achimenes longiflora (Gesneriaceae), 255

Acorns: stored by woodpeckers, 13–15; ripening of, 228; sizes of, 271

Acquarone, Paul, 40, 47

Aguacatán, Guatemala, 268

Agustín, Silverio, 266, 268, 270, 273

Akee, *Blighia sapida* (Sapindaceae), 47

Alaria esculenta (Laminariaceae), 27, 29

Alder, *Alnus arguta* (Betulaceae), 244, 255, 260, 269, 280; flowering and leaf renewal of, 220–21

Algae: on rockbound coast, 26–30; on hairs of sloths, 71

Almirante, Panamá, 52, 53, 56–57, 58, 67, 77, 92

Almirante Bay, 56, 92

Alsacia Plantation: author's arrival at, 154; disputed nationality of, 154–55; laborers at, 155–56; view from, 156; second-growth thickets at, 156–58; birds at, 157, 158–64, 166–80; departure from, 180–81. Picture, 151

Alta Verapaz, Guatemala, 191–92, 196

"Amazilia." *See* Hummingbird, Rufous-tailed

Amazon River, 140

American Museum of Natural History, 289, 294

Ames, Oakes, 248

Amphipods, 29–33

Amphithöe rubricata (amphipod), 29–33

Andes, 210, 211

Anhinga, *Anhinga anhinga*, 67

Ani, Groove-billed, *Crotophaga sulcirostris*, 110; nesting of, 131, 313, 314–15. Picture, 311

—Smooth-billed, *Crotophaga ani*, 49, 315

Animals: on making pets of, 115–17; inner lives of, 315–16

Antigua, Guatemala, 93, 198, 202–3

Antpitta, Scaled, *Grallaria guatimalensis*, 267; habits of, 236–37

Ant(s): inhabit cecropia trees, 70–71; attack nestlings, 80–81, 86, 303

—Fire, *Solenopsis geminata*, 80

—Leaf-cutting, *Atta* sp., 54, 90

Antshrike, Barred, *Thamnophilus doliatus*, 157

—Slaty, *Thamnophilus punctatus*, 110

Apel, Rodolfo, 265, 273

Aphelandra aurantiaca (Acanthaceae), 98

Apple of Peru, *Nicandra Physalodes* (Solanaceae), 137

Araçari, Collared, *Pteroglossus torquatus*, 297–99

Arbutus, Trailing, *Epigaea repens* (Ericaceae), 4

Arbutus donnell-smithii (Ericaceae), 212, 218, 245, 258

Armadillo, Nine-banded, *Dasypus novemcinctus*, 111–13

Arnold Arboretum, 248, 283

Ascophyllum nodosum (Fucaceae), 28, 29

Ash tree, *Fraxinus uhlei* (Oleaceae), 269

Aspen, American, *Populus tremuloides* (Salicaceae), regeneration of, after fire, 36

Aster exilis (Compositae), 238

Atitlán, Lake, Guatemala, 212–13, 242–46, 253

Ayutla, Guatemala, 249, 253

Bagworms, 54

Balboa, Canal Zone, 285

319

REPUBLIC OF GUATEMALA
(excluding northern Petén)

00 10   30   50 kilometers

00 10   30   50 miles

MEXICO

Río Usumacinta

Río Copón

SIERRA DE LOS CUCHUMATANES

San Miguel Acatán•
•Soloma
□ CHAILÁ

•Putul
•Cotzal
•Nebaj
•Chiantla
HUEHUETENANGO ○
•Aguacatán

○ COB

•Tac

▲ Vol. Tacaná
▲ Vol. Tajumulco
•Momostenango

SIERRA DE CHU

Río Motagua

QUEZALTENANGO ○

▲ Vol. Zunil   SOLOLÁ
□ SANTA ELENA
□ CHICHAVAC
•Tecpán

Vol. Santa María ▲ ▲
Colomba•   •Santa María de Jesus   Lago Panajachel
□ SAN DIEGO   Atitlán   ▲ Vol. San Lucas   ○ CHIMALTENANGO
San Felipe•   Vol. San Pedro   □   Vol. Atitlán   ○ ANTIGUA   ⊕ GUATEMA
○ RETALHULEU   MOCÁ   Vol. de Fuego ▲ ▲ Vol. de Agua
Lago de Amatitlán

•Ayutla

•Champerico

○ ESCUINTLA

OCEANO PACIFICO

San José•